Time Machines

Time Machines
Telegraphic Images in Nineteenth-Century France

Richard Taws

THE MIT PRESS
Cambridge, Massachusetts London, England

To the west I now see the cathedral and, it appears, the Strasbourg telegraph, whose index of death is sinister and almost sublime; it moves its scissors like a Fate—the needle of the balance of people inclines and declines the compass of time.

Jean Paul[1]

This work, in its discontinuity, proceeds by means of two movements: the straight line (advance, increase, insistence of an idea, a position, a preference, an image) and the zigzag (reversal, contradiction, reactive energy, denial, contrariety, the movement of a Z, the letter of deviance).

Roland Barthes[2]

1 *The Compass of Time*

The dream of reason produced mechanical monsters, angular and winged, secretive and strange. On July 12, 1793, one such creature made its debut at the top of a hill on the outskirts of Paris, on a street that now bears its name.[1] Two concrete water towers, topped with a lofty communications antenna, occupy its place today, looming above Belleville cemetery and indeed the whole of Paris; as the architects of the device knew, this is one of the city's highest points.[2] A jointed wooden zigzag, silhouetted against the sky, it stood at first still and inscrutable, before slowly assuming a series of static poses, contorting itself into one after another until the sequence was complete and it could resume its sphinxlike resting pose (fig. 1.1). Right away, it seemed clear to passersby that the strange mechanism was speaking, although not in a language they understood. While, squinting a little, some of the shapes thrown by the spasmodically rotating limbs looked a lot like conventional letters—a giant *H* here, an oversized *Z* there—these similarities revealed only a desire to extract meaning where there was none, for the substance of the signs remained obscure: at best they were hieroglyphics, at worst gibberish. Conducting its conversation in full view, the apparatus was visible for miles around, but it wasn't talking to casual observers. The instrument's interlocutor was nowhere to be seen and its arms appeared to move of their own accord. Far in the distance, though, the glint of

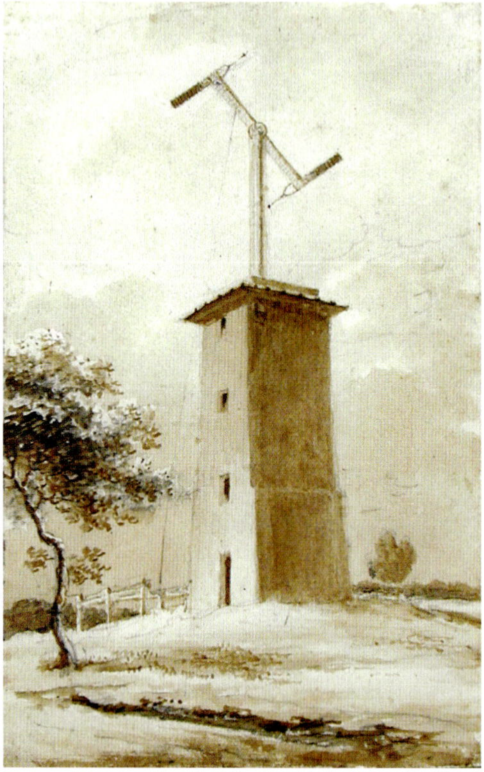

Figure 1.1
Bernard Gaillot, *Le Télégraphe du Trou d'Enfer*, 1842, pencil, sepia, and gouache on paper, 16.5 × 10.5 cm. Musée du Domaine royal de Marly, Marly-le-Roi.

a telescope betrayed the presence of an unseen viewer who understood the significance of these cryptic signals, which they received with revolutionary speed and clarity.

During the French Revolution, many novel inventions such as this one were proposed. Most of them never saw the light of day and of those that did, few survived an experimental phase. In contrast, this peculiar contraption would, against the odds, take hold and remain in operation for over fifty years. In that time, it altered radically the mental as well as physical landscapes of nineteenth-century France, shaping how people encountered and understood events both local and global, transforming perceptions of their place in time and space, and, in subtle but substantial ways, shifting and conditioning their apprehension of the relationship between past and present. However, it was also fully a product of its time; it did not act on a passive social body, or determine straightforwardly the course of events, but was the outcome of numerous individual interventions as well as broader political and cultural conditions, which in due course it also mediated.

At its introduction no one could anticipate the invention's future success, for like the Revolution, it only seemed inevitable or necessary in retrospect. Yet from the beginning, the *machine*—that was the term used to describe it then—carried a powerful visual charge. It was made to be seen, and it conveyed news of things seen elsewhere.

Unsurprisingly, the machine's intrusion into Parisian space, as well as into towns, villages, fields, and mountains across the nation, often atop familiar and cherished buildings, did not go unremarked upon by eyewitnesses. The previous year, in September 1792, an experimental version of the machine had been installed at the same site in the parc Saint-Fargeau. Its baffling appearance on the scene was documented in a gruff police report:

> On the Belleville Road, many citizens noticed a type of obelisk in the parc Saint-Fargeau, raised to the height of a windmill; everyone asked why it had been put there. One citizen says it's to receive the new word signals, from Lille to Paris, in a quarter of an hour. Others ask how it works; some say it's by the voice, others that it's by way of brass lines between each one. Another said that in the Vendée, brigands had a relay from one windmill to another and knew all the movements of our troops. We must define the device, and find out if the committees, who need to know about it, should be instructed on this invention.[3]

Reactions to the startling materialization of the alien monolith at Belleville give some sense of the astonishment and fear that accompanied its arrival, before the device and its protocols were naturalized by use.[4] The onlookers described in the anonymous policeman's report had smashed that one to pieces, convinced it was a nefarious agent of counterrevolution.[5] Threats of destruction or extinction accompanied the "terrifying logogriph" from the start.[6]

As this account reveals, the windmill-obelisk didn't yet have a name. Even so, it was already received in aesthetic terms, pitched somewhere between the windmill's familiar rotational technology and the unfathomable monuments of ancient Egypt. Such responses that exoticized or racialized the machine and its operators appeared early on. Over four decades later, from the roof of the Ministère de la Marine, one device would indeed face the

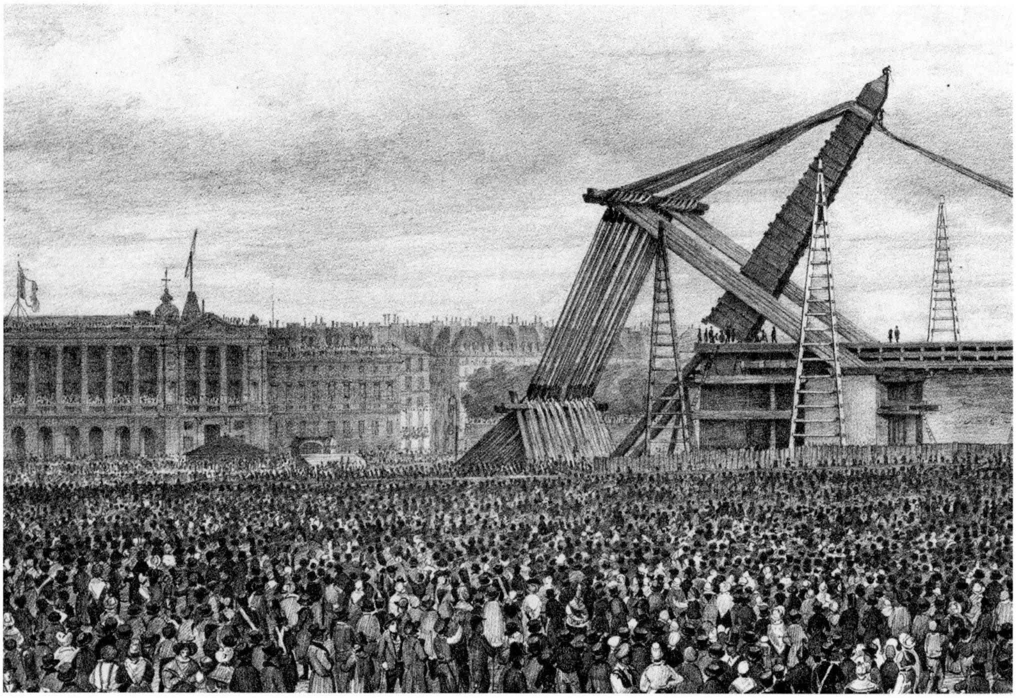

Figure 1.2
Detail of Théodore Jung, *Erection de l'obélisque de Louxor, place de la Concorde, le 25 octobre 1836*, 1836, lithograph, 36.1 × 55.7 cm (full image). Musée Carnavalet, Paris.

Luxor obelisk, following its 1836 installation on the place de la Concorde, while for many years another nestled among the windmills of Montmartre (fig. 1.2).

Suspicion of the machine's purpose never really abated, and although in time it became commonplace, hiding in plain sight as other affairs prevailed, it never ceased to be an object of considerable perplexity; the "enigma of our age," one contemporary called it.[7] Echoing the impression of the bizarre new apparatus as an uncanny spy, an unwelcome or threatening interloper, later observers, reaching for an appropriately diabolical language, described it as abject, anti-aesthetic, broken-down, or, perhaps grasping for terms, as a witchy familiar: a spider, a snake, or a chrysalis.[8] Or, taking their cue more directly from its appearance, they pictured it as something malevolent and winged: a beetle, a bird, or a bat. The machine demanded new words, before it spoke itself. For a while it was known as a "tachygraphe."[9] Soon after, though, it answered to one name only: "telegraph."

T Is for Telegraph

A new coinage, in an era when the malleability of language was intensely meaningful, "telegraph" brought together "that which writes—or draws—at a distance"—from the Greek *tele* (far off) and *graphos* (writer/drawer). This chilly archaism, an inherited vocabulary sent from a distant time, soon functioned as shorthand for the presence of modernity, and in due course it eclipsed its original meaning to describe quite different objects and actions. By 1833, such was the familiarity with the telegraph's appearance that an image of a machine, its limbs extended, headed a lithograph representing the letter *T* in Victor Adam's *Nouvel abécédaire en enigmes*, preceding vignettes of other everyday objects ("un tonneau," "un tambour," "des tentes," "une table," and so on).[10] In time, the device lent its name to roads, hills, fields, and mountains around the world. It didn't take long for the telegraph to become miniaturized and domesticated as one more fragment of commercial culture, in everything from ceramics to hallmarks, vineyards to cafés, pamphlets to children's games (fig. 1.3). Demonstrating its overlap with other media networks, the naming of newspapers after the telegraph began early—Pierre-Charles-François Mithois published *Le Télégraphe politique* in Paris in 1794, immediately after the advent

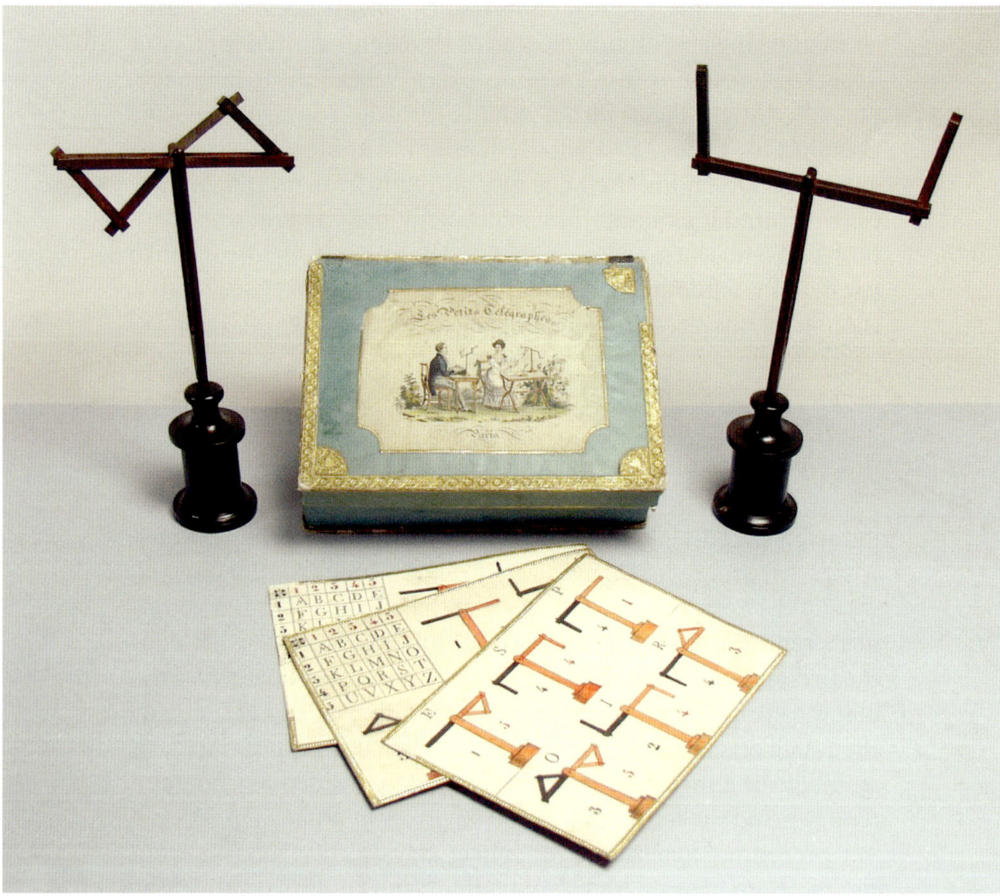

Figure 1.3
Les petits télégraphes game, c. 1845, cardboard, paper, and wood, 21.5 × 7 × 16.5 cm. Museumsstiftung Post und Telekommunikation.

of the system, while in Port-au-Prince *Le Télégraphe* was from 1813 until 1843 the official Haitian government gazette.[11]

The public encounters with the telegraph in 1792 and 1793, when crowds either tore it apart or marveled at its peculiarity, marked the coming into visibility of something both mundane and extraordinary in a world that had, in the preceding three years, already transformed beyond belief. In this early iteration, before the widespread introduction of electromagnetic systems at mid-century, telegraphy was a predominantly visual medium. In essence a simple mechanical semaphore viewed through a telescope, sight was intrinsic to its operation, and it tested the limits of vision. Operating in public but conveying secret messages, this technology, variously termed "semaphoric telegraphy," "aerial telegraphy," or more emphatically "optical telegraphy," was brought into official service at the height of the revolutionary Terror, when the legibility of signs and the use of images for political ends were pressing issues.[12] It saw its last major use during the Crimean War, at the Siege of Sevastopol in 1855.[13] However, such bookends tell only part of the story, for the telegraph's demise was predicted long beforehand, and like a sharp-elbowed revenant, it made repeated encores in the decades that followed its official replacement.

The telegraph did not operate autonomously. As with any new technology, it was a product of diverse cultural forces, among which, perhaps unexpectedly, art should be counted. Reproducing and reinforcing the telegraph's visual orientation, visual images were best poised to make sense of the telegraph and its effects. This book argues that the optical telegraph was in unique dialogue with its visual representation—certainly to a far greater extent than its electrical successors—and that the confluence of line-of-sight telegraphy and a wildly varied range of images revealed anew the temporal dynamics of both art and communications technology at a crucial juncture in modernity. In the nearly sixty years of its use and in its persistence thereafter as an object of memory, the telegraph inflected visual culture, and how people experienced it, in ways both straightforward and insidious. But more than this, the relationship between the optical telegraph and the proliferating and promiscuous universe of images into which it emerged was such that it might more properly be considered a vital component of the nineteenth century's effervescent visual culture, rather than adjacent to it, or merely its subject. For some, it was itself an object of aesthetic wonder; as the correspondent for the literary and dramatic newspaper *Vert-Vert* put it in 1835, "nothing is as beautiful as the telegraph, when it declaims in the street, when its black rulers break in the void and stenograph the azure of the sky."[14]

Longer histories of information transmission have tended to focus on modern media's exclusion of the visual in favor of increasingly virtual, dematerialized forms of communication. Meanwhile, visual culture in our digital age has been susceptible to a tendentious downplaying of its materiality, masking its dependence on exploitative labor and its environmental harm.[15] Countering these tendencies, there is another history of telegraphy—that of a visual and material phenomenon that operated in tandem with, and at times against, more exclusively artistic objects. An increasingly eclectic range of images and objects produced in diverse media for different audiences articulated responses to life under this new technological regime, and more importantly, played a role in producing the regime itself.

This relationship was not straightforward, however, and sometimes it was antagonistic. In a departure from art's familiar frontal address, the telegraph made margins and correlates meaningful in new ways, at times challenging received ideas about art's communicative function. This book traces how the new technology altered not only the transmission of state and military information but the way those on the ground perceived it, or failed to do so, and it examines how both the technology and its reception were registered, anticipated, historicized, or rejected in the visual and material culture of the nineteenth century. It asks that we shift our historical gaze from the streets to the sky, from faces and facades to rooftops and horizon lines, and from the coherence of the whole to nagging details, to consider how people living through a time of immense political and social turbulence processed a—sometimes aloof, sometimes intimate—relationship with the politics, aesthetics, and affective capacities of infrastructure.

Jerking and Fugling

> It is possible that I will be called and questioned about the telegraph machine. I beg you to tell me at once what I should answer and to send me instructions on this subject. . . . The town is very quiet. Today is the anniversary of the death of the Tyrant.[16]

But perhaps we should begin by asking the same question as the Belleville policeman. Or Baron, here, in January 1794, writing in nervous anticipation of a visit to Lille by the fearsome duo of Philippe-François-Joseph Le Bas and Louis-Antoine de Saint-Just, sent by the Committee of Public Safety to reorganize the armies of the north. What *was* the telegraph? (And what was it not?) The system that interests me was developed by former priest Claude Chappe and his four brothers in the early 1790s, and brought into official service just after Thermidor, in the fraught summer of 1794. It consisted of a network of relays, each one comprising three metal "wings," "sails," or "arms," joined together in an articulated T shape atop a static pole, and manipulated via pulleys by an operator, or *stationnaire*, housed below (fig. 1.4). Chappe described his stationnaires as "a sedentary army," although when it came to the act of transmission, they were anything but, the movements of their hands on the control assembly mirrored by those of the telegraph's signaling limbs above.[17]

Seen from outside, the telegraph was unavoidably anthropomorphic—Thomas Carlyle later described its elbow joints "jerking and fugling in the air"—while the operation of the machine policed the bodies of those who animated it (an instruction manual for stationnaires published in 1827 urged that "a good stationnaire should never work in fits and starts. He must have soft movements").[18] The telegraph's internal and external mechanisms—its gears and handles, pivots and telescopes—align with Helmut Müller-Sievers's argument for the centrality of cylindrical objects and processes in the nineteenth century, and for the vital role played by mechanisms of rotation and translation in the epistemology of the age.[19] Müller-Sievers observes that electrical telegraphy was inseparable from the kinematics of the cylinder—rolled wires and cables, railway lines and steamships, and the rotating drums of telegraphic transmitters and receivers—but in fact similar dynamics were already present in the movements of its semaphoric precursor.[20] "On the whole, is it not, O Reader," Carlyle asked, linking the telegraph to a more terrifying revolutionary contraption that also exploited rapid shifts between the seen and the unseen, "one of the strangest Flame-Pictures that ever painted itself; flaming off there, on its ground of Guillotine-black?"[21]

The external part of the machine comprised a long cross beam, or regulator, set atop a fixed mast, rotatable through 180 degrees around a central axle. At the end of each arm a smaller,

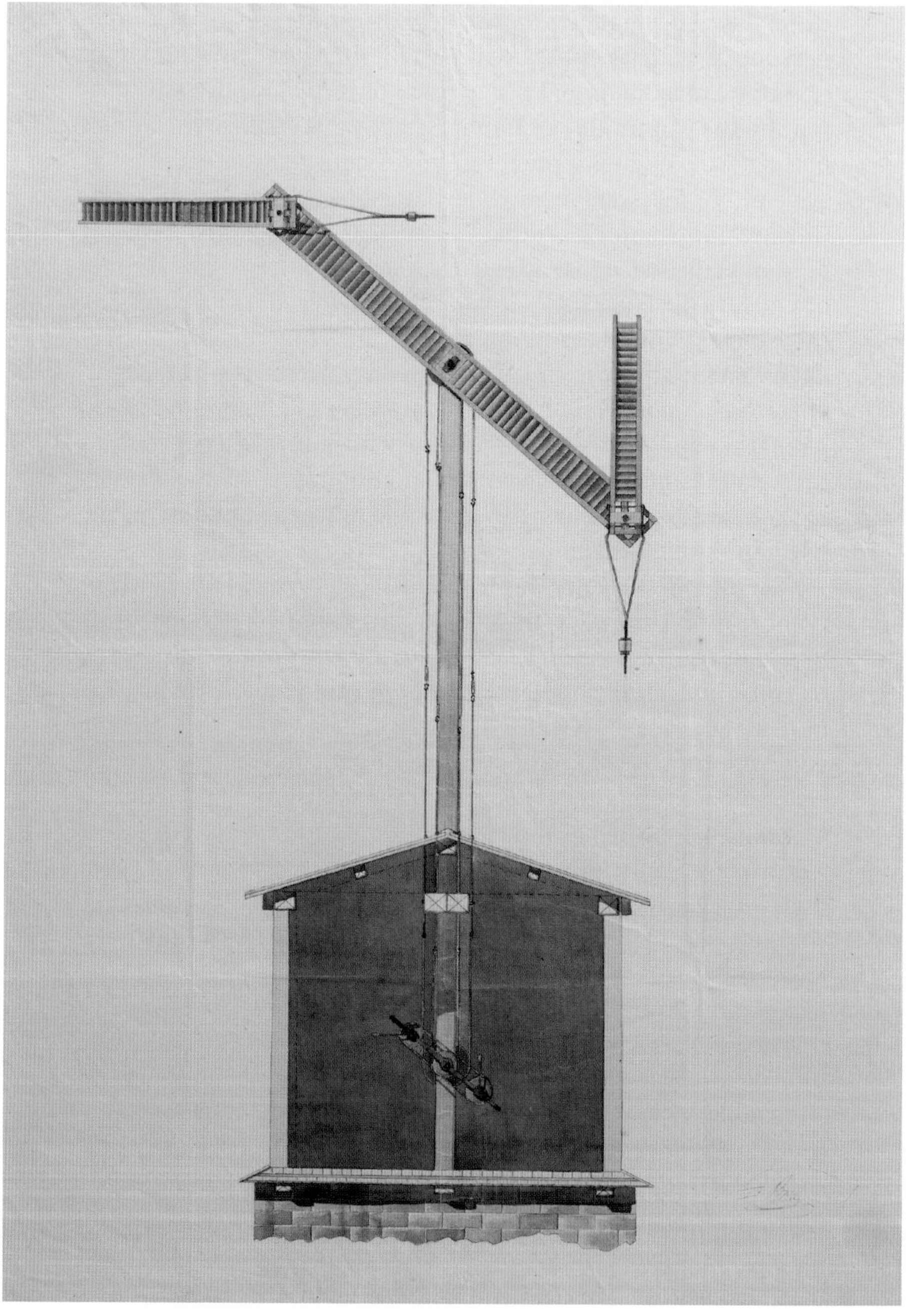

Figure 1.4
Plan du télégraphe de Chappe, early nineteenth century, ink, watercolor, and gouache on paper, 66 × 51 cm. Musée de La Poste, Paris.

counterbalanced pivoting arm, or indicator, was fixed.[22] This mobile, jointed mechanism could be moved into static positions, each one precoded with a letter of the alphabet, or more commonly, a word, phrase, or standard instruction (end of message, suspension due to fog, and so on). Providing for the first time a standardized apparatus, a stable and largely reliable infrastructure, and more particularly, a widely accepted code, after many false starts the telegraph became a uniquely far-reaching method to send messages at previously unheard-of velocities. By 1840, in ideal conditions, a message could be sent the 250 kilometers from Lille to Paris in two minutes (twenty-two stations), and the 600 kilometers from Brest in eight, passing through fifty-four stations. The code, which was refined on several occasions, was in many ways Chappe's primary innovation, making his telegraph much more successful than the many similar projects being developed at the same time.[23] Each indicator could be rotated into seven different positions, while the regulator could be moved between horizontal and vertical. A deceptively simple mechanism, this made for ninety-eight initial combinations, although this range was expanded to include more positions (fig. 1.5). By transmitting paired symbols, it became possible to expand the code to accommodate 8,464 words, phrases, and control signals.

Telegraph apparatuses were placed on tall buildings or on purpose-built structures, and their signals were viewed through a long telescope by another operator at the next station, approximately six to eight miles away. Telescopes were set in fixed boxes to ensure consistency and correct orientation, and they pointed in both directions along the line. While English optical instruments were used at first, the telescopes were soon manufactured in France (fig. 1.6). A civil engineer, Elisabeth-Théodore de La Poix de Fréminville, proposed that lenses be made of high-quality crystal ground down from captured English luxury goods, a reverse engineering of aesthetic objects into technical ones that cached a powerful symbolic appropriation in a practical solution.[24] Later, Fréminville supplied scientific lenses and optical instruments for use in the French occupation of Egypt.[25] Stationnaires initially operated in groups of two, one to handle the apparatus, and the other to keep an eye on the horizon. When the lookout noted the distant symbols via the telescope, he passed them to his colleague, who replicated these characters and passed the secret messages along the channel until they reached their final destination,

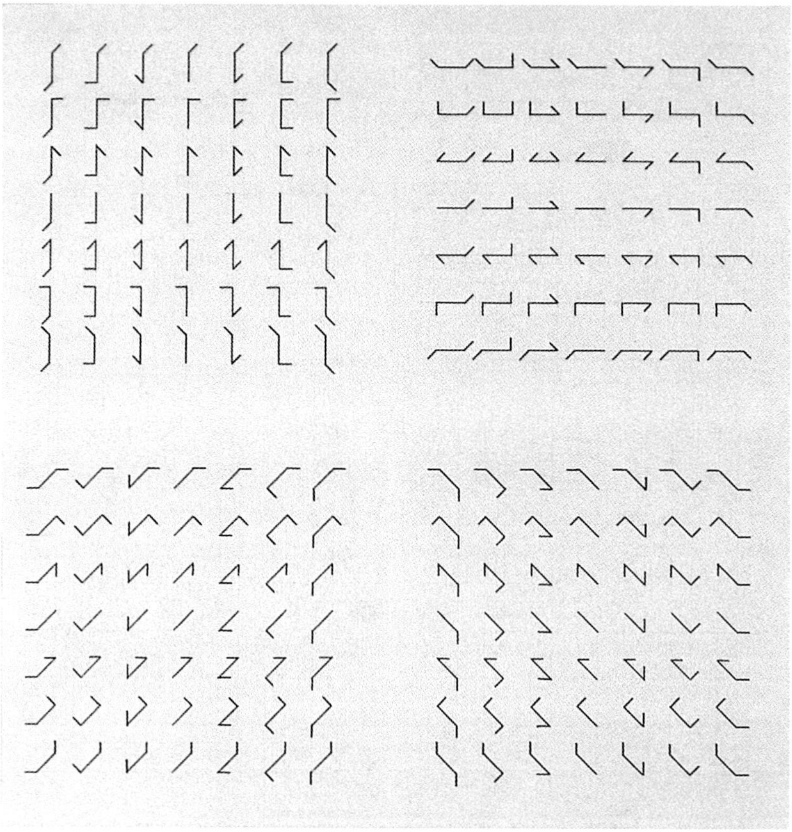

Figure 1.5
Charles Dupuis [engraver], *Graphic Showing the 196 Characters That Can Be Transmitted by the Optical Telegraph. Each of the four sections corresponds to the four possible positions of the regulator,* after 1793, engraving, 28.4 × 27.4 cm. Museumsstiftung Post und Telekommunikation.

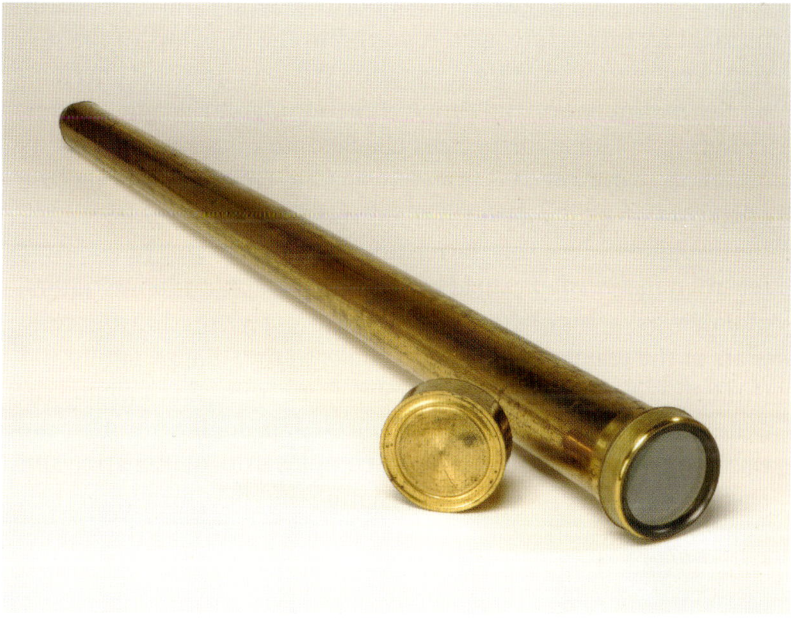

Figure 1.6
Lunette d'approche pour le télégraphe optique système Chappe, 1792, brass and glass, 96 × 7.7 cm.
Musée des Arts et Métiers, Paris.

where they were decoded and transcribed. Beginning in 1809, the paired stationnaires alternated shifts, changing over at midday, meaning that for most of the time they worked alone (fig. 1.7).

To ensure security, stationnaires did not have access to the code, and they were innocent as to the content of the information they passed back and forth. A shorthand language, giving the degree and direction of elevation of the regulator and indicators, oriented to either *ciel* or *terre*, allowed operators to describe the signals verbally to one another. Using this minimal vocabulary, stationnaires only required a combination of three to five words and numbers to share a signal.[26] A hierarchy of telegraphic administrators was arranged to guarantee smooth operation and trustworthy messaging. Inspectors, who reported to the directors of the system, were responsible for paying the stationnaires and checking on the maintenance and proper functioning of the apparatuses. Traveling by horse, inspectors checked nonsequential outposts on a line at unexpected intervals, having covertly surveyed the operation of a telegraph from a distance. They would examine the machine and consult the record of transmissions. The code was recorded in *vocabulaires* held by senior telegraphic administrators, or *directeurs*, across France. The directors, recruited in the early years of the telegraph via familial links to the Chappe family and through trusted recommendations, were from 1824 former students of the École Polytechnique selected through a competition; they were responsible for the more complex work of preparing and encoding efficient messages, translating the code, managing their immediate network, and when necessary, modifying the *vocabulaires*. Decoded messages were often not delivered verbatim to a wider audience but were finessed for public consumption.

Notwithstanding multiple experiments with illuminated machines, signals could only be transmitted feasibly during daylight hours, and the apparatus was painted dark colors—usually black—to increase visibility from a distance. The signaling parts of the machine were louvered to reduce wind resistance, and smoke from the stoves installed to keep stationnaires warm was evacuated to the side of the building to avoid interference with transmissions. The air was the medium in which the optical telegraph operated and bad weather its greatest obstacle.[27]

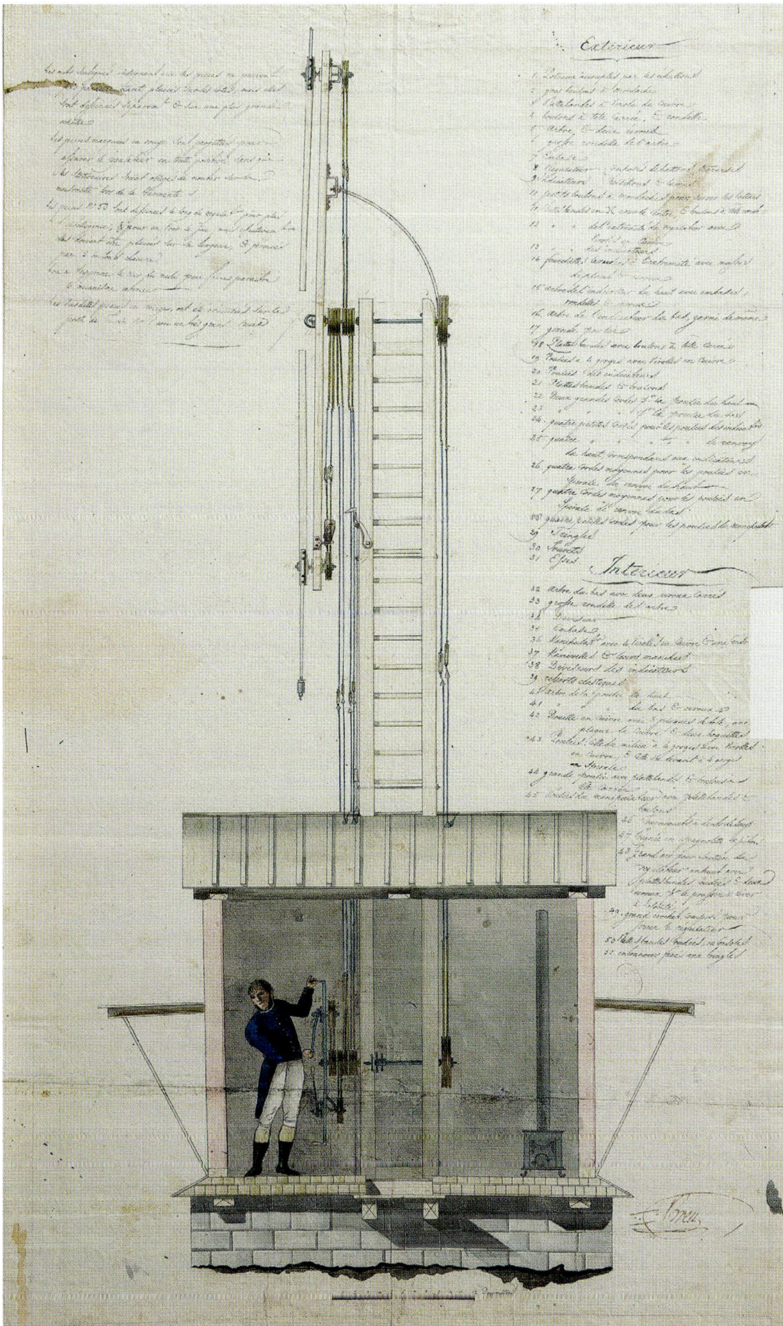

Figure 1.7
Stationnaire dans un tour de télégraphie Chappe, 1809, ink and watercolor drawing on paper, 65 × 35 cm. Musée de La Poste, Paris.

In France in the first half of the nineteenth century, this innovative but deceptively simple technology achieved an unprecedented range. The optical telegraph's oblivion was abrupt. In 1868, popular science writer Louis Figuier remarked on the swiftness of its being forgotten, and well before then references in art and literature to obsolete or decaying telegraph towers, typically isolated on romantic outcrops, had coalesced as a genre in their own right; celebrating the optical telegraph's extinction paradoxically kept its memory alive (fig. 1.8).[28] But in its day, it seemed to be everywhere. For Lewis Mumford, Chappe's telegraph was, following the cannon, "the first of the modern space-annihilating devices."[29] Between the revolutions of 1789 and 1848, the optical telegraphic network expanded to reach 5,000 kilometers across France, the five-fingered spread of its lines to each point of the nation echoing the waving hand from which the machine derived its "natural" authority (fig. 1.9).[30] While there were certainly blind spots (a big chunk of central France, for instance), these areas too were transformed by the telegraph's rapid dissemination of news, and if necessary they could contribute to the system by sending messages via other means to the nearest telegraphic outpost, from where they could be accelerated.

The network eventually extended into Germany, the Netherlands, and northern Italy, as far as Venice. It inspired variations on the system worldwide: in England, Poland, Russia, Sweden, Norway, Finland, and Denmark, and across the Americas and in northern and South Africa. James Gillray's 1795 caricature of Whig leader Charles James Fox as a colossal telegraph traitorously beckoning an invading revolutionary horde indicates the network's international impact, its shadowy alignment with Republican politics, and its terrifying efficiency as a component of the French war machine (fig. 1.10).[31] In Gillray's etching, the telegraph points to the spires of London and the dome of St Paul's, but it has become unmoored from conventional architectural supports, standing with its own feet on the ground as an autonomous android enemy. The telegraph's centrifugal expansion did not mean that its channels of communication, winding inexorably out of Paris, moved only in one direction. It prompted new fears about what kind of information—in this case, what kind of invitation—might be sent back.

Visible from the ground, the telegraph's opaque gestures transformed how public space was experienced and represented. Advertising the specter of the unknown, the awareness of imminent

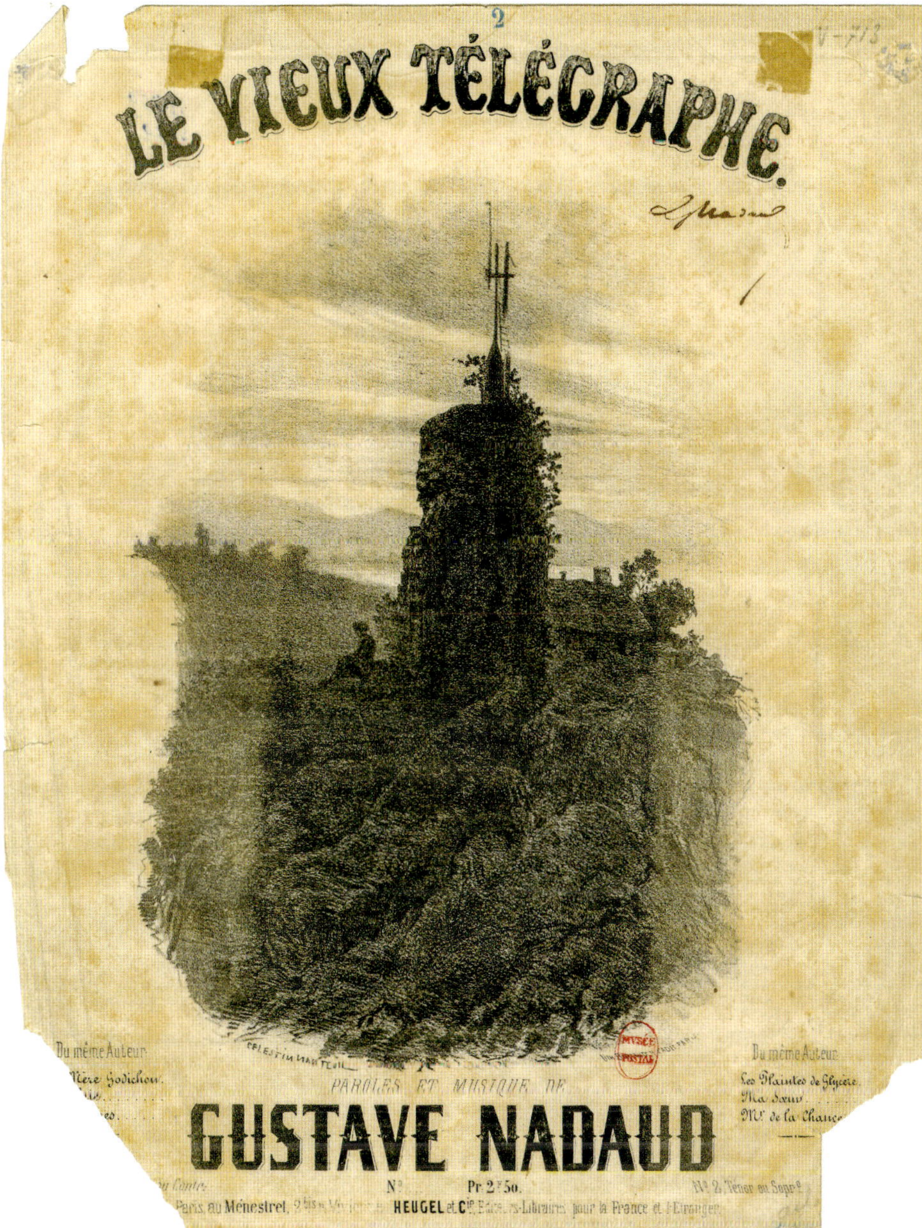

Figure 1.8
Bertauts [lithographer], after Célestin Nanteuil-Leboeuf [draftsman], *Le Vieux Télégraphe. Paroles et musique de Gustave Nadaud* (Paris: Ménestrel, Heugel et Cie, c. 1878), lithograph, 33 × 25 cm. Musée de La Poste, Paris.

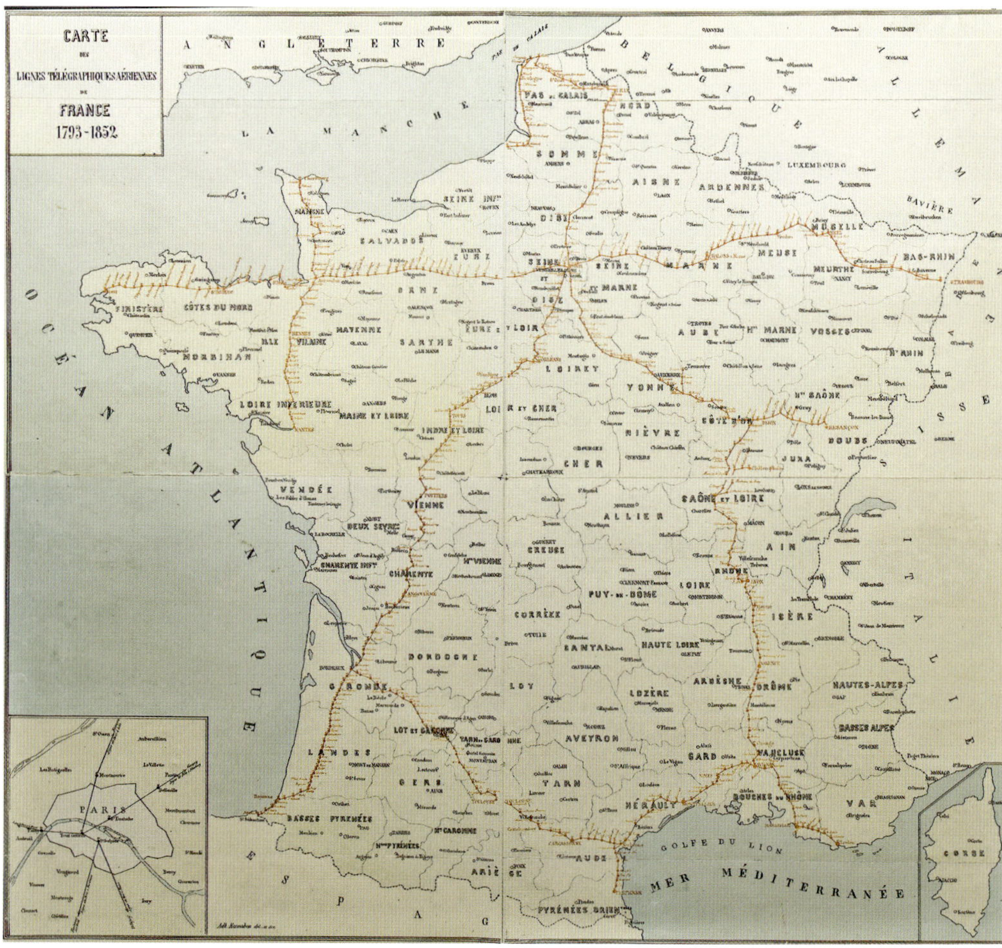

Figure 1.9
Ernest Jacquez [editor] and Adhémar Kermabon [designer], *Atlas des lignes télégraphiques aériennes, construites en France de 1793 à 1853* (Paris, 1892), 75 × 85 cm. Musée de La Poste, Paris.

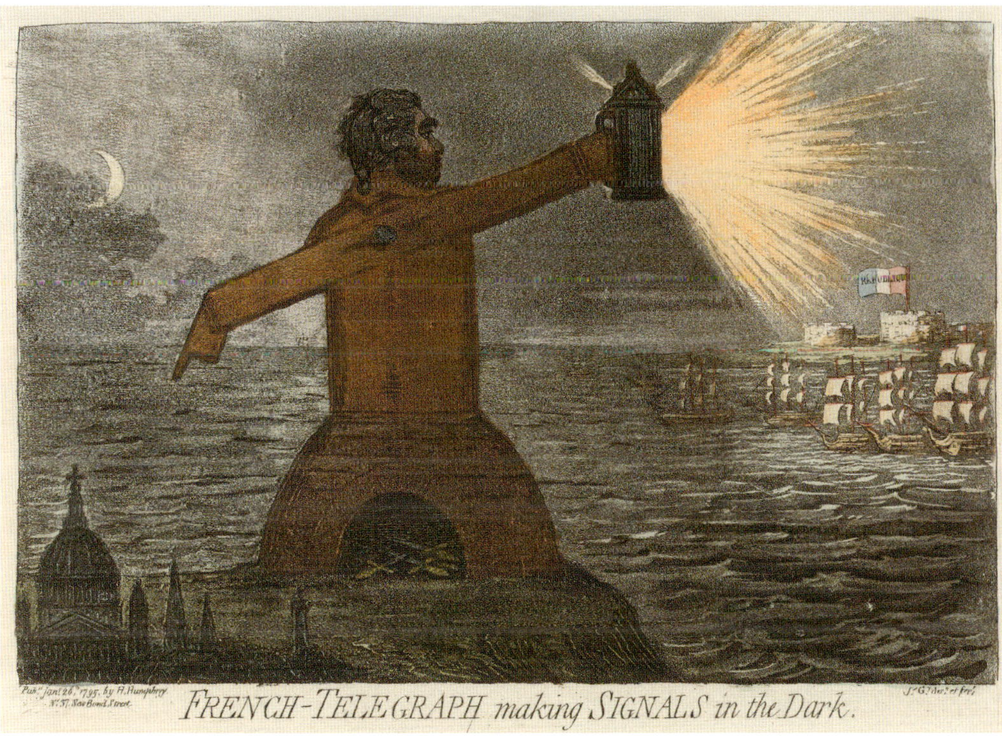

Figure 1.10
James Gillray, *French-Telegraph Making Signals in the Dark*, 1795, hand-colored etching and aquatint, 22 × 33.5 cm. Yale Center for British Art, Paul Mellon Collection.

news whose content—good or bad—was not yet revealed, its presence suffused routine environments with the charge of an uncharted future and a new alertness to the nation beyond. Telegraphs appeared on many of the most prominent buildings in Paris—among other places, at Montmartre, Saint-Sulpice, the hôtel de la Marine, and for several years on top of the Louvre—as well as on church spires, modified monuments, and specially constructed towers across France (fig. 1.11). The network expanded to include over five hundred telegraph stations. In time, Chappe could effectively put a telegraph wherever he wanted. Trees that blocked the view between relays were felled, and familiar buildings were rendered strange, even grotesque.

The case of Notre-Dame-de-l'Épine, near Châlons-en-Champagne, Marne, is a noteworthy example: in 1798 Chappe's workers lopped off one of the two spires of this masterpiece of Gothic architecture to install a telegraph on the Paris-Strasbourg line. In 1838, Victor Hugo described its fifteenth-century tower, "worked like lace," being "accosted" by the neighboring telegraph, which the surviving tower considered "very disdainfully, like the great lady she is."[32] The symbolic connection between such decapitations and that of the king whose execution in 1793 anticipated by months the arrival of the telegraph was plain to see; as David Charles has described, Hugo rhymed Notre-Dame-de-l'Épine with "guillotine."[33] The spire of the basilica was rebuilt at the end of the Second Empire but can be seen in amputated form in a lithograph from Baron Taylor's *Voyages pittoresques et romantiques dans l'ancienne France*, where despite the antiquarian context for the image's publication, no attempt is made to conceal the severity of this intervention (fig. 1.12). If telegraphy was, at one level, presented in *Voyages pittoresques* as an agent of unpardonable vandalism (the text is unequivocal on this issue), the plate representing Notre-Dame-de-l'Épine's truncated facade also expresses, and is noticeably inquisitive about, the tension between different kinds of knowledge transmission and the era's rapid media change.

The broader contours of this development were summarized most concisely by Hugo in *Notre Dame de Paris*, published in 1831 at the telegraph's peak: "this will kill that. The book will kill the edifice."[34] "The tower," Hugo added, "will crumble." As a serial print publication comprising twenty-four volumes issued over fifty-eight years, *Voyages pittoresques*' documentation of the national patrimony was itself implicated in these processes; yet, rather than

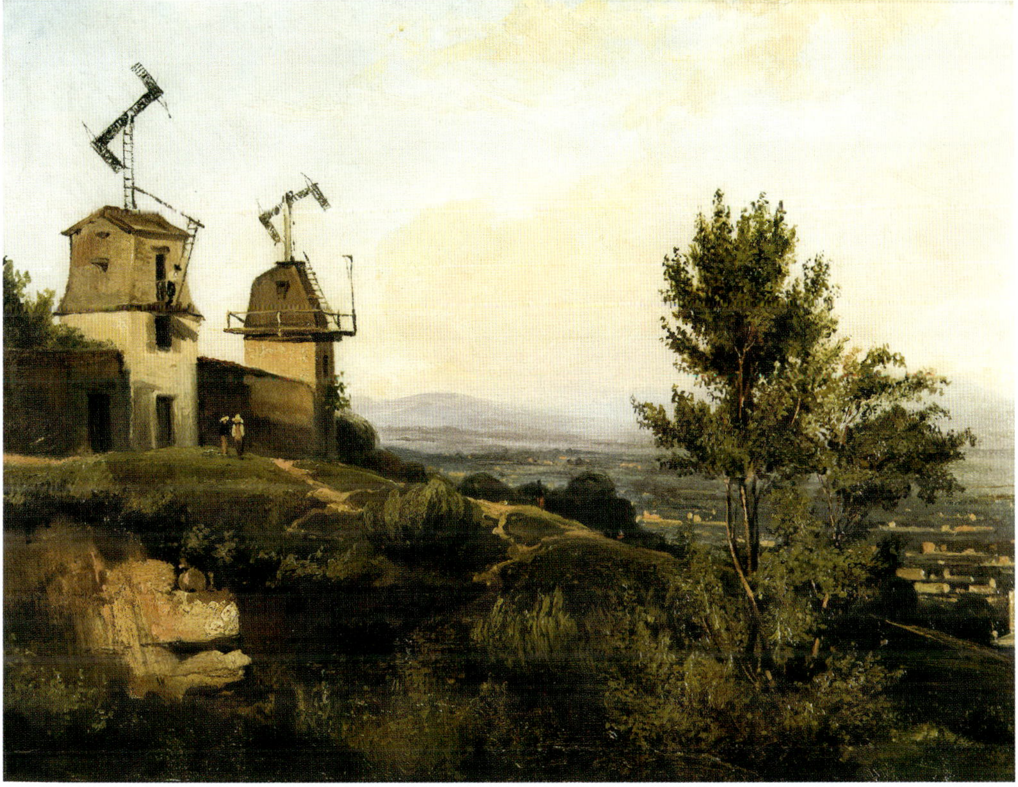

Figure 1.11
Chappe Telegraphs at St Just (Near Lyon), c. 1840, oil on paper, 26 × 34 cm.
Musée de La Poste, Paris.

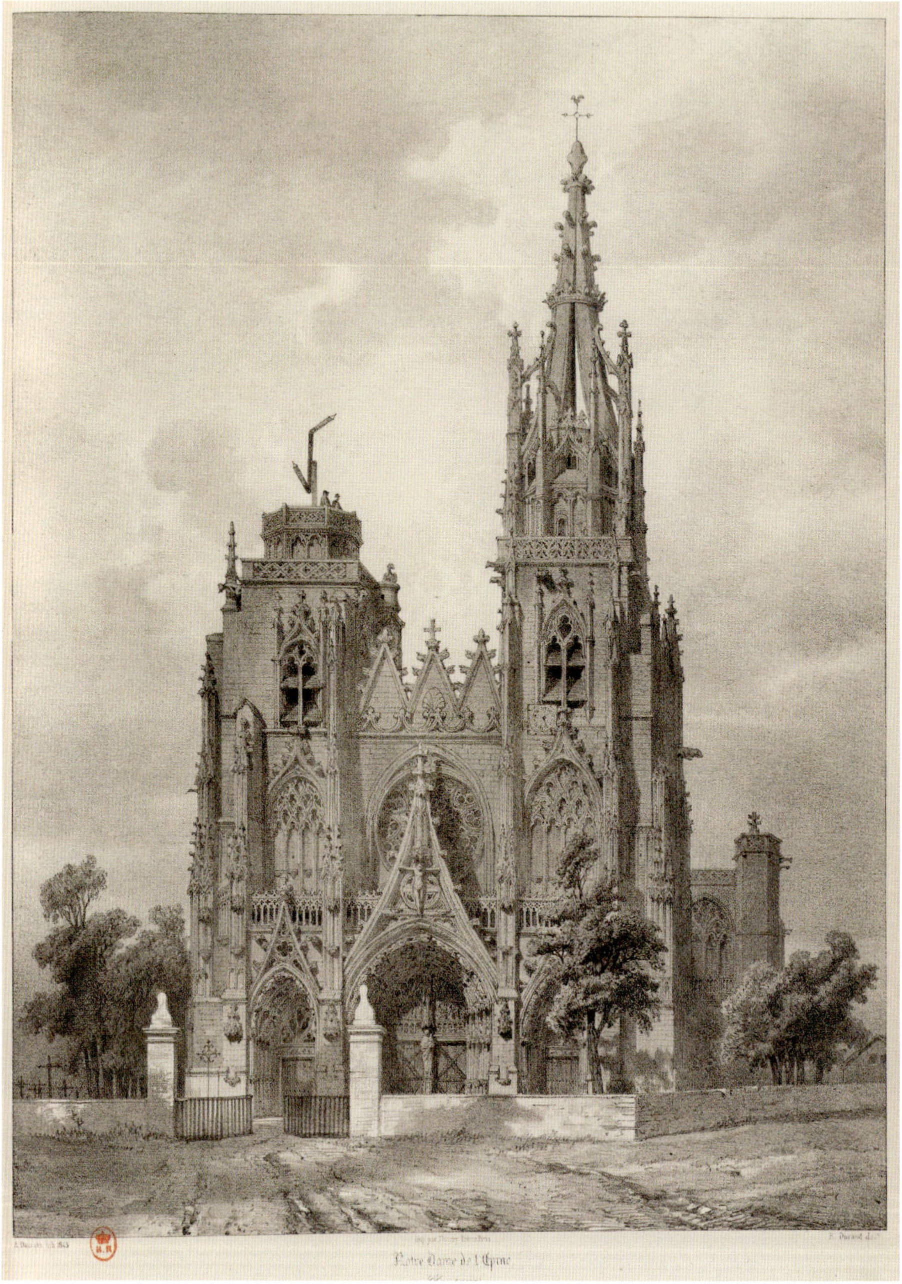

Figure 1.12
Adrien Dauzats [lithographer] and H. Durand [draftsman], "Notre-Dame de l'Épine, Champagne" (1845), from Charles Nodier, Isidore Justin Séverin Taylor, and Alphonse de Cailleux et al., *Voyages pittoresques et romantiques dans l'ancienne France. Champagne. Atlas vol. 1* (Paris: A.-F. Lemaître, 1857). Bibliothèque nationale de France, Paris

replicating Hugo's model of one medium replacing, or "killing," the one that preceded it, *Voyages pittoresques* attempted the juggling act of preserving the past while at the same time embodying the novel reproductive technologies that were displacing older truths. Despite the series' smooth aestheticization of these historical tensions, the telegraph's shocking presence in this plate made them impossible to ignore. The encoded words of the telegraph, dutifully transcribed in logs, represent a bureaucratic version of the book that pushed beyond the limits of the codex. Augmenting further this temporal and material complexity, most such machines had been withdrawn from use by the time this volume of *Voyages pittoresques* was published, after which point the crumbled towers prophesied by Hugo came to stand in the popular imagination for the once omnipresent telegraph.[35]

Synchronize Clocks

Chappe's system replaced—and overlapped with—earlier experiments with a shutter telegraph, based on blocking out sections of light within a sectioned framework, and a telegraph in which a clock face was encoded with the alphabet. The horological sensibility evident in these tentative steps pervaded the practice of telegraphy, as well as its representation, long after Chappe's experimental systems were superseded. Synchronicity was always the goal, and often the method. The famous clockmaker Abraham-Louis Breguet collaborated with Chappe on the design for the optical telegraph, and his grandson Louis was partly responsible for one of its electromechanical successors, the Foy-Breguet telegraph, which continued to use the Chappe code; its tiny arms, animated by electricity, were scrutinized from close up by a lone operator rather than at a distance through a telescope.[36]

The French Revolution had transformed concepts of historical time, accelerating the present and forging a profound breach with the past while at the same time dredging it for antique moral archetypes.[37] As Colin Jones notes, with the Revolution, "Parisians had become keenly aware of time. . . . They politely apologized for not giving an exact time if they had forgotten their timepiece."[38] Breguet's intricate clockwork marvels of the 1790s, his *tourbillon* (whirlwind) mechanism of 1795, or his anti-shock devices seem saturated with postrevolutionary sentiment; they are practical meditations on how to comprehend and regulate time in the wake of the storm. During the Restoration there was a fashion for timepieces incorporating miniature telegraphs, the arms of which moved on the hour as a crystal torrent fell from the front (fig. 1.13). The precedent for these had been set by a much more elaborate marble and bronze telegraph clock, made by Jean-André Lepaute for Napoleon's bedroom at the palais Rohan in Strasbourg. The stylized "telegraph" at the top of Lepaute's clock is composed of an articulated bow and arrow, operated by a miniature Cupid, aided by Mercury, and the design includes references to Napoleon's correspondence with Joséphine, which at times was conducted by telegraph. From the bedroom window at the palais Rohan, the emperor could see the real telegraph turning above the transept of Strasbourg cathedral.

Later, after midcentury, the telegraph appeared as a three-dimensional component of a painted *tableau mécanique* alongside other symbols of synchronicity such as trains and a clocktower, all of which would be set into motion at a given time (fig. 1.14). The ostensible modernity of the scene jars with the painting's fussy genre stylings and with the antiquated form of the *tableau mécanique*, an elaborate and expensive philosophical toy of the kind

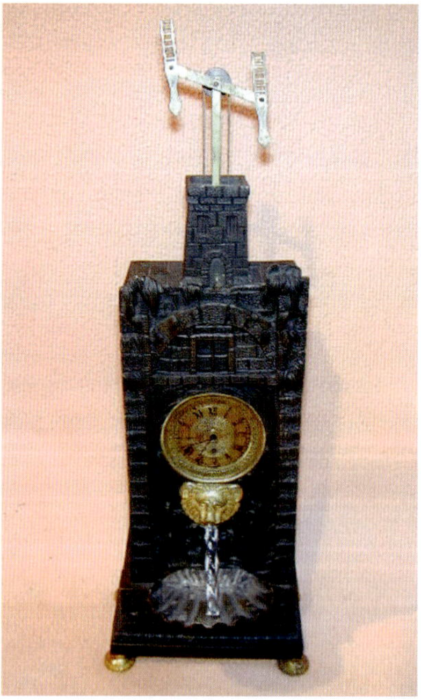

Figure 1.13
Tavernier-Cadet, *Telegraph Clock*, c. 1820, patinated bronze and glass, 40 × 14 × 8 cm. Musée de La Poste, Paris.

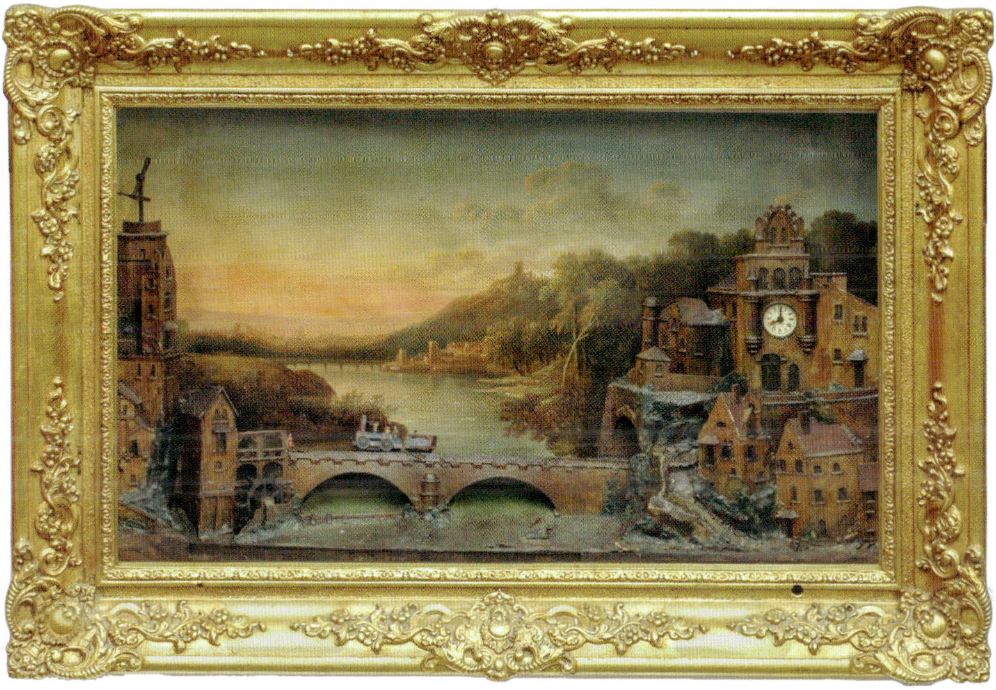

Figure 1.14
Anton Olbrich [clockmaker] and anonymous painter, *Tableau-horloge animé en relief*, third quarter of nineteenth century, oil on wood and iron, cardboard, metal, 70 × 103 cm. Musée de La Poste, Paris.

that had largely vanished with the Revolution.[39] The Chappe telegraph, by the time this railway automaton was made, also existed only in vestigial form, but its presence here communicated a complex responsiveness to the past that had accompanied it from its earliest days. The appearance of Chappe's machine in the *tableau* also gave tangible form to contemporary electrical iterations that were less visible. By the middle of the nineteenth century, electrical telegraphy began to play a key role in railway signaling, and by the 1880s it was implicated closely in the drive to standardize global time zones (with, again, rail travel as a key imperative; the need to regulate timetables across national borders).[40]

The 1790s and early 1800s had witnessed numerous ambitious projects aimed at transmitting revolutionary ideas, from printed images of diverse quality and audience to modifications in clothing and speech.[41] The telegraph may be considered among these initiatives; it was one of many technologies and protocols invented or modified during the revolutionary decade that promised to transform the everyday experience of space and time, from developments in geographic science to the bureaucratic reorganization of the *départements*, and from the standardization of the meter to the Republican Calendar.[42] While these were intended to homogenize units of measurement or centralize political and administrative units, they often produced forms of doubling, even double consciousness, as citizens retained former behaviors and beliefs while outwardly living within the new structures. Time systems, especially, overlay one another to vertiginous effect. Year IX could, several years after the new calendar's introduction, still be experienced as 1800, and the valence of which calendar one used, when, and in which combination, was politically freighted.

Among these novel systems, telegraphy was accorded a unique, deterministic place, for especially in the phase immediately following its introduction, it promised a universal means of transparent communication that chimed with revolutionary ideals, giving form to principles that were otherwise abstract. The Revolution's widespread rethinking of time was ultimately realized more concretely here than it ever was in the decimalization of time itself—telegraphs were, it transpired, far more potent and commonly encountered time machines than ten-hour clocks. Producing a fantasy image of simultaneity (the idea of a message sent from one place being received in almost synchronous time), telegraphy appealed to a political future in which the often-irresolvable fractures of the Revolution might be smoothly overcome. However, in its attempt to clarify and accelerate knowledge, the telegraph also threatened to produce dizzying breakdowns, lags, asymmetries, and reversals in signification, crumpling the stability of established timeframes.

Perhaps to offset such risks, legislator and educationalist Joseph Lakanal, in a crucial 1794 speech to the National Convention advocating for the telegraph's use, pressed home its consequence and singularity. For Lakanal, the two greatest inventions of the eighteenth century were the telegraph and the hot-air balloon.[43] Both were French, or at least they appeared to be (in fact, the telegraph emerged from a lengthy period of international experimentation: the likes of Robert Hooke and Richard Lovell Edgeworth in Britain and Ireland or, later, Abraham Niclas Edelcrantz in Sweden), and between them they promised to command vertical and horizontal space.[44] The development of the network was far from seamless or straightforward, as Chappe often struggled for political and financial support, yet the telegraph's utility and world-historical significance were acknowledged early on, and in Vendémiaire Year VII (September to October 1798), when the Institut National offered a description of all the occupations that might be counted among the "arts," "télégraphie" was allotted a place among them.[45] Lakanal's evangelism for the telegraph was shared by many of his contemporaries, who considered it a distinctively revolutionary invention whose impact would be comparable to that of printing, the compass, and gunpowder.[46] As a print by Étienne-Claude Voysard after Claude-Louis Desrais shows, the telegraph was swiftly assimilated into a narrative of French technological superiority and colonial mastery that was presented as deriving from natural laws—the lightning striking the conductor adjacent to the telegraph in the left-hand background of this bizarrely crowded scene seems to suggest the machine's animation by supra-human energy, as well as forecasting its electric future (fig. 1.15).

Desrais's image, which depicts several loincloth-clad Black figures in a landscape populated by the Montgolfiers' balloon and hydraulic ram, Chappe's telegraph, and the parachute pioneered by Louis-Sébastien Lenormand and subsequently André-Jacques Garnerin, offers a troubling vision of telegraphy's potential. These figures bear witness to the industrious performance of Enlightenment knowledge but play no role in it. Gesturing in wonder at

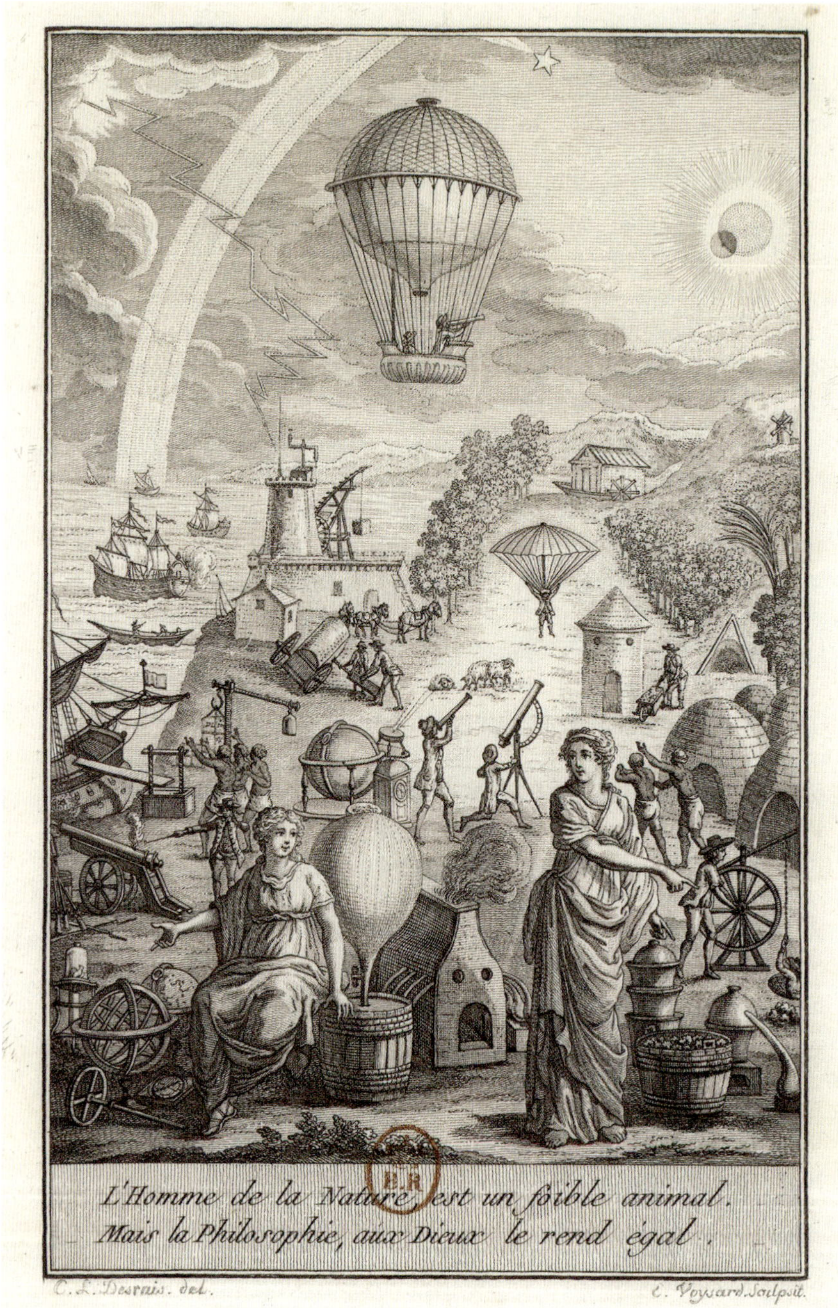

Figure 1.15
Claude-Louis Desrais [draftsman] and Étienne-Claude Voysard [engraver], *L'Homme de nature est un faible animal. Mais la Philosophie, aux Dieux le rend égal*, 1799, etching and engraving, 15.7 × 9.7 cm. Bibliothèque nationale de France, Paris.

the technical marvels that appear before them, with their arms raised, oathlike, they are depicted as unaffected counterparts to the telegraph's limbs and the slanted telescopes hoisted by white Europeans: humans, but also machines. At the same time, they are ciphers for the "man of nature," the "weak and feeble animal" of the print's legend, which it is claimed only enlightened philosophy has the capacity to improve. Muddied by allegory, the scene—set in a space that is both France and not-France—downplays the horrors of colonialism and enslavement. The plank pitched on the ship is presented as the destiny of these men, toward which they move inexorably and seemingly willingly.

The French optical telegraph may have allowed messages to be sent at a speed "equal to that of thought," but the terrain that thought encompassed had its limits.[47] As Simone Müller has observed, in writing on telegraphy there is a tacit assumption that "telegraphy=globalization."[48] Here, the distinction between the Chappe telegraph and the electrical cables that later spanned continents is sharp. While the French optical system eventually spread beyond France's borders, it was primarily confined to the hexagon. The optical telegraph was undoubtedly vital to France's conception of its place in the world, and as such it was intrinsic to a broader colonial frame within which telegraphy came to play an important role. However, this was not achieved by a hardware that linked the globe (at least not at first). Its colonial ramifications found more complex expression instead in representations such as Desrais's print, a reproductive image that was designed for dispersal and synchronous comprehension, but whose complacent triumphalism and legitimization of slavery could, as revolutions in France and the Caribbean advanced, no longer be guaranteed a universally positive reception. The "Frenchness" of the telegraph was in tension with another set of possibilities: those aligned with the telegraph's potential as a means of universal communication among global citizens. Telegraphy transcended borders in theory, even as it affirmed them in practice.

Müller rightly points to the Atlantic telegraph's relative exclusivity—the expense of sending a telegram—as a caution against overreading its democratic potential. In the context of the earlier optical telegraph, however, no such access was possible.

Although other applications, particularly the transmission of news for print publication, were proposed from the moment of its invention, repeated attempts to turn the French telegraph toward commercial communication were resisted, and it was used almost exclusively for the transmission of state and military information, operating in what Eric Wertheimer calls telegraphy's "executive" mode.[49] As *Le Charivari* observed in 1840, granting the machine human agency, "the telegraph does not correct himself; this is all the more unfortunate because, given his high position, he is more likely to set a bad example to society."[50] The device's lofty taciturnity sent as powerful a message as its loquaciousness, for "when the telegraph needs to speak, when thousands of families, for the sake of their children, anxiously fix their gaze on his great arms, when the attention of France is suspended on his strings, the scoundrel will not move, and we could not wring the smallest sentence from him. He remains in complete silence."[51]

During the First Empire, the only other kind of telegraphic message permitted was the dissemination of winning lottery numbers, a use grounded in fears about illicit print; it was hoped that telegraphic transmission would prevent the delay that allowed counterfeit tickets to be produced (it was the emperor's nephew, as Napoléon III, who would in 1851 begin the process of opening electrical lines to public use). Corruptions of the system were possible, most famously in the case of the brothers François and Joseph Blanc, bankers employed at the Bordeaux stock exchange, who for two years from 1834 carried out a successful scam that allowed them to receive early knowledge of the movements of the Paris Bourse. A co-conspirator in Paris passed on news of the Paris stock exchange to a telegraph operator in Tours, on the Bordeaux line, who added this information to the message as it moved forward. Making clever strategic use of the signal for transmission errors, which "deleted" the previous signal, the brothers hid a code of their own—based on a pattern of mistakes—in the already encoded messages, which arrived unaffected. The Blancs were ultimately released, for despite governmental monopoly, there was at the time no clear law making such interventions illegal.[52] Nonetheless, interference of this kind helped erode faith in the optical telegraph's seemingly unassailable authority. As a contemporary noted, "If the Egyptians had understood the hieroglyphics invented by their priests, they would have lost all religion. Well! The telegraph has found its Champollion."[53]

The highly regulated state control of the telegraph made a considerable difference in how people responded to the machine: as something to be watched rather than used. For this reason, too, anachronistic descriptions of the telegraph as a "nineteenth-century internet" are wide of the mark in discussions of the optical network—not just for their presentism, but for the fact that the Chappe network was not truly rhizomatic in its essential structure, as a linear system, or user-oriented in practice.[54] Despite a shared origin in military hardware, in most ways the optical telegraph couldn't be further from how we use the internet now. The Chappe telegraph was a closed system, even to those who operated it, although anyone could see it at work. Forcing reflection on the nation and its borders, the telegraph participated in broader processes of centralization that drew all information to Paris (a specific control signal gave priority to messages passing to and from the capital). Yet by turning things inward, the telegraph offered unexpected opportunities for considering aspects of modern selfhood, at a time when such a concept—the cataloguing of its potential beneficiaries as well as the classification of those who were denied access to it—was subject to radical revision and reformulation.[55]

In a Narrow Circle Trapped by Glass

One of the signal characteristics of nineteenth-century modernity, as it seems to us now and as it appeared to contemporaries, was a reduction in time and space between people and things. New technologies—the railroad, photography, the telegraph—made it ever easier to connect people across distance. Meanwhile, mechanical reproduction separated images from their origins and brought them to meet their viewers directly. The world that resulted was at once more knowable and, by virtue of an increase in information, simultaneously harder to grasp. Much of this is a familiar story, in which an inexorable quickening, often associated with vexed notions of technical or political "progress," takes precedence. However, in the long, politically tumultuous afterlife of the Revolution, the effects of these temporal fractures and recursions found complex expression at the interface of art and science, passing into everyday life in unexpected ways and accruing histories of their own.

Although attempting to disentangle telegraphy's temporal and spatial dynamics is a fruitless task, in many ways the telegraph's temporal dimension takes priority in my account. "Farwriter" may have edged out "fast-writer" as the machine's name, but only because it described a shift from transcription *within* a homogeneous time and, presumably, space (that of the *logographe* speedily recording events in the legislative chamber, for instance) to inscription and transmission that *traversed* time as well as space. Without exception, the telegraph's velocity was crucial to its purpose. Yet by performing the work of homogenizing temporal and spatial intervals, the "distance" of the new apparatus also called attention to disparities between divergent points and different histories; their unevenness and their incommensurability.

Already by 1798, when politician, poet, and art critic Pierre-Jean-Baptiste Chaussard referred to the "winged symbol" of the optical telegraph as a "bearer of the fates," his allusion was not solely to the telegraph's ability to transmit important information, annexed to familiar temporal mechanisms ("see, in a narrow circle trapped by glass, time marching on").[56] Rather, Chaussard suggested that telegraphy might also drive the progress of history. Like Walter Benjamin's "angel of history," the telegraph moved into the future with a face to the debris of the past, its arms similarly outstretched in the wind.[57] Yet the optical telegraph has often been understood as contributing to an acceleration of life characteristic of the nation-state, of modernity, and of a defining technological shift; what media theorist Friedrich Kittler, comparing the Chappe telegraph and the daguerreotype, described as the "fundamental trend of modern media technologies to replace static values with dynamic values and to replace steadfastness with speed," or what Paul Virilio characterized more locally as "the increase in the penetration speed of means of transport and communication (such as the Chappe telegraph at Austerlitz)."[58] For Kittler, the Chappe telegraph, which "remote-controlled France's armies with secret codes," marked the point at which this accelerationist tendency was allied for the first time to optical image storage, enabling the telegraph's vital role in the success of the Napoleonic armies.[59]

Such approaches, in which technology is the motor of its own forward march, minimize the extent to which technological artifacts are products of culture and disregard the agency of individual subjects in determining their meaning. The Chappe telegraph emerged not as an inevitable, independent technological object, but as what David F. Bell calls an "emblematic artifact created at the conjunction between the technological and the social."[60] With this in mind, it is important to note too that Kittler's and Virilio's militarism helps perpetuate the gendering of such technologies as uniquely male; yet the telegraph's history was closely intertwined with the lives of those who lived alongside it and, in later iterations, used it, including of course many women.[61] Moreover, the telegraph's wider effects were not always anticipated by those who brought it into being, and significantly for my purposes, teleological or deterministic accounts of telegraphy obscure the fact that this was a technology that looked backward as well as forward. Telegraphy enabled new ways of thinking about time, but not all of these associations had to do with the new speed of communication it brought to bear; rather, many had to do with more far-reaching insights into the relation between technology, time, and revolutionary process. Certainly, we can see both "steadfastness" and "speed"—Kittler's opposed terms—coming together in the many depictions of the telegraph apparatus on the crumbling church tower of Saint-Pierre de Montmartre, a dialectical image if there ever were one (fig. 1.16).

In a landmark essay, historian and theorist of communication James W. Carey outlined how, at the time of writing, in 1983, telegraphy remained "one of the least studied technologies, certainly the least studied communications technology."[62] Carey's essay is notable for his sustained attention to telegraphy's profound reshaping of structures of language and knowledge, and for his anthropological account of its transformation of "the practical consciousness of everyday life."[63] "A thing to think with, an agency for the alteration of ideas," the telegraph, for Carey, decoupled communication from both transportation and geography, allowing it to control physical objects and spaces rather than be controlled by them.[64] It ultimately came to serve as a means to coordinate time itself. These effects were, Carey argued, inseparable from new ways of thinking about monopoly capitalism and empire, as the telegraph leveled markets and moved speculation from arbitrage to futures, the domain of times not spaces. The broader implications of telegraphy were felt at all levels of society and culture, even where the telegraph was not explicitly invoked, from the advance in organic metaphors that accompanied the rise of such communications systems to new forms of writing that were molded to fit, or mimicked, the staccato brevity of the telegraphic mode.[65]

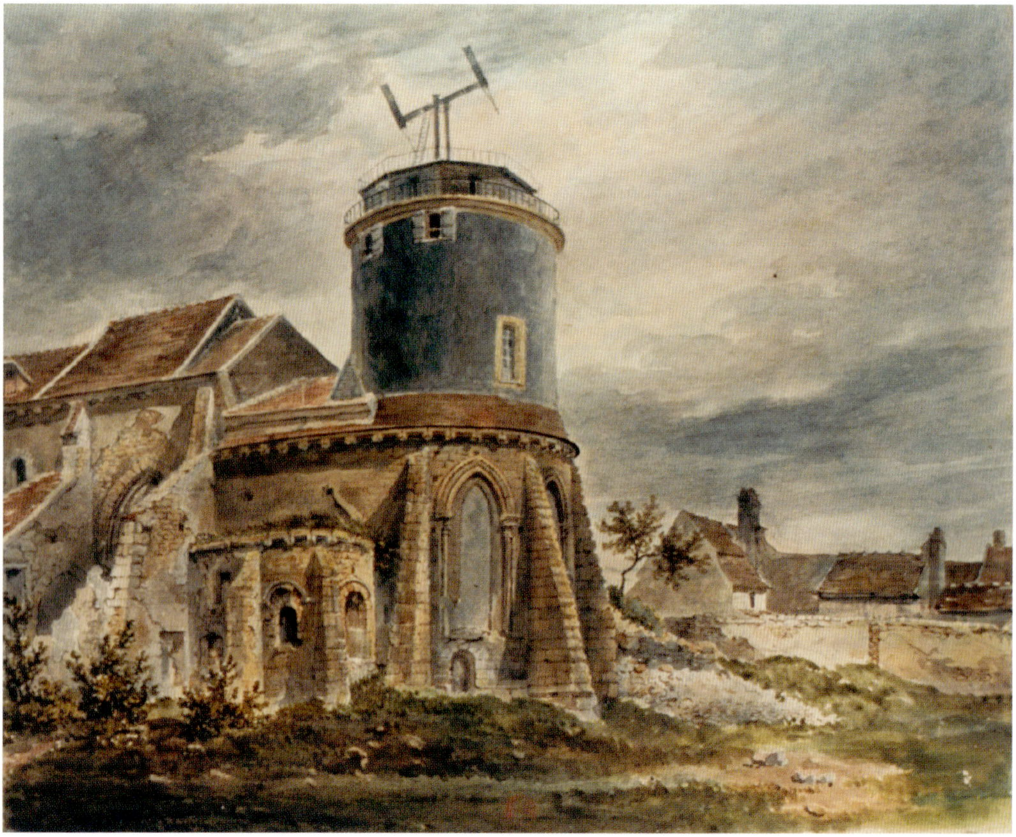

Figure 1.16
Ancienne église et la tour du télégraphe à Montmartre, c. 1815, watercolor, 24 × 29.7 cm.
Bibliothèque nationale de France, Paris.

Powerful as Carey's argument is, the cultural examples he invokes are, as is often the case, predominantly textual, as well as secondary to his argument, and his focus is on the electric telegraph in America, where, prior to Morse, there was a less integrated and far less extensive telegraphic system than existed in Europe. While experimentation with optical telegraphy was widespread in the late eighteenth century, only in France was it developed to form a truly national network, and it was perceived as a singularly French phenomenon. In passing over this period before electrical networks, when telegraphy nonetheless achieved great success, Carey is not alone. Modern histories of communication have prioritized the electromagnetic system whose claims to universal significance skyrocketed with successive, and finally successful, attempts to lay a telegraph cable across the Atlantic in 1857–1858 and 1865–1866 (an undersea cable was installed between France and Britain in 1850).[66] The trial of the Gauss-Weber telegraph in 1833, the Cooke and Wheatstone system patented in 1837, or Morse's first message in 1844, are the haloed origin points for modern communications.[67] For most, in fact, that *is* the telegraph. Carey does mention earlier semaphoric or line-of-sight systems, which operated briefly on sections of America's east coast as well as in France. But he saw these as mere "premonitions," of interest only as precursors to what followed. Electrical telegraphy, Carey argues, is what provides the "decisive and cumulative break," because it simulated fully and functioned as a control mechanism for "what has been left behind"—namely, telegraphy's clunky semaphoric predecessors and the railroad.[68]

Yet as a "thing to think with," telegraphy was not bound absolutely to the hardware with which it was realized, which was, in any case, subject to constant experimentation and adaptation. And, as it is a central claim of this book that the French optical telegraph became integral to ways of thinking historically, not only simulating prior media but haunting those which followed, the "left behind" cannot help but fascinate. "Decisive breaks" are perhaps best avoided. In contrast to Carey, then, and to the scholars for whom the history of telegraphy is inseparable from its electrical iterations, my focus remains largely on the semaphoric system that came before and at times overlapped with them (the word *sémaphore*—"carrier of symbols"—also appeared in France in the early nineteenth century, moving swiftly into English). While electromagnetic communication enters the picture at a number of points—for telegraphy was, in truth, always plural, "telegraphies"—I have deliberately set its heroic narrative, well described elsewhere, to one side, or at least I have tried to complicate this trajectory. The study of this subject has frequently been treated as a series of triumphal discoveries and ruptures wherein one technology leads to another, often with a vision of human perfectibility as the end goal. As a corrective to such deterministic narratives, the French *machine* that was introduced for the first time during the French Revolution, lasting until the early years of the Second Empire, appears here with its modernity and its archaism embracing one another, as they did from the start. It makes a great deal of sense that most early accounts of the development of the telegraph and descriptions of its wondrous future utility, from Lakanal onward, relayed an eclectic prehistory of this novel machine that proceeded from smoke signals and running messengers. Such accounts did not merely celebrate the culmination of a linear sequence of innovations in the present day, but described an archaeology of telegraphy that looked for its contemporary iterations in the past.[69]

The optical telegraph's eventual extinction made it a useful repository for nostalgic visions of historical time, but in its curious relays, compressions, and delays it was also a conduit for rethinking the very category of obsolescence. Vitally, however, while the nonsequential quality of technological change certainly affected the hardware and operation of the telegraph, it found more complex expression in a panoply of images and objects that in ways direct and indirect invoked a telegraphic imagination. Those questions of obsolescence were picked up here too, as the timeliness of images and their sufficiency to contemporary life were intelligible in new ways in a world alert to the telegraph's appearance. At the same time, images outlived—as is their way—both the technology they represented and those who brought it into being. Ignace Chappe, Claude's older brother, recognized the limitations of a unique concentration on the machine, writing in 1807 to Claude-Pierre Molard, the first director of the Conservatoire des Arts et Métiers, that "the telegraph is undoubtedly worth a great deal because of its strikingly simple mechanism; but this object is only secondary."[70] By turning to artworks that deployed vocabularies and techniques developed for scientific and military purposes, as well as to technological objects that articulated new aesthetic possibilities, we might track how experiments with the future also shaped an understanding of that which came before. Produced at one remove from the telegraph, artistic works were frequently able to articulate more fully the future-oriented metaphorical meaning of "to telegraph" as anticipation or prophesy than was the machine itself, while also possessing a unique potential to appraise and represent the past.

Art under the Sign of Telegraphy

Literary scholars have been attentive to the ways in which telegraphy forced a reconsideration of effects of space, plot, and immediacy across different kinds of writing in this period, from novels to newspapers.[71] Yet the optical telegraph—despite its lingering presence in some of the era's most notable images, its operation by visual means, and its residence for over half a century on rooftops across France—has struggled to find a place in histories of art and architecture. It has been proposed that "Chappe telegraphy hardly inspired painters, apart from some local artists who represented monuments adorned by the famous machine."[72] Such a claim rather misses the point, for the interference that telegraphy caused in the domain of visual art was not restricted to its distinctive iconography. At times the telegraph was present even when it was, ostensibly, not. Clouds, skies, landscapes, bodies, buildings of all kinds: all signified differently in the era of telegraphy.

The telegraph was never likely to be absorbed into advanced painting's preferred repertoire of sources, but found a more regular home in urban topographies, caricature, games, photographs, and commercial objects. Yet while images of the telegraph were, on the surface of things, a more straightforward inclusion in genre or landscape works than in large-format history paintings, we would do well to avoid reinforcing value judgments about "local artists," many of whom take their turn in the spotlight here. As Jacques Rancière has put it recently, "inferior genres, such as paintings of landscapes, flowers, or animals, usually offer more material for the work of the imagination than the more elevated genre of history painting."[73] Moreover, the system *was*, in fact, depicted and described by a wide range of illustrious artists—from Grandville to Pissarro, Courbet to Daumier—as well as numerous lesser-known figures, who were variously fascinated, amused, and horrified by the telegraph's intractable coded messages and its uncanny environmental and perceptual disruption. We get a different picture of nineteenth-century modernity if we switch our point of reference. At the same time that Paris's famous arcades began to crisscross the city at ground level, this more enigmatic network was operating above.[74] Modern France took shape, quite literally, under the telegraph's sign.

But if history painting, still the most ambitious and politically relevant genre of artistic practice at the telegraph's inception, was required to legitimize or undermine the interpenetration of telegraphy and art, it could be summoned. In *The Paris Sketch Book*, published in 1840, William Makepeace Thackeray offered an idiosyncratic reading of some paintings of the French school. Strangest among these was the novelist's interpretation of Jacques-Louis David's 1784 *Oath of the Horatii*, which he illustrated with a rendering of the Horatii as ideographic stickmen:

> Of the great pictures of David, the defunct, we need not, then, say much. Romulus is a mighty fine young fellow, no doubt; and if he has come out to battle stark naked (except a very fine helmet), it is because the costume became him, and shews off his figure to advantage. But was there ever anything so absurd as this passion for the nude, which was followed by all the painters of the Davidian epoch? And how are we to suppose yonder straddle to be the true characteristic of the heroic and the sublime? Romulus stretches his legs as far as ever nature will allow; the Horatii, in receiving their swords, think proper to stretch their legs too, and to thrust forward their arms thus,—
>
> 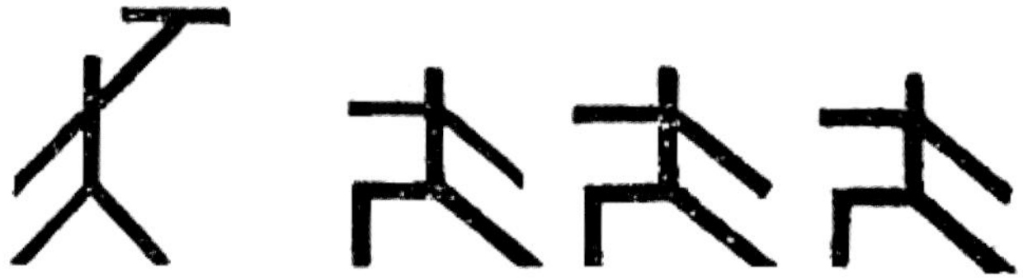
>
> Romulus's is the exact action of the telegraph; and the Horatii are all in the position of the lunge. Is this the sublime? Mr. Angelo, of Bond-street, might admire the attitude; his namesake, Michael, I don't think would.[75]

Thackeray's mocking destabilization of the heroic gestures of late eighteenth-century history painting is intriguing on many levels.

Misreading the code—deliberately perhaps—Thackeray reversed the position of the figures in the painting and misrecognized the elder Horatius as Romulus, the subject of David's 1799 *Sabines*, hanging close by in the Louvre, and another work that could not be more telegraphic (fig. 1.17). First displayed in front of a mirror, so that viewers could imagine themselves part of the painting, its slick surface examined carefully by visitors with *lorgnettes*, Hersilia's outstretched arms in the *Sabines* were, like those of the telegraph, made to be seen through glass.[76] Robert de La Sizeranne probably had this painting in mind when he described David's "pink statues, maneuvering their arms like Chappe's telegraph, or doing the splits."[77] Such associations were not wholly derived from the work's formal properties, for David had in fact witnessed a celebrated moment in the telegraph's genesis. Miot de Mélito reported in his memoirs that when Ignace Chappe visited him at the Ministry of War in 1793—a pivotal meeting at which Miot suggested the machine be named *télégraphe* instead of *tachygraphe*—it was "the famous painter David" who brought him there.[78]

David was close to the action in 1793, but several decades later Thackeray's account of "David, the defunct" polemicized the artist's obsolescence, obtaining its force from his juxtaposition of David's prerevolutionary *grande machine* and this far more prosaic technology that in 1840 remained a regular sight above the rooftops of Paris. Imagining painting as telegraphy rendered David's painting both declarative and incomprehensible, drawing attention to its seemingly arcane codes and the discrepancy between its advanced aesthetic claims and an unintentionally mundane effect. It also suggested easy affiliations between such a technology of state and this most technocratic of artists. But Thackeray's breezy image of the Horatii-as-telegraph was a more complex analogy than he allowed. Not only did it accentuate the inherent visuality of telegraphy at this time, connecting it to a wider image world that included elite academic art, but it also exposed the complicated temporal effects of both painting and telegraphy, showing them to be integral to the French Revolution's protracted afterlives. Rather than suggesting that telegraphy adumbrated a future of smooth technological progress, Thackeray's sketch indicated that it might also provide a language with which to comprehend images from the recent past, such as the *Horatii*, whose production and display in the mid-1780s preceded the machine's invention.

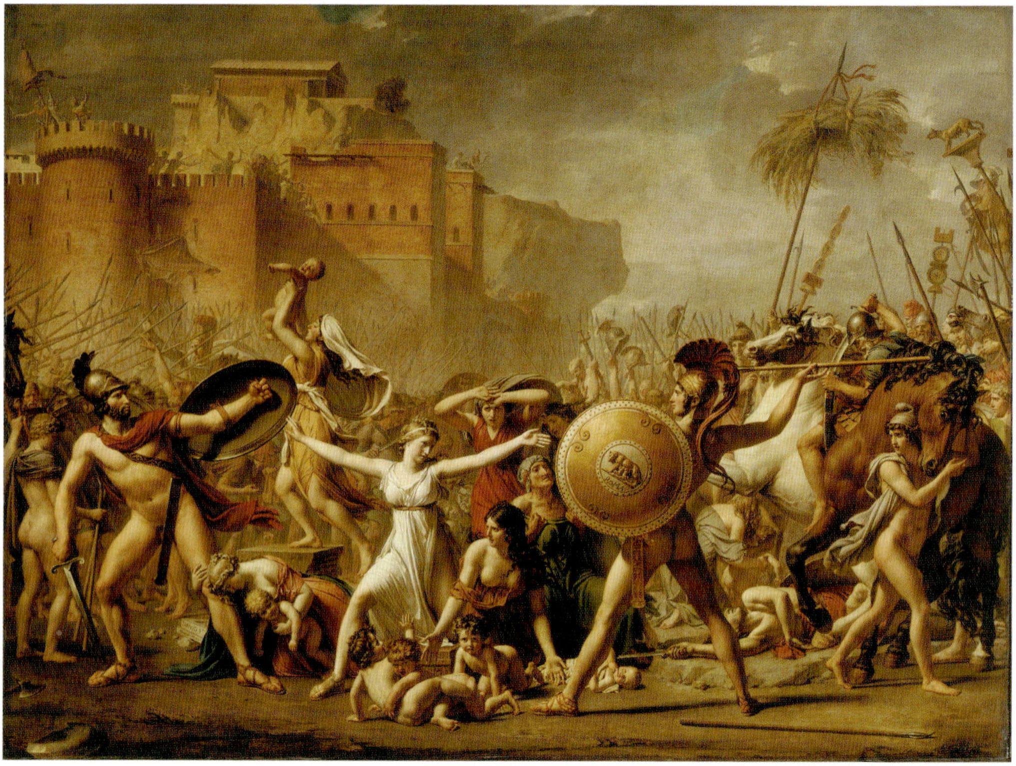

Figure 1.17
Jacques-Louis David, *Les Sabines*, 1799, oil on canvas, 3.85 × 5.22 m. Musée du Louvre, Paris.

In Paris in 1840, Thackeray would not have had to walk far outside the Louvre to encounter an optical telegraph. For several years a telegraph had been positioned atop the newly established museum, until it was relocated to Saint-Sulpice in 1805. However, given the secrecy that surrounded the codes and protocols of the telegraph, knowledge of what occurred in the process of telegraphy was limited. Sometimes, visitors to France, such as Thackeray, were more clear-sighted on this issue than those who lived with the machine. For foreigners, the telegraph expressed clearly the divergent timescales the French Revolution brought to bear, and many were fascinated by its acceleration of history as well as its agitated geometries.[79] Images of the telegraph appeared across Europe shortly after its introduction, prompting curiosity about its operation (fig. 1.18). Lawyer and travel writer Friedrich Meyer, in his *Fragments sur Paris*, published in German in 1797 and translated into French a year later by the traitorous exile General Dumouriez, observed that "this excellent invention by Chappe is better known for its exterior properties than for its interior layout, which is as simple as it is effective; I've never seen a report made by an eyewitness. To ascend to the telegraphic observatory of the Louvre, you need to obtain an express governmental permission from the Minister of the Interior or be taken there by the inventor himself."[80] (Needless to say, Meyer managed to get in, and offered a detailed description of the interior arrangement of a telegraphic transmission room.)

A series of prints by architect and engraver Louis-Pierre Baltard, who later contributed designs to the *Description de l'Égypte*, corroborate Meyer's impression of the Louvre telegraph's proximate detachment. Published in 1803 as part of *Paris et ses monumens*, Baltard's volume comprised meticulous renderings of the architecture of the Louvre, Saint-Cloud, Fontainebleau, and the Château d'Écouen, with accompanying texts by Amaury Duval.[81] Three prints present the telegraph at the Louvre to dramatic effect, viewed from different angles (fig. 1.19). Another, however, is titled simply *Vestibule du Louvre sous le télégraphe* and shows us the space underneath the device. The telegraph's presence is felt even when it cannot be seen, with clouds substituting for its aerial presence. With the apparatus identified but concealed from view, we can discern in Baltard's prints a premonition of those points—in images and in everyday life—when the telegraph stopped being noticed, when it invaded other patterns of visual experience, or when, as

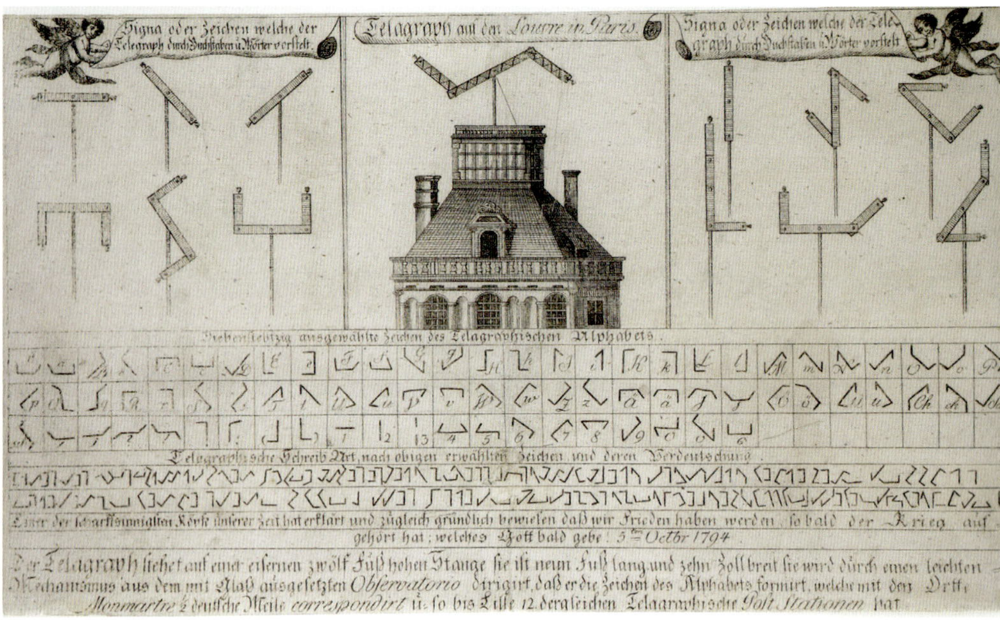

Figure 1.18
Telegraph auf den Louvre in Paris, 1794, etching, 19 × 33 cm. Museumsstiftung Post und Telekommunikation.

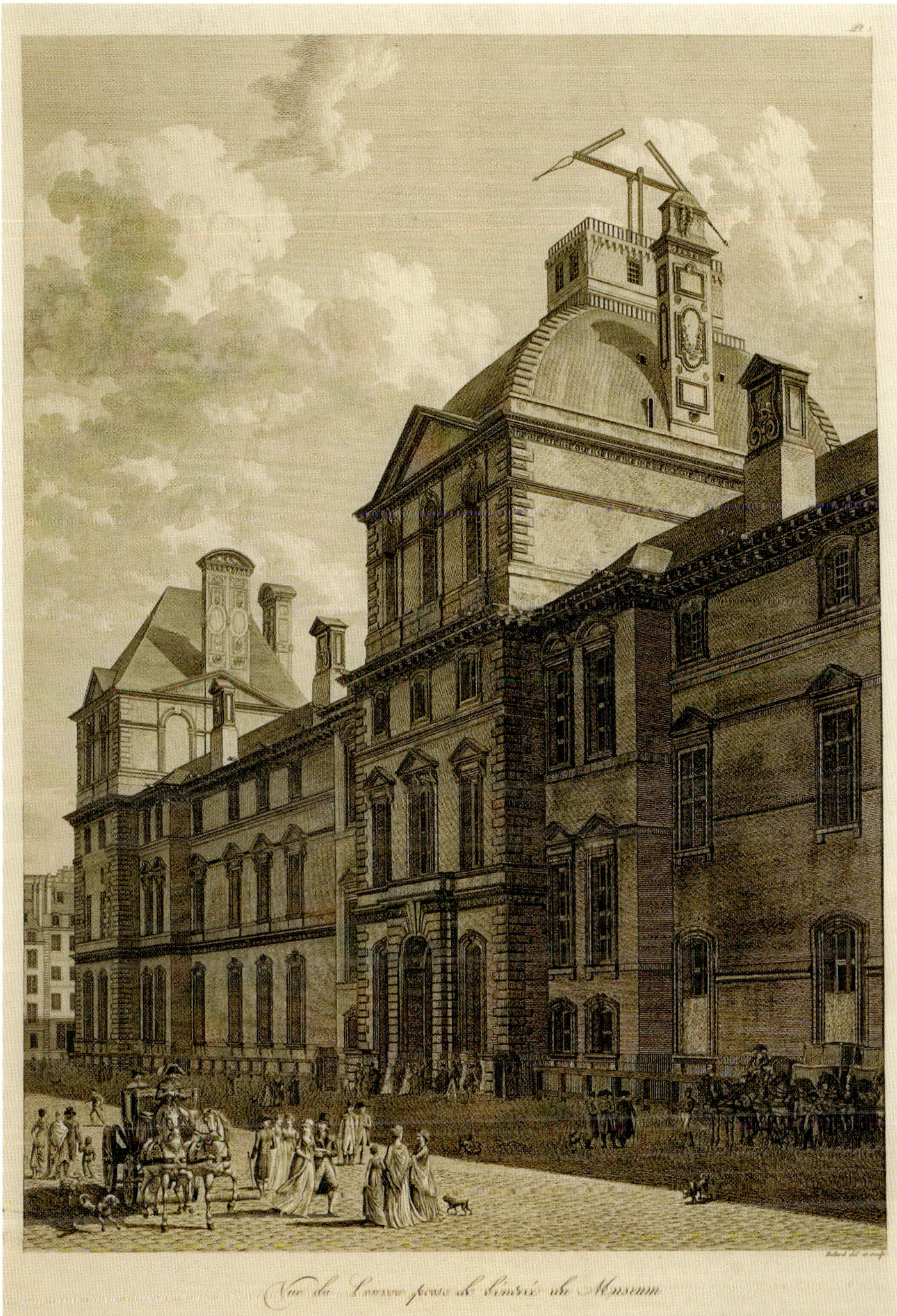

Figure 1.19
Louis Pierre Baltard, *Vue du Louvre prise de l'entrée du museum*, 1803, etching and engraving, 54 × 38 cm. Harvard Art Museums.

with the networks of wires and pipes that enable modern communications infrastructure, its ubiquity rendered it invisible before, unexpectedly, it came abruptly back into view.[82]

In Baltard's engravings, dogs fight and frolic, carriages trundle by, and families clad in Empire fashions stroll past the new museum. Mentioned only once in passing in the text, the telegraph is naturalized as part of the building's architectural fabric. But unlike other aspects of the building, it is something to be seen only from without, a shape lacking interior form, its animation a mystery. The cognitive and imaginative flights the telegraph produced were not related to the detail of its operation but were triggered by its visual presence alone. Years later, Ignace Chappe, in his 1824 *Histoire de la télégraphie*, quoted Johann Bergsträsser, inventor of a rival device, who had perceptively recognized the power of the telegraph's silhouette: "I fear that the French will not use their telegraph for anything other than a political purpose: they use it to amuse the Parisians, who, with their eyes constantly fixed on the machine, say: *it works, it doesn't work*."[83]

Bergsträsser was wrong in some respects, for the telegraph was much more than an amusement, but he was right in diagnosing the fascination exerted by the machine, the "eyes constantly fixed" on it, and the palpable start that heralded the arrival of as-yet-undeciphered news. When, in 1844, Ennemond Gonon published his own design for a competitor telegraph that would work at night as well as during the day, he was explicit that it was the product of years of watching Chappe telegraphs in action, as an outsider to the network.[84] Telescope-assisted vision was inherent to the telegraph's use, but equally significant was the way it was seen, from nearby, from afar, glimpsed in part or while passing in both urban and rural environments. Or, as in Baltard's prints, ignored, as people went about their business. The telegraph as seeing machine was inseparable from the telegraph as an apparatus to be seen, and its inscrutability did not prevent significance being attributed to it. While the meaning of the telegraph's jarring visibility shifted over its long operational life, visual questions were always at the forefront of discussions of the system.

Thackeray, in the guise of Mr. Titmarsh, identified something else which, like Baltard's prints, exploited the telegraph's slippage between visibility and invisibility, although in a quite different way. Neoclassical painting—from sketch to *académie* to finished work—prioritized the contours of expressive bodies, accentuating the ways in which their codified gestures might indicate intense emotional states. Images such as Ingres's studies for *The Martyrdom of Saint-Symphorien* demonstrate this in especially concentrated form and are representative of a broader aesthetic tendency with a long history of its own (fig. 1.20). Legibility was vital, and as has been well described elsewhere, in the 1780s and 1790s the presence of the awkward semaphoring bodies in such paintings was deliberately intended to achieve a strange, inharmonious effect, cutting through the niceties of established artistic form; a strategy that soon operated as a signifier of radical politics.[85] Giorgio Agamben, in his "Notes on Gesture," claims that by the end of the nineteenth century the Western bourgeoisie had "lost" their gestures, becoming obsessed with them as a result and sublimating them in cinema.[86] The Chappe telegraph might be considered a significant missing step toward the path outlined by Agamben: a hysterical last flourish before the retreat of semaphoric pantomime. Or perhaps it shows that gesture was never really lost, it just went elsewhere, outsourced to proxy machines.[87] Either way, comparing David's work of the revolutionary period to the telegraph that emerged around the same time was not as powerful a takedown as Thackeray imagined: these are indeed telegraphic images. One could pick any one of this artist's major works, from *The Death of Socrates* to *Brutus* to *The Death of Marat* and find "telegraphic" a useful descriptor for the way it stages communicative gestures between figures within the painting, and between ancient and contemporary subjects and their viewers. As Philippe-Alain Michaud writes, apropos of Warburg, "Sign language, which allows the *istoria* of painting to be transmitted independently of speech, opens up access to the past for the modern viewer."[88] This strategy was not limited to neoclassicism, as many artists working in different genres and media used telegraphic gestures to articulate a sense of history.

Analogies such as these risk accusations of pseudomorphism. Seldom used as a term of approbation, pseudomorphic connections are based on nothing more than accidental similarities of shape, structure, or color, often linking works made in different times and places. The anthropomorphism of the optical telegraph encourages this kind of leap. Yet as Amy Knight Powell has argued, "in

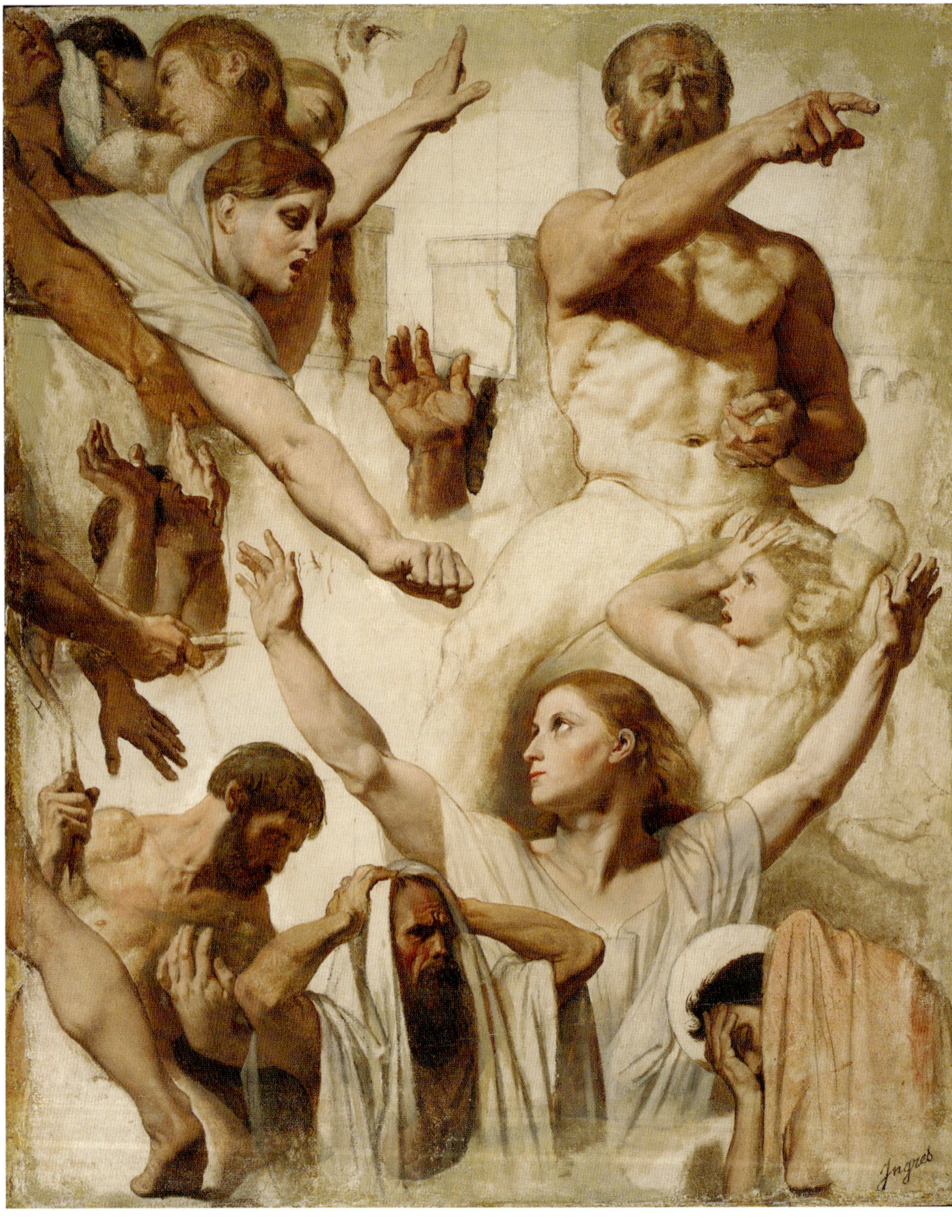

Figure 1.20
Jean-Auguste-Dominique Ingres, *Studies for "The Martyrdom of Saint Symphorien" (Saint, Mother, and Proconsul)*, 1833, oil over graphite and red chalk (stylus transfer), squared in graphite, on paper mounted on canvas, backed by a wood panel, 60 × 47.6 cm. Harvard Art Museums.

its promiscuity, form has a way of detaching works of art from the people who worked them and from the time and place in which they were made, not by transcending history—that would be the metaphysical understanding of form as 'idea' or 'essence'—but rather by transgressing history, at least, our linear conception of it."[89] Although the period of the optical telegraph's use was, in historical terms, relatively short—certainly shorter than the one Powell has in mind—these processes are not restricted to interactions between works of art but might also take in other visual objects and media that shared their time, as well as those that did not. The telegraph's reshaping of temporal experience lent itself to the "transgression" of history in the manner described by Powell; slipping into and out of view, it posed questions about the relationship between artistic form, time, and politics in ways that were not limited to the representation of the apparatus or the intention of the artist.

Take, for instance, a work such as Hubert Robert's 1794 image of his fellow inmates at Saint-Lazare prison being brought milk (fig. 1.21). The painting depicts a moment of connection in an environment that would seem to occlude such a possibility, capturing fleeting communication and relays across time and space. Is it just me, or does the foreground figure, and the balustrade for that matter, take up the Z-shaped angle characteristic of a telegraph? Once one starts noticing telegraphs, one sees them everywhere (like the figures in the painting, I am probably reaching here). Having been incarcerated since October 1793, Robert most likely had not encountered a telegraph at this point, but the telegraph-as-hardware gave form to questions of transmission and contact that long preexisted it, in all areas of life, which were thrown into uniquely sharp focus with the Revolution.

Or, better, consider Louis-Léopold Boilly's *A Girl at a Window*, made around 1799, in which the dual telescopes, one facing in, one projecting out of the window, seem to allegorize the telegrapher's task at a time when telegraphic outposts were being constructed across the nation and with greatest density in Boilly's adopted Paris, as well as, for many years, in his native Lille, the terminus of the first telegraph line (fig. 1.22). Boilly seems to have never painted a telegraph, a notable—and presumably deliberate—omission by an artist who was so attentive to the specificities of urban space. But perhaps he just had more nuanced ways of representing its main idea. My suggestion here is not that this is an image "about"

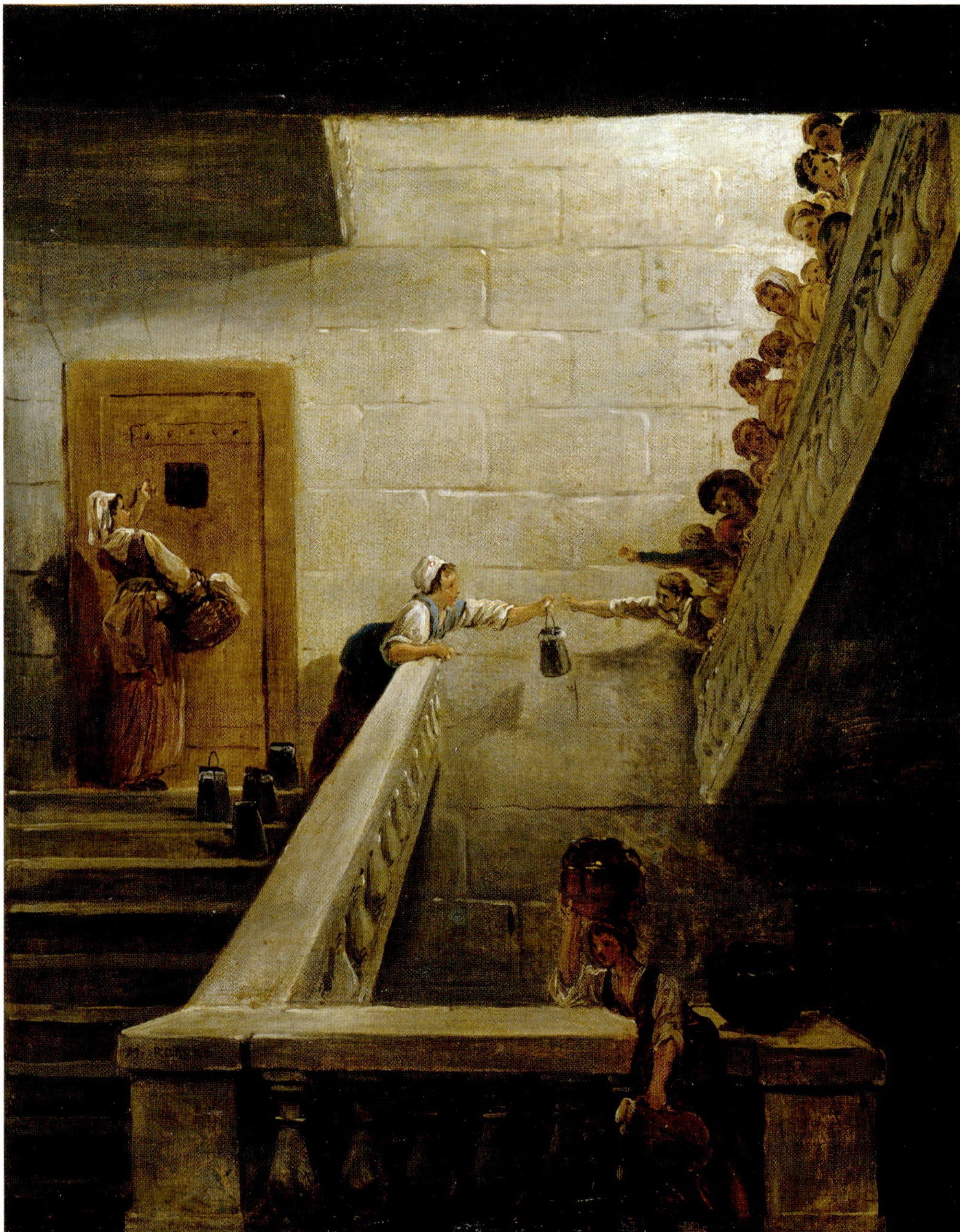

Figure 1.21
Hubert Robert, *Le Ravitaillement des prisonniers à la prison de Saint-Lazare*, c. 1794, oil on canvas, 40.5 × 32.5 cm. Musée Carnavalet, Paris.

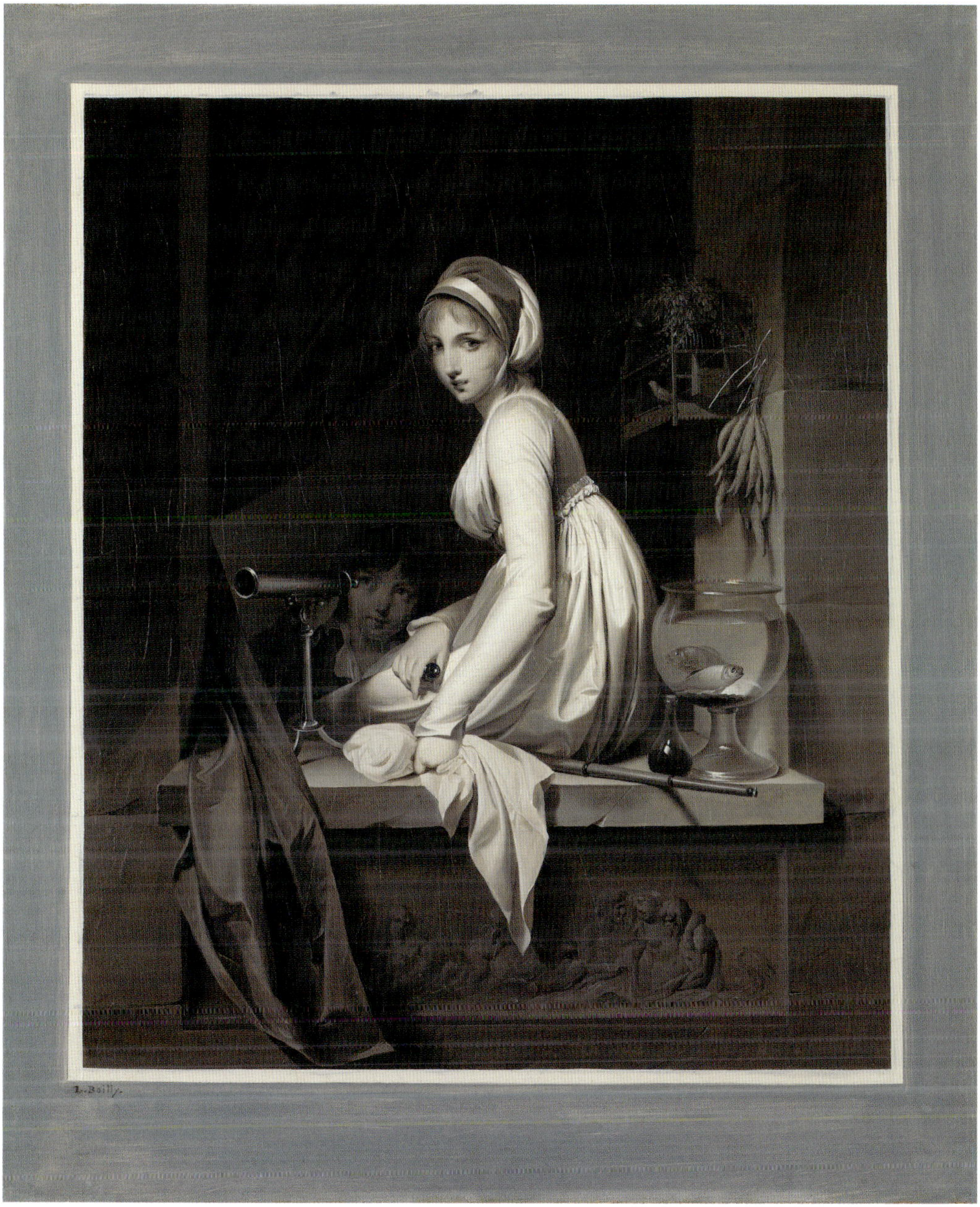

Figure 1.22
Louis-Léopold Boilly, *A Girl at a Window*, c. 1799, oil on canvas, 55.2 × 45.7 cm. The National Gallery, London

telegraphy, as such, but that it is indubitably a telegraphic image, and that such images were legible in new ways following the introduction of telegraphy's abstruse signing, temporal-spatial compressions, and mysterious codes. Best understood as an intersection of cultural, technological, sensory, and, crucially, aesthetic phenomena grounded in the particularities of its use in France in the first half of the nineteenth century, the meanings of optical telegraphy extended beyond the development of the technology or its immediate application. To take another example, Louise-Adéone Drölling's image of an artist at work, the church of Saint-Eustache framed by the window behind her, could not fail to be seen differently after 1834, when the telegraph was installed on the building (fig. 1.23). Such an association could be made only retrospectively, animating an image from the past with knowledge of its future appearance. In its direct window-to-window view and its meditation on inscriptive processes, both musical and artistic, Drölling's painting, which she made around 1815 when telegraphs were already a feature of several Parisian churches, seems to anticipate these changes.

"Telegraphic images" are thus not merely works representing telegraphy or produced in direct connection to the telegraph itself—its codebooks, maps, and letterheads—although such things do of course repay careful attention. Sometimes the absence or excision of the telegraph from a scene speaks loudest. At other times, artists got the telegraph wrong, mangling its appearance, which given the simplicity and familiarity of its form seems symptomatic of a more profound displacement. Neither are "telegraphic images" simply those that invoke, in one way or another, the machine's gestural choreography—although again, the presence of this tendency in several images requires explication. As Jennifer Roberts notes, in assessing the visual impact of telegraphy, "the best way . . . is not to locate some direct causal impact of telegraphy on picture making, and not to limit analysis to the iconography of telegraphs as they were depicted in paintings, but rather to delineate concerns about coding, replication, and transmission shared by both visual media."[90] Telegraphic images were those in which issues of mediation, transmission, and time were foregrounded, their historical meaning illuminated by qualities intrinsic to telegraphy. Returning the message, representations of telegraphy, and even some that foreshadowed telegraphy's conceptual foundations regularly made sense of telegraphy itself, as well as the other way

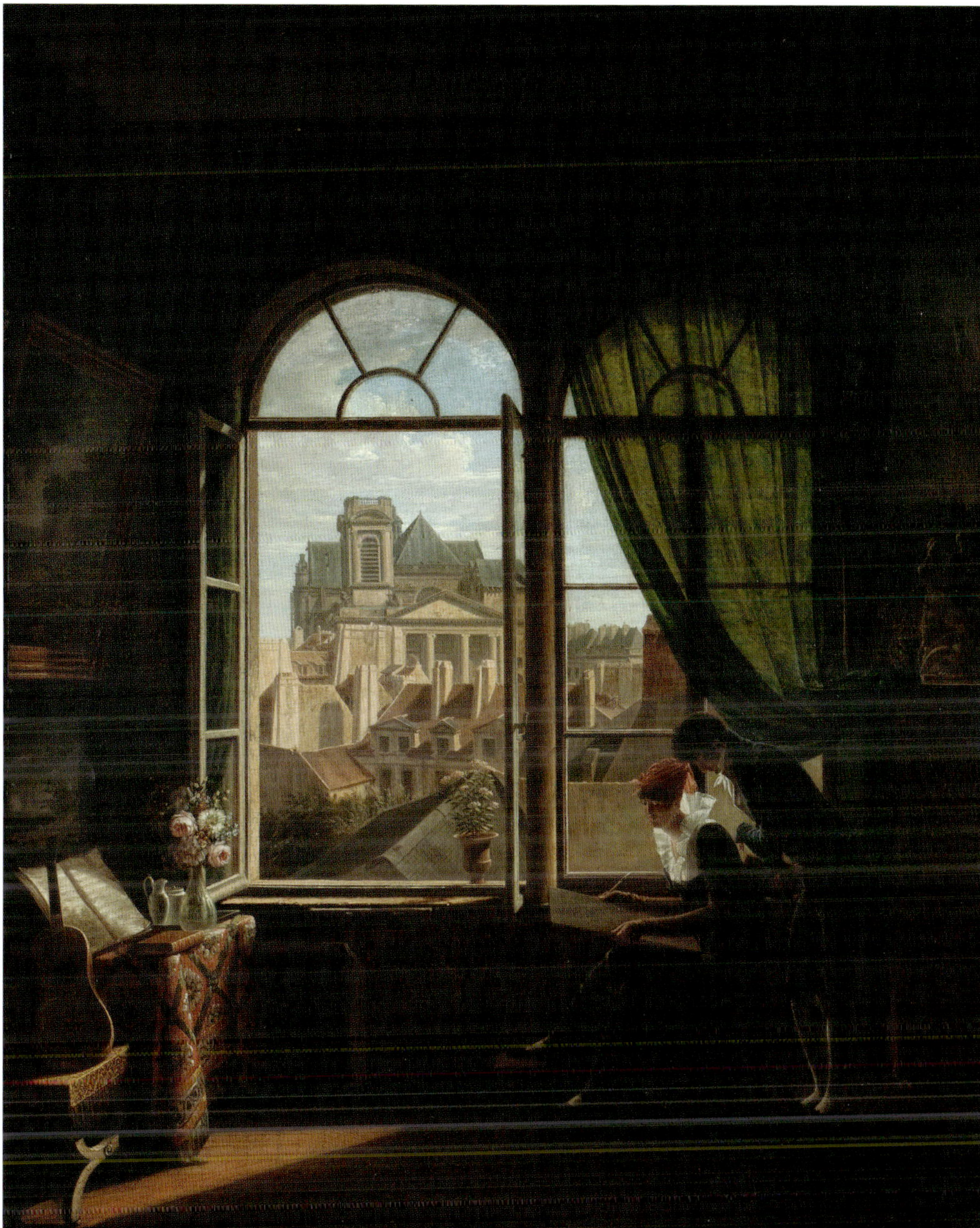

Figure 1.23
Louise-Adéone Drölling, *Intérieur d'artiste, avec vue sur la façade de l'église de Saint-Eustache*, c. 1815, oil on canvas, 73 × 59 cm. Musée Carnavalet, Paris.

around. However, the manner in which this took place was not uniform. Telegraphy sometimes correlated closely with artworks' transmission of information or a work's own movement through space, but at other times it offered a way of thinking about fixed points and local phenomena; it spoke to the urgent and polemical as well as the overlooked and the mundane. Broadly put, art in the time of telegraphy was connected in intricate ways to the network's reconfiguration of the sensible. If I risk overstating my case by implying that all art's properties might be housed under the optical telegraph's rickety canopy, it is only because telegraphy seemed at times to promise this kind of total intelligence. Such vaunting claims for the telegraph's capabilities were exacerbated by the fact that most who encountered the machines were ignorant of their messages' content, projecting a range of desires and interpretations on them as a result.

Returning briefly to Boilly's *A Girl at a Window*, we might consider how, with its lattice of deflected and direct gazes, telescopes, fishbowls, lenses, handheld *lorgnette*, and of course, the fact that it is a trompe l'oeil *grisaille* made to mimic a print, the painting plays with viewing conventions and shifting optical impressions. Boilly was an enthusiastic collector of optical devices and his interest in their effects can be discerned in many of his works. At first sight, optical telegraphy would seem to find a ready place among the panoramas, phenakistiscopes, stereoscopes, peepshows, and lanterns that made up spectacular culture in the early nineteenth century. But a note of caution: that "optical" may be misleading. The Chappe telegraph did not, unlike many of these devices, participate in a wholesale recalibration of perceptual knowledge, an epistemic shift in which interiorized vision was allied to new forms of power.[91] Or at least, such effects cannot be squarely attributed to its optical hardware but were a secondary outcome. Telescopes were, for the most part, still just telescopes, and were hardly a new technology in 1794. For Louis-Sébastien Mercier, writing in *Le Nouveau Paris* in 1798, "Everything is optical. Or an optical game." "From close up," Mercier added, "things are different from how they might be judged from afar!"[92] Part of his discussion of how revolutionary newspapers could never hope to keep up with the speed of events, thereby distorting perception, Mercier's remark was farsighted. In the post-telegraphic world everything was, in one sense, optical. Yet the telegraph's broader significance was also discerned in subtle gestures and other expressive characteristics that were not reducible to, or that exceeded, opticality. Central here were the many images and objects that in manifold ways emulated, anticipated, or refused the telegraph's austere mannerisms and runic signs. The visual character of the telegraph—its opticality—was not contained within, or by, telegraphic hardware but proliferated where one might least expect it to do so.

Straight Lines and Zigzags

In November 1971 a group of dancers, led by choreographer Trisha Brown, silently performed a relay of corporeal semaphores across the skyline of New York's SoHo. At first private in its execution and unrecorded, *Roof Piece* was staged two further times, in June and July 1973, and was documented in photographs by Babette Mangolte and Peter Moore (fig. 1.24). *Roof Piece* was reperformed in 2011 on New York's High Line, and again in 2020 during the Covid-19 pandemic, via Zoom, as *Room/Roof Piece*.[93] In the first iterations of the work, Brown's dancers threw shapes from water tower to water tower, loft to loft, reproducing each set of movements until the signal reached the end of the line. The scarlet costumes that stand out in Moore's 1973 photograph echo those worn by photographer Félix Nadar a century earlier, when—as I will discuss—he mimicked the telegraph's jagged gyrations. While *Roof Piece* took place above normal lines of sight, it was not invisible: one could see it if one knew where, and when, to look. Complementing, even competing with, the crisscross of preexisting informational networks which it overlaid, *Roof Piece* had the effect of rendering visible that which is normally taken for granted. As with Chappe's optical telegraph, Brown's work called attention to gesture, and made rooftops and the things they held come into focus in new and unexpected ways.

Such comparisons are not arbitrary, for as Thackeray's *Paris Sketch Book* showed, the early nineteenth-century telegraphic imagination was sensitive to this kind of connection over time, across distances, and between objects as well as people. Like Brown's piece, in which the precise articulation of the original message inevitably deteriorated slightly as it was passed on, fidelity was not always the end of such connections, for they accrued new meanings along the way. Perhaps, rather than thinking about the telegraph solely in terms of its "visual" or "optical" qualities, we might also consider it productively in terms of performance.

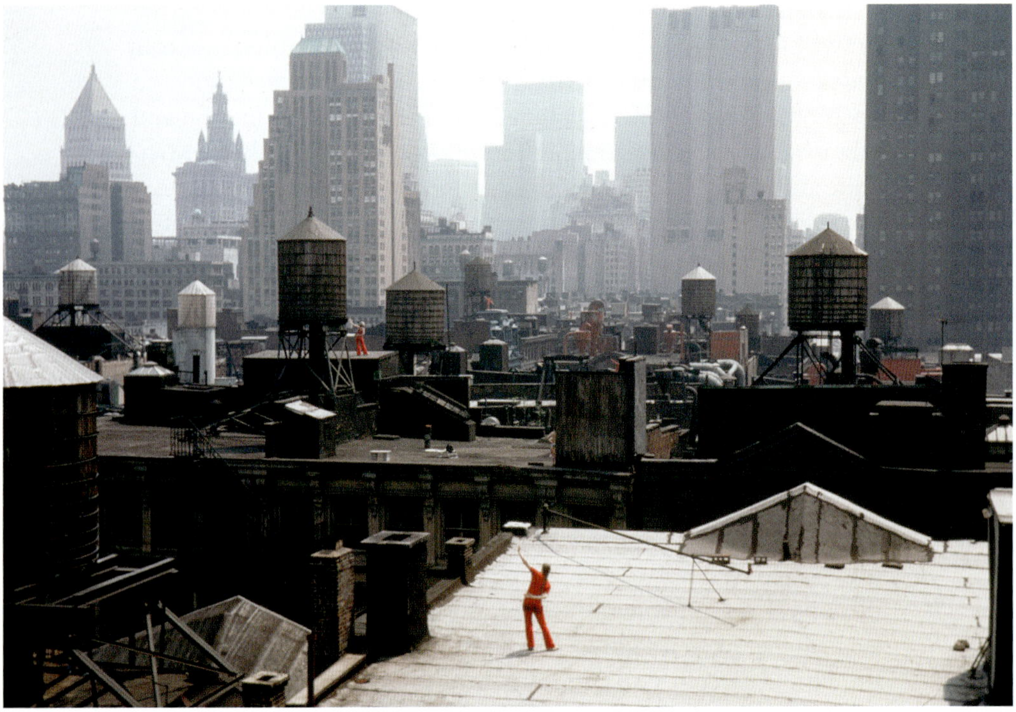

Figure 1.24
Performance view of Trisha Brown's *Roof Piece*, New York, 1973. Photograph by Peter Moore. © Northwestern University. Peter Moore Photography Archive, Charles Deering McCormick Library of Special Collections, Northwestern University Libraries.

Telegraphs were a frequent plot device in *vaudevilles* during the Empire and Restoration, and in some such productions the telegraph even took on a character role.[94] The telegraph was, without doubt, an old ham. Yet, more significantly, reorienting our attention toward the telegraph's performative qualities rather than its ostensible opticality or its hardware allows greater access to its temporal effects. The telegraph not only followed a series of scripts, but it was time-limited in its use. Telegraphy was ephemeral, even if its hardware and the representations that followed were adamantly material. Following the completion of each short message, the only record remained in the far-away transcription of the signal, each of its shapes drawn by the stationnaire and decoded at the end of the line, and later in a different way in press reports and visual images. In other words, much like Brown's *Roof Piece*, telegraphy survived only in documentation, in the archive, and in the code.[95]

Yet as Rebecca Schneider has argued, to think of performance as always vanishing and the antithesis of "saving" restricts our understanding of it to a dominant Eurocentric logic of the archive and to the persistent disavowal of performance as historical practice. As Schneider proposes, "when we approach the archive not as that which disappears (as the archive expects), but as both the *act* of remaining and a means of reappearance (though not a metaphysics of presence), we almost immediately are forced to admit that remains do not have to be isolated to the document, to the object, to bone versus flesh."[96] The contexts for Schneider's work and mine here are evidently different, but her argument aligns with my contention that the telegraph's returns, as well as its combination of intrusive materiality and dematerialized code, were part of its performativity and its historicity, not antithetical to them. Survivals, of one kind and another—whether as a kind of Warburgian *Nachleben* or as something more mundane—were as intrinsic to the telegraph's relationship to artistic practice as the disappearance of its visual acts.

Seen from this perspective, the hardware of the telegraph acquires new meaning, even when reduced to rust and rubble. We like this kind of thing these days. Put cynically, modish nostalgia for an earlier—now, from a safe distance, cuddlier-seeming—military or industrial culture, stripped of its economic and social context and the horrors it may have produced, is accompanied by a fetishization of ruins and an easily commodified desire for the remnant, for the artfully distressed, for the mournful or belated. Scholarly and quasi-scholarly turns have amplified, and in different ways questioned, this phenomenon. Both elite artistic culture and more explicitly popular or commercial practices understand its logic and replicate its signs. It seems that in one way or another infrastructure doomed to fail tells us something we want to know about ourselves.

Folding present into past, approaches that subject such issues to scrutiny have also done much to undo simplistic teleologies of technical progress, not just digging up old things but theorizing a subtle challenge to established historical epistemologies.[97] Nonetheless, there remains the risk that prioritizing obsolescence might have outcomes that are just as pernicious as the narratives of heroic invention destabilized by the "traveling discipline" of media archaeology.[98] In the context of a postrevolutionary history, foregrounding extinct things can serve reactionary ends, highlighting loss and failure without acknowledging transformations and gains. Worse, as Michelle Henning points out, purportedly obsolete material objects and the cultural practices that accompany them can be too readily aligned with a tacit perception of the communities who continue to use them as equally outmoded or marginal.[99] While no one uses the optical telegraph nowadays, people tend to fall out of the historical picture in close readings of individual technologies. This is particularly the case when we are in the company of military hardware that makes space for humans only begrudgingly, as metaphorical part-objects, all listening ears and gesticulating arms. In the shadow of these machines, we experience a kind of moderated sublime, domesticating the strangeness of military and state information systems, and turning unknowable information into something more manageable.

The situation is yet more complicated when one considers a technology that was spectacularly successful during its lifetime, such as the optical telegraph. Here we are faced with a different set of historical concerns than those accompanying quaintly thwarted systems, as we encounter machines whose extinction is a frame applied only retrospectively. By focusing on a unique technology that blossomed, one risks remaining blind to the many other systems and apparatuses that operated alongside it for varying periods of time, or which never had a chance to do so. In such cases the tendency to conceive of the operational time of a particular technology as static, rather than as subject to continuous flux, is particularly acute. This book is hardly immune to such hazards, but I hope to confront at least some of them by showing how these exact questions of time and history were integral to the telegraph's use and reception during its lifespan, beginning with the accelerating quickstep of revolution, but also detectable before its slow invention and long after its protracted demise. Focusing on a single technology, even one with ragged historical edges, has its pitfalls, but in truth the telegraph never really stood alone.

It should be clear from this introduction that I do not pretend to offer a comprehensive technical or social history of telegraphy in the nineteenth century. Just as the image of the telegraph recurs again and again at the limits of vision, a nagging, irritable trickster at the edge of perception, so it appears in the pages of this book. While the chapters follow a loose (if occasionally digressive) chronological order, from the revolutionary prehistory and origins of the telegraph, passing through its use during the Empire, Restoration, July Monarchy, Second Republic, and Second Empire, and continuing to its afterlife in the Third Republic, this book is not just about the telegraph. Rather, a history of early nineteenth-century French visual culture written not through the history of individual artists or of institutions, but by way of the telegraph—the ultimate connecting system of the era, and one that for more than half a century operated visually—prompts the question: what happens if we look at art telegraphically? This is not a story that can be told evenly, for it trades in more oblique glances, including those of contemporary viewers who in the twenty-first century encounter "telegraphic images" sent from distant times and places. Telegraphy encompassed both straight lines and zigzags, clarity and direction, alongside interference, detours, and delays, conflating visuality with a pronounced historical dynamic. How might an attention to the shared terrain between artworks and communication technologies help us see both differently, in a split second and over time, from up close and far away?

In the next chapter, I will have more to say about the visual world of the Chappe optical telegraph before, during, and after the French and Haitian revolutions. I consider how telegraphy produced new ways of thinking critically about history and revolution, raising the possibility of communication between all liberated peoples, as well as the thwarting of that ambition. Chapter 3 steps away from the immediate context of Chappe's system to focus on the work of artist-engineer Nicolas-Jacques Conté during the Consulate and Empire, particularly his "invention" of the modern pencil, and his challenge to the autonomy of drawing via his deployment of a mechanized "engraving machine" to produce regulated lines in the skies and backgrounds of plates in the *Description de l'Égypte*. These processes, I suggest, are linked to Conté's contemporaneous work in telegraphy and to related forms of representing the movement and control of information in colonized space.

In chapter 4, I focus on the work of three little-studied genre painters, Jacques-François-Joseph Swebach-Desfontaines, Étienne Bouhot, and Georges Michel, to consider the permutations of the optical telegraph network at its center, in Paris. Examining urban reconstruction—fountains, canals, and quarries—and other overlapping "worlds of flow," from the movement of money, people, and information to the passing of clouds, these artists registered telegraphy's diffuse effects in the context of their local environment, assessing its material realities as well as its historical, political, and even ecological impact. The penultimate chapter turns to the posthumous image of optical telegraphy, via discussion of Giovanni Caselli's "pantelegraph," the first device capable of sending graphic images down telegraphic lines. The pantelegraph, which swiftly became obsolete, features unexpectedly in the writing of Nadar, who mobilized it—alongside the Chappe network—to think about the technological past and photography's future. A concluding chapter examines the death and afterlife of the optical telegraph in its posthistory as both absurd monument and tragic promise.

Time Machines departs from the principle that no technology operates outside representation, and that such representations must be treated as both dynamic and historically and culturally specific. Optical telegraphy's complex encodings within culture were produced by, as well as reflected in, the rapidly changing artistic and extra-artistic landscape of postrevolutionary France, at a time when new visual media, from lithography to photography, were reshaping the parameters of what images might do. Yet although the conditions of both scientific and artistic practice were transformed in the telegraph's operational lifespan, their effects were not limited to their era but were the subject of continual reflection, incorporation, and return through the century or so after its introduction. In its staging of visibility and invisibility, transparency and disinformation, novelty and superannuation, optical telegraphy brokered a range of issues regarding technology, citizenship, and the legitimacy of media that remain urgent contemporary concerns. Like art itself, the telegraph promised to defeat time, and indeed this promise remains its most remarkable feature: to be at once wildly futuristic and always already belated, an emissary—sometimes mournful, sometimes spry—moving between past, present, and future.

2 *Revolutionary Horizons*

In revolutions, it pays to keep an ear to the ground and, what's more, an eye on the horizon. On October 30, 1789, an anonymous writer published a short text, accompanied by simple line diagrams, investigating new ways of transmitting the message of the French Revolution to the widest possible audience.[1] Written in the tense period between the king's return to Paris on October 6 and the National Constituent Assembly's expropriation of church property on November 2, this *Mémoire* presented four inventions that promised to disseminate revolutionary ideas across space and time. As the text explained: "We must find a machine that allows them to hear the voice of a man of bearing. It's necessary to find another one that will make visible, in an instant, the result of reflections that limited experience or nervousness prevent them from making. We should also find a means to give the greatest intensity to the voice of an orator in a grand assembly, because often, as they are poorly heard, their words are misinterpreted."[2] The author of these words was prescient; faltering voices, as well as strident ones, came to play a significant narrative role in the French Revolution, from Robespierre's famously quiet oratory to the muffling of Louis XVI's final speech on the scaffold. Error, misapprehension, or confusion could be deadly. Rapid and unambiguous communication of ideas and information was vital. Consequently, the first of the devices proposed in this pamphlet was a *Porte-voix du peuple*, a serpent-like trumpet that amplified the speaker's voice and could be used to communicate to large crowds. There were precedents here: Ignace Chappe later described a similar instrument, a copper trumpet invented by Morland and presented to Charles II in 1670, in the process incorporating such amplifiers into a long history of telegraphy.[3] Although the proposals in the *Mémoire* preceded the Chappe brothers' device, and although their solutions were different, the questions to which they responded were essentially the same as those faced by the telegraphers. Chappe's earliest telegraphic experiments in 1790 had involved beating casserole pans to mark a code based on precisely syncopated clocks, but the limitations of human hearing and the interference of other sounds swiftly made this method unpracticable.

The second invention from the 1789 *Mémoire*, a *Tableau populaire*, was visual in character. It consisted of paired wooden drums, mounted on spindles and placed high on a prominent building. Between these turning cylinders stretched a screen of transparent canvas, manipulated by a winch. Perhaps to diffuse its strangeness, the author compared this spectacle to more relatable scenarios, such as the division of skeins of yarn on louvered drums, the latter sold by *tabletiers* (medieval vendors of wax tablets for inscription who, by the eighteenth century, had diversified their business into the sale of small objects in wood, enamel, gold, and ivory).[4] The French Revolution had transformed almost overnight the implications of such mass communication methods, but the language with which to comprehend them had not yet caught up, harking back to familiar forms of artisanal knowledge.

Clear lines can be drawn between the *Tableau populaire* and devices that followed closely on its heels, such as the moving panoramas of the nineteenth century, which similarly set a world of information in motion.[5] Yet from our historical vantage point, the *Tableau populaire* seems closer to the mechanisms of cinema. It was, at the most basic level, a filmic apparatus, albeit one grounded in the materiality of the printed word. Organized in cartons next to the device, like a giant typesetter's table, were letters and punctuation marks, each around eight inches high. When the time came, female operators were to pin these figures to a sheet of canvas, which had been premarked with lines, before attaching it to the rotating screen. The advantages of this system were both corporeal and medium-specific; eradicating strain on an orator's voice, it obviated the need for printed bulletins, thrown from windows.

The diagram in the *Mémoire* gives a sense of the all-caps tone imagined of its messages: "Wait, we are going to reveal the black aristocratic plot."[6] The device was not limited to daytime use but could be illuminated by lanterns situated behind the screen. A parabolic mirror projected its paranoid communiqués far into the night sky.

The third and fourth inventions proposed in the *Mémoire* recognized the shortcomings that accompanied the successful elimination of the body in the previous two apparatuses, and they were reproduced alongside one another with some minor changes in Louis-Marie Prudhomme's newspaper *Révolutions de Paris* (fig. 2.1).[7] Both the *Porte-voix du peuple* and the *Tableau populaire* had obscured the facial movements, gestures, and affective nuances necessary to move the people. The *Siége orale mobile*, on the other hand, allowed a speaker to reach his required audience with minimal effort (the images make clear that the orator was imagined as male), while returning to the scene of communication the everyday pleasures of one human semaphoring to another. This movable stage comprised a giant parabolic screen that bounced the speaker's voice outward through a latticed floor. Considerations of distance were crucial. The audience had to be placed at least 118 feet from the reflective surface, to prevent the spoken words being drowned out by their echo.[8] In Jonathan Crary's influential account, the early nineteenth century witnessed a profound epistemic shift, with optical experience and a disciplining of spectatorship taking precedence over an earlier sensorium in which sound, smell, and vision had competed for attention.[9] The devices from the *Mémoire*, which combine different modes of address, seem to hang on the cusp of these worlds.

Citing traditions of Roman oratory, theatrical stagecraft, and Rousseau's description, in his dictionary of music, of a wooden box used to amplify orchestral volume, the author developed their invention, moving it indoors. The similar *Siége orale fixe* was intended for use at the heart of government, in the meeting hall of the National Assembly, which had not, at the time of writing, been constructed yet in Paris. Situated on each lateral side of the oval chamber, sound mirrors would allow speaker and president to converse intimately and in public: they could be heard by the whole Assembly without needing to raise their voices. The image representing this project envisions the words of both orators as dotted lines of pristine information, traversing the empty space of the hall. In the years following the French Revolution,

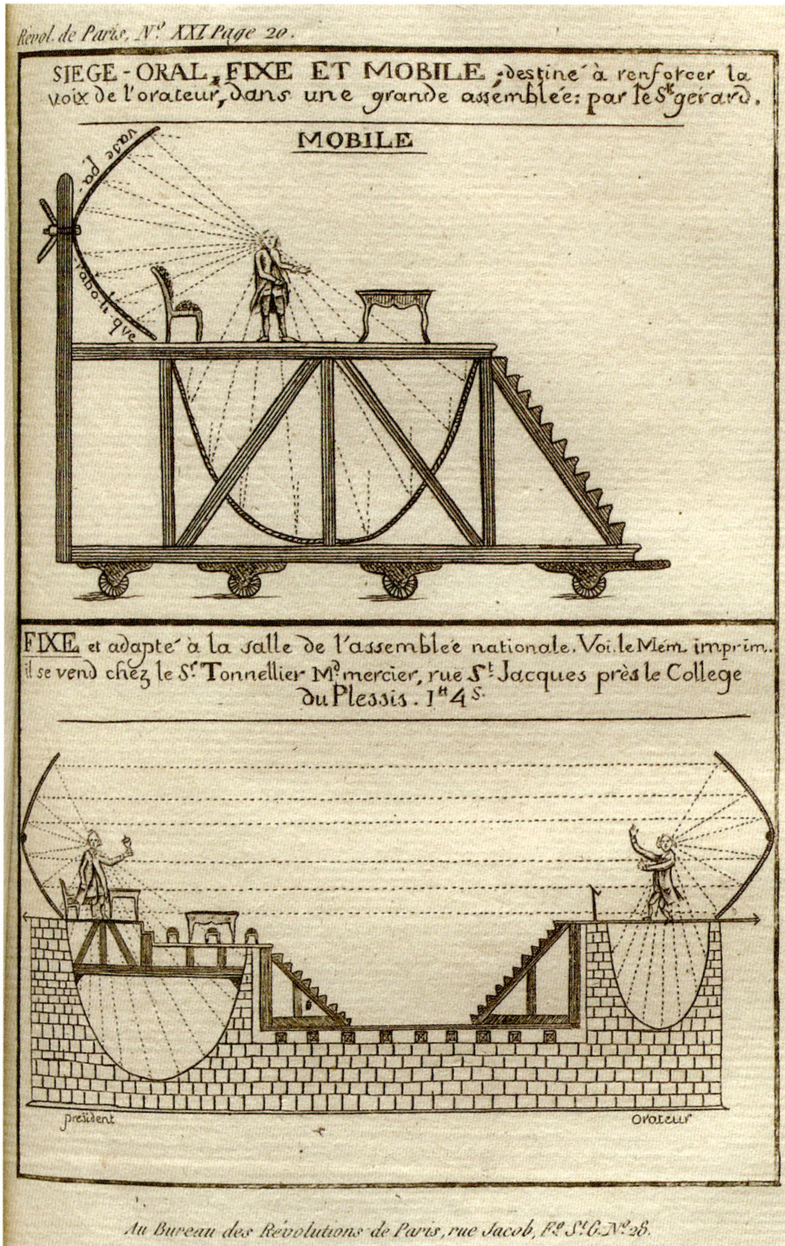

Figure 2.1
Siège-oral, fixe et mobile, from *Révolutions de Paris, dédiées à la Nation*, no. 21, November 28, 1789. Bibliothèque nationale de France, Paris.

making visible such ostensibly invisible content became a matter of increasing consequence. Seeing communication processes in action, as well as hearing them, was necessary, in a political culture in which transparency came to be prized above other virtues, and was also a source of anxiety, desire, and speculation, when the meaning of signals could not be understood by those who saw them passing back and forth.

The prototypes of the *Mémoire* did not appear in a vacuum, emerging alongside devices that were outside or adjacent to state communications technology. To take an example that followed in the Revolution's wake, in 1816 Parisian doctor René Laënnec was walking in the courtyard of the Louvre when he spotted two children signaling to each other by scratching one end of a long piece of wood, the amplified sound of which could be picked up by putting one's ear to the other end. Laënnec recalled this passing encounter some months later, when treating a woman with heart disease—rolled paper, later refined as a wooden tube, composed what became the first stethoscope. Amplification, compression, and diagnosis were here aural in nature, but they might be visual too. As the *Mémoire* made clear, sound and gesture offered different but overlapping ways of coding a signal. And it is notable that the origin story for Laënnec's auscultation device took place at the Louvre, which shortly before had been the site of Paris's most prominent machine—telegraphy was by then the most eloquent expression of a broader cultural imperative toward clear communication.

In its emphasis on precision and limpidity, Chappe's semaphoric code shared much with the "reformed" sign language recommended for the deaf; indeed, at the time the telegraph took hold, an uncluttered, transparent sign language was being proposed as a realistic alternative to speech, for the hearing too, as a revolutionary form of communication between individuals that promised to eliminate contingent meaning (as evident in regional accents, patois, or slang).[10] These issues were culturally pervasive through the Revolution, and inflected political discussion of all kinds. Belprey's "télélogue" or "speaking cylinder," described in 1799 (and published as *De l'optilogue* two years later), was an updated version of the *Tableau populaire*.[11] Having developed the device for the use of deaf people, Belprey suggested it might also transmit messages to vast crowds on the Champ-de-Mars, or indeed to an entire city; such investment of sign language with grand political purpose shaped the context into which telegraphy too emerged. As one contemporary commentator put it, "the discovery of the telegraph isn't just an ingenious speculation; its results leave no uncertainty in the literal transmission of the different characters appropriate to the language of signs."[12] The neologism under which the system labored—"far writing"—testified to the same desire for economical expression.

Glancing sideways, we can discern a shadow of the *Siège orale fixe* in the contemporaneous work of the most illustrious painter of the period. David's *Tennis Court Oath*, its figures organized in a concave arrangement to mirror the semicircular seating plan of the National Assembly, where it would provide a model of emulation for France's legislators, might be thought a two-dimensional manifestation of the *Siège orale fixe*'s parabolic mirror (fig. 2.2).[13] Collapsing the auditory into the visible, the curvilinear arrangement of representatives of the Third Estate in the *Tennis Court Oath* was, like the inventions proposed in the *Mémoire*, a means of transmission, and like them, it drew on theatrical precedents.[14] They speak an oath, but they semaphore it too. The similarity did not end there. For neither David's painting nor the devices described in the *Mémoire* were completed. The production of the painting could not keep up with the accelerating pace of events and the transforming reputations of those depicted. As Philippe Bordes writes, "David was faced with an unavoidable dilemma: he could finish his painting and ignore the course of the Revolution or abandon it and somehow catch up with the Revolution."[15] Revolutionary representation was always both charging ahead and lagging behind the times. Equally, the amplifiers, screens, and sound reflectors of the *Mémoire* belonged to an early phase of the Revolution, when reaching an audience took precedence over the content of the message, the clarity of which was soon muddied by the contingent matter of politics.

Something about all this seems familiar. Revolutions have a habit of throwing their voice. Consider, for example, the acoustic mirrors at Dungeness, on the Kent coast in southeast England, three monuments to the technological past that mark the landscape. These parabolic reflectors were developed between 1928 and 1930 to detect incoming aircraft.[16] As it transpired, the reflectors

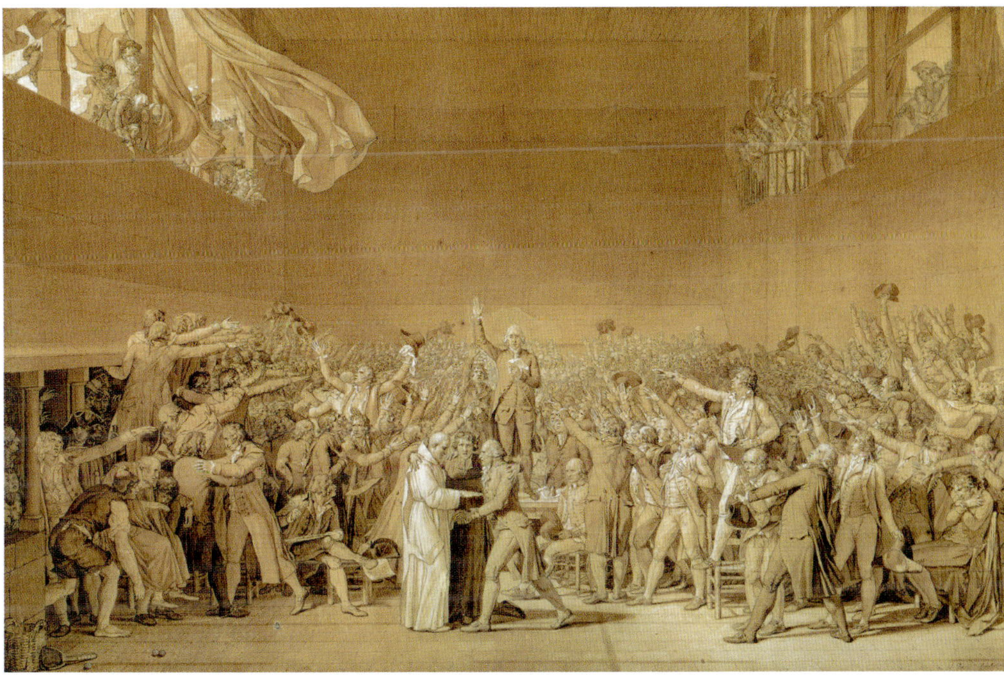

Figure 2.2
Jacques-Louis David, *Le Serment du Jeu de Paume à Versailles le 20 juin 1789*, 1791, pen drawing with white highlights, 66 × 101 cm. Châteaux de Versailles et de Trianon.

confused different sounds and thus were only briefly useful for their intended purpose. These concrete mirrors—receptive rather than projective versions of the *Siège orale*—were soon rendered obsolete by the introduction of radar. Yet they remain, crumbling on the marshes. Unlike the devices of the 1789 *Mémoire*, which were never constructed, the Dungeness listening machines now fulfill a different role as haloed artifacts of a future gone awry.[17] Despite their differences, these kinds of things can be worth thinking about together—the now quirky-seeming contrivances of the anonymous *Mémoire*, David's projected *grande machine*, and aborted technologies that echoed them felicitously many years later. They illuminate not only the enjoined histories of different kinds of sensory knowledge and the devices that communicated it, but also the ways in which the transmission of revolutionary information was susceptible to interference and transformation while *en route*: across distance, over time, in history.

The painted sketch and technical diagram also show how a seemingly transparent gestural language operated alongside numerous subtler codes and, crucially, both betrayed an attempt to intervene directly in the practice of political representation. Yet there is a sense in the *Mémoire*'s designs and David's preparatory image that both their authors are grasping for something that could not yet be realized. How to represent rapid movements of information from center to periphery and back again? How to bring together visual communication and speech acts in one dynamic process, rather than considering them separately? How to reconcile universal aims with specific circumstances, to allow for both ideal abstractions and human contingencies? As in so many other areas, the French Revolution changed how all of this might be understood. But the story does not end with this beginning. The fullest realization of these issues—the end of the revolutionary Terror on 9 Thermidor Year II, often taken as a terminus of sorts—was only really the start. Seen the other way, the official introduction of the optical telegraph that same month, a signpost for the beginning of a new regime of communication, concentrated a range of concerns—spanning science and the arts and exemplified by the proposals in the *Mémoire*—that had been current since at least the start of the Revolution, if not before. Its echoes were felt for many years afterward and were heard in other revolutions, the roar of which merged with and transformed the original signal.

The telegraph—in the form that claimed that name—was inconceivable without the political and social transformations of the revolutionary period which produced it and which it enabled. However, as with the workings of the telegraph itself, to move forward, we need first to look backward. Histories of technology are not reducible to inventors, but the Chappe family network was vital to the shape the telegraph eventually took. Claude Chappe was born in Brûlon, in the Sarthe department of northwestern France, on Christmas Day 1763. A twin, his sister Marie-Marthe, followed the next day.[18] Inheriting his surname from his maternal grandfather (he was, otherwise, de Vert), Chappe had a famous scientific uncle, the astronomer Jean-Baptiste Chappe d'Auteroche, who was celebrated for his observations of the transit of Venus in 1761 and 1769, and for his account of his travels in Russia, published with engravings by Jean-Baptiste Le Prince. Chappe d'Auteroche's *Voyage en Sibérie* was allegedly the first book his nephew read.[19] He died of yellow fever twenty years before the start of the Revolution, while studying the passage of Venus over the sun in California.[20]

After studies in Rouen and at the Collège Royal de La Flèche, Claude was named *abbé commendataire* (without religious obligations). Having moved to Paris, he opened a "cabinet de physique," where he conducted a range of experiments on static electricity, silkworm rearing, telecommunications, and sound effects.[21] With the outbreak of the Revolution and the suspension of priestly benefits by the Constituent Assembly on November 2, 1789, Chappe returned to Brûlon, where he continued his scientific research alongside his four brothers, René, Abraham, Ignace-Urbain-Jean, and Pierre-François, who had also lost their positions. He was twenty-six. According to Abraham, he began to think about a new form of government communication almost immediately, early the following year.[22]

Although Claude is now credited with the telegraph's invention, it was a family affair. Lateral connections between siblings correlated with and empowered the telegraph's horizontal duplications of information and its branching structure, and their collaboration was essential to the management of the network's intricate bureaucracy. After Claude's death in 1805, his brothers controlled the telegraph system until 1830, acting as prospectors, directors, and administrators, and surviving ten successive political regimes. None of the brothers outlived the system they had invented and developed: René was the last of the brothers to die, at Brûlon in 1854, just as their network was replaced by electrical signals. The Chappe brothers' first experiments with a telegraph took place at Brûlon on March 2, 1791, with a message sent between there and Parcé, fourteen kilometers away. Claude Chappe approached the National Assembly and received permission to establish a telegraph on one of Claude-Nicolas Ledoux's pavilions of the barrière des Champs-Élysées (also called barrière de l'Étoile or de Neuilly) in spring 1791. The machine was set up, but the experiment could not proceed because wood for the machine was stolen. A new machine was installed—this time with five wings, as screens were judged less effective—in the parc de Ménilmontant. Other systems were tried, including sound signals, as well as electrical communications (abandoned because of difficulties insulating lines).[23] Ignace Chappe's election to the Legislative Assembly, representing Sarthe, eased the presentation of the results to the Assembly, who permitted construction of three machines to conduct more significant experiments.

From March 1792, Claude Chappe installed his machines, soon to be destroyed by suspicious onlookers. A few days after his abortive trial in September that year, the National Convention replaced the Assembly, and the projects were adjourned. Within a few months, Chappe developed—with the essential expertise of Abraham-Louis Breguet—a new machine that replaced his earlier shutter systems: this was the distinctive machine that became known as the telegraph, the form of which remained relatively stable for half a century. As ever, during the Revolution, timing was crucial. Lakanal's support was especially vital. His *rapport* to the Convention in support of the experiments on July 26, 1793, elaborated on the technicalities of the system and argued that were it not taken up, "fanaticism would soon raise its altars, and servitude would cover the earth."[24] Successful demonstrations of the system that month between the parc Saint-Fargeau at Ménilmontant, Écouen, and Saint-Martin-du-Tertre (a distance of twenty-five kilometers) were convincing—Lakanal had been there to see it, based at the final post with Abraham Chappe, while his colleague Daunou was installed with Claude in the parc Saint-Fargeau. The day after Lakanal's speech, Chappe was granted the title of "ingénieur thélégraphe." Charles-Gilbert Romme, best known as the author of the Republican calendar, had already noted the telegraph's potential usefulness to the French military, arguing in a report to the Committee of Public Safety of April 1, 1793 that the machine offered—uniquely among many similar proposals—a reliable method of commanding troops from a distance. "An ingenious means of writing in the air" was how Romme described it.[25] With Valenciennes falling into enemy hands to the north and Dunkerque threatened, members of the Convention were eager for military advantage, and decided that the first line to be built should run from Paris to Lille. The order having been signed by Georges Couthon, Bertrand Barère, Marie-Jean Hérault de Séchelles, Louis-Antoine de Saint-Just, Jacques-Alexis Thuriot, and Maximilien Robespierre, it began operation in July 1794.

Although it was the first line to be developed, the Paris-Lille line was also the last to be retired, in 1847. The line originally ran from Ménilmontant, but to reduce the distance between the terminal device and the National Convention, the first outpost was relocated to the Louvre, linking from there to Ménilmontant, and onward to Écouen. Its construction ordered on 24 Prairial Year II (June 12, 1794) by the Committee of Public Safety, the Louvre telegraph was placed on Lemercier's *pavillon de l'horloge* at the western end of the *cour carrée* (the *pavillon* was built from 1624 to 1645, but the clock for which it was later named was added after the telegraph's time). During its residency there, it was the system's most visible node. Proximity to the Convention meant that the machine was subject to frequent security concerns; the royalist uprising of 13 Vendémiaire Year IV (October 5, 1795) prompted fears that enemies of the Republic would gain control of the telegraph.[26] However, additional guards posted at the site posed problems of their own. As Auguste Hubert, architect of the Louvre, wrote to Minister of the Interior Pierre Benezech: "I observe that this post is becoming all the more useless as the majority of the soldiers who spend their time on guard duty there, not wanting to take the trouble to go downstairs for their natural needs, allow themselves to do so below their room, which carries an infectious odor into the artists' workshops below and rots the upper boards of these workshops."[27]

Already fed up with being peed on by guardsmen, sixty-one resident artists signed a petition against the telegraph in 1800 when for security reasons a restaurant in the Louvre run by Swiss "militaires-portiers" had been closed, a move they considered "insensitively prejudicial." The Louvre telegraph ultimately fell afoul of Napoleon's desire to fully complete the transformation of the palace into a national museum, and the head of the Lille line was finally moved to Saint-Sulpice in 1806. Despite the Chappe brothers' protestations, the demolition of the telegraph above the museum began in May that year. Fifteen workers, supervised by neoclassical architect Pierre-François-Léonard Fontaine, took the machine apart and restored the dome of the tower.[28] Photographs of the *pavillon de l'horloge* for the *Missions héliographiques* around the 1850s, shortly after the retirement of the system, taken by Gustave Le Gray and Édouard Baldus, and at other times by Charles Nègre and Charles Soulier, still seem haunted by the absent figure of the telegraph, a half-century after its removal. Ruination and reconstruction shaped the other end of the line too: the tower of Saint-Pierre de Lille, where the telegraph stood until it was moved to the church of Sainte-Catherine in 1795, was the remainder of a Roman structure destroyed by revolutionaries in 1793.[29]

With the telegraph functioning as of July 1794, the following month Abraham Chappe sent news of the taking of Quesnoy. Bertrand Barère praised the invention, insisting that "we should trace a few lines today on this telegraph machine." It was, Barère contended, up to legislators "to put an end to the clamor of ignorance or the anxieties of curiosity," and to encourage the arts and sciences in the name of freedom and Enlightenment.[30] The reaction was muted, for the Convention was not sitting at the time. When, however, Lazare Carnot announced to the Convention on August 30 that Condé had been taken, and that the news had been received by telegraph, the reception was rapturous. It was immediately decreed that Condé's name should be changed to Nord-Libre, and that this information should be sent back via telegraph. This success prompted the development of other lines. A line from Paris to Strasbourg followed the Lille connection in 1798 and one from Paris to Brest a year later. The latter line, departing from the place de la Concorde, was primarily designed to serve the navy, with a view to a prospective invasion of England—the architecture supporting the machines on this line was particularly well reinforced. These telegraphs play a key role in a remarkable fantasy of aerial, seaborne, and subaquatic invasion published in 1803, where they seem to be conducting the balloons above as well as boats and soldiers below (fig. 2.3). This line also served as a main avenue for news from France's Caribbean colonies.

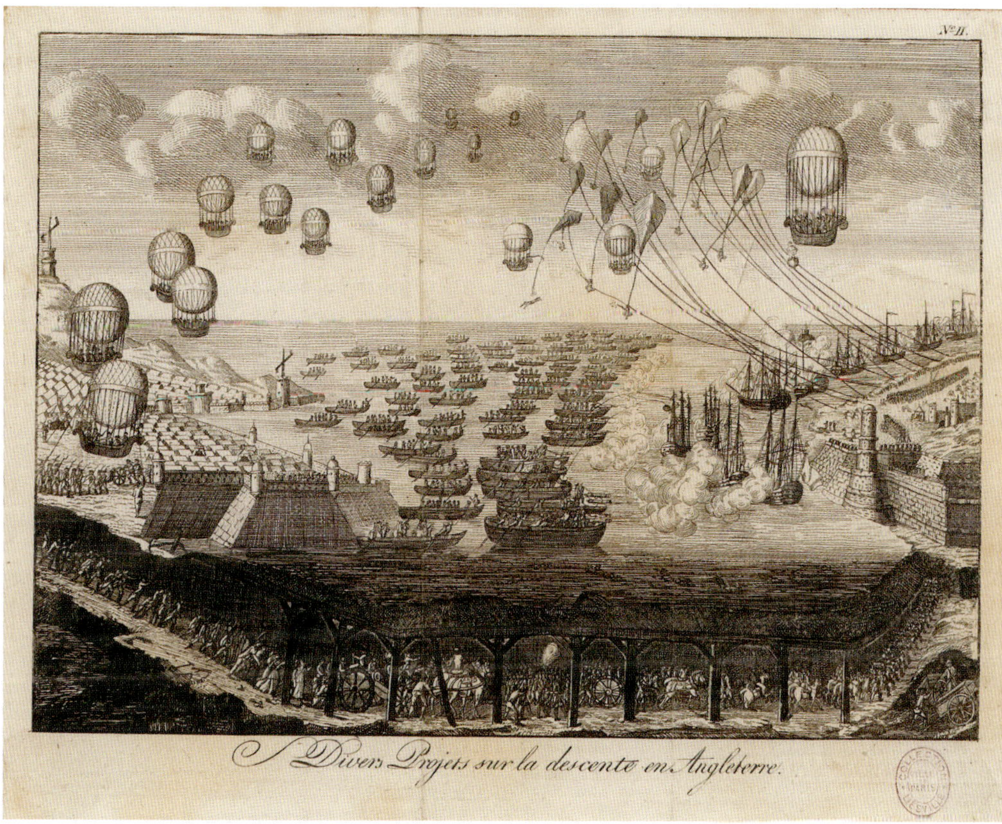

Figure 2.3
Divers projets sur la descente en Angleterre, c. 1803, etching and engraving, 19 × 24.2 cm. Musée Carnavalet, Paris.

While the development of the network seemed in retrospect a sure-fire success, it did not seem so for those involved. Not only were telegraphs frequently placed near the front, risking direct assault, but access to financial support was a constant concern ("funds, funds, funds, once more, funds," pleaded Claude Chappe in Ventôse Year VI [February–March 1798]), and anxiety and exhaustion plagued those tasked with the many obstacles that appeared as the lines were, station by station, built and staffed.[31] Ignace Chappe remembered how the "death [of the project] hung over everyone's head."[32] However, by the time the Directory was replaced by the Consulate following Napoleon's seizure of power in 1799, there were three lines: north to Lille then Dunkerque, west to Brest via Saint-Malo, and east to Huningue via Metz and Strasbourg.

Although the Treaty of Amiens in 1802 caused a brief shutdown of all but the Paris-Brest line (again partly due to shortage of funds), renewed hostilities with England from May 1803 led to the construction of new lines from Lille to Brussels and Boulogne, and the following year saw Paris connected to Dijon and onward to Lyon in 1805, with additional lines from Lyon to Turin, and from there to Milan. A Lyon to Toulon line, via Marseille, was built in 1821, and Paris reached Bayonne a year later. The network continued to expand until 1846, with connections spreading out to Antwerp, Amsterdam, Venice, Mainz, Calais, Bordeaux, Bayonne, Nantes, Perpignan, Avignon, Montpellier, Besançon, Agen, Eu, Behobie, and Cherbourg. New lines were established on an almost annual basis, each one passing through many towns and villages on its route. The telegraph's appearance astonished onlookers: when used in the battles of the north, "the soldiers of the coalition were struck with stupor; they began to believe that the French had something gigantic and supernatural."[33]

Revolution in the Air

As with all media, the telegraph's visibility was proportional to its "newness," and numerous responses to it appeared in the interregnum between the introduction of the network and its acceptance. The telegraph's most obvious order of representation, the shapes thrown by its arms, was paralleled by the storage and translation of this information in the official registers of telegraphic clerks. Its more insidious, accidental, or disruptive appearances, however, can be discerned in images—artistic and otherwise—whose variety was congruent with the divergent ways in which the telegraph's presence invaded the public sphere. Optical telegraphy was connected to diverse forms of cultural expression in ways that began with, but were not limited to, the iconography of the telegraph's angular convulsions.

For some, the telegraph promised a source of visual pleasure. Arguing for the placement of the telegraph on the "Pavilion of Unity" at the Palais National (the Tuileries Palace), Jacques-Antoine Rabaut-Pommier suggested that the telegraph would offer aesthetic enhancement to the building's decoration, claiming that "by means of an optical illusion, the poles designed to support the telegraph gallery will disappear, so that it will appear suspended and without support."[34] Echoing the enthusiasm for the phantasmagoria and other optical illusions in post-Terror France, the telegraph's secret language promised further magic (the visual trickery foreseen by Rabaut-Pommier also anticipated the mirage effects discussed in the next chapter). Claude Chappe himself was more than aware of the arresting visual and symbolic effects of the telegraph, over and above its operational function. He designed most features of the system, and although his brothers soon took over its administration, he was involved in this too, finding sites, sourcing materials, and directing construction.[35] His signature appears on numerous passport applications for workers to travel the country to construct and facilitate the network. He designed uniforms (the copper buttons for the coats of telegraph engineers sported a miniature telegraph), and he made a set of watercolors representing the apparatus.[36]

As a printed identity card issued to a telegraph operator demonstrates, the Chappes exploited the potential for the telegraph's arms at rest to mirror visually the Masonic triangles, pyramids, and scales of justice of Republican iconography (fig. 2.4). In this example, the telegraph replaces the pike or axe at the center of a fasces inscribed on the right-hand side of the pyramid. Meanwhile, the masthead for a schematic map of the inaugural Paris-to-Lille telegraph line combines telegraph arms with an eye of surveillance and *bonnet rouge* to create a compound vignette that is at once innovative and effective (fig. 2.5). There is porosity between such images and more self-consciously artistic works, such as Regnault's 1794 *Liberty or Death* (fig. 2.6), where several of these emblematic elements recur, and which in its turn echoes

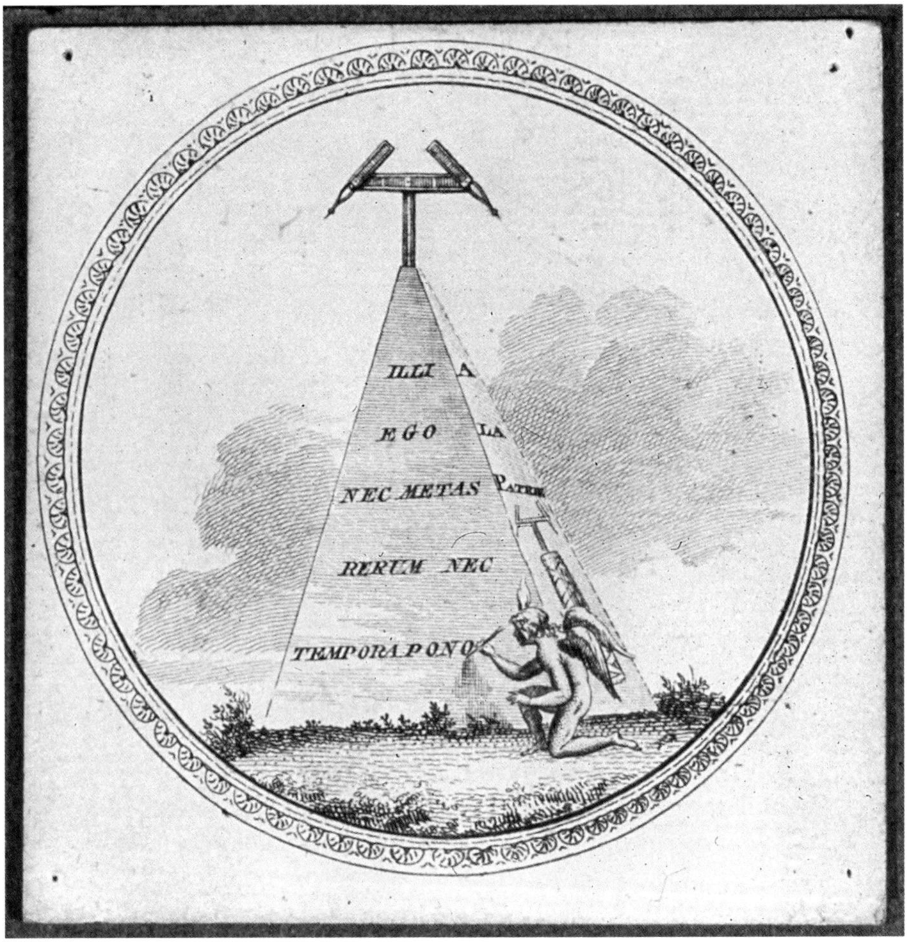

Figure 2.4
Carte d'admission, c. 1794, etching on cardboard, 7.5 × 7 cm. Musée de La Poste, Paris.

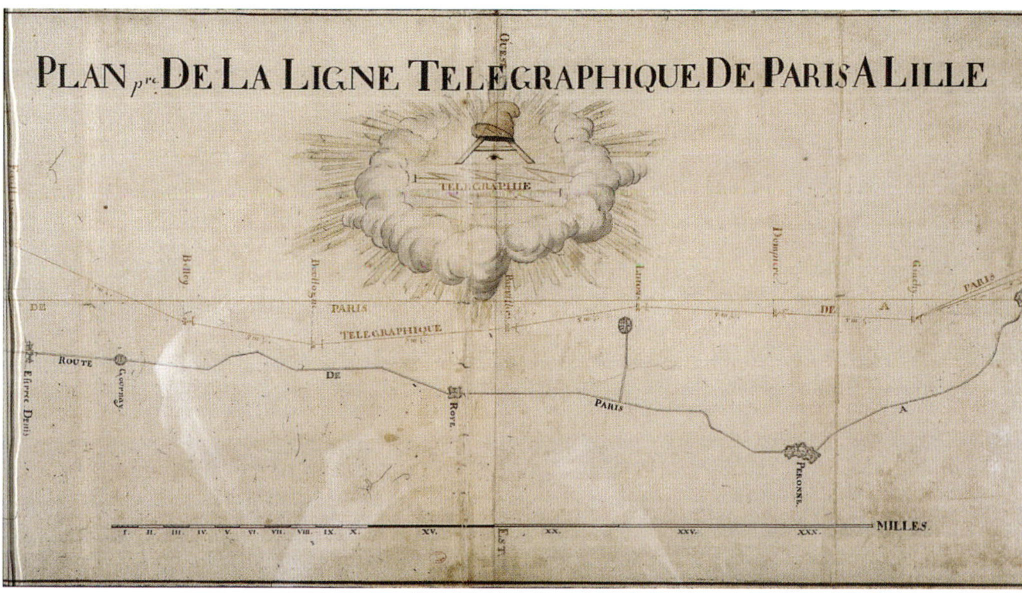

Figure 2.5
Plan de la ligne télégraphique de Paris à Lille, 1794, drawing and watercolor on paper, 36 × 170 cm. Musée de La Poste, Paris.

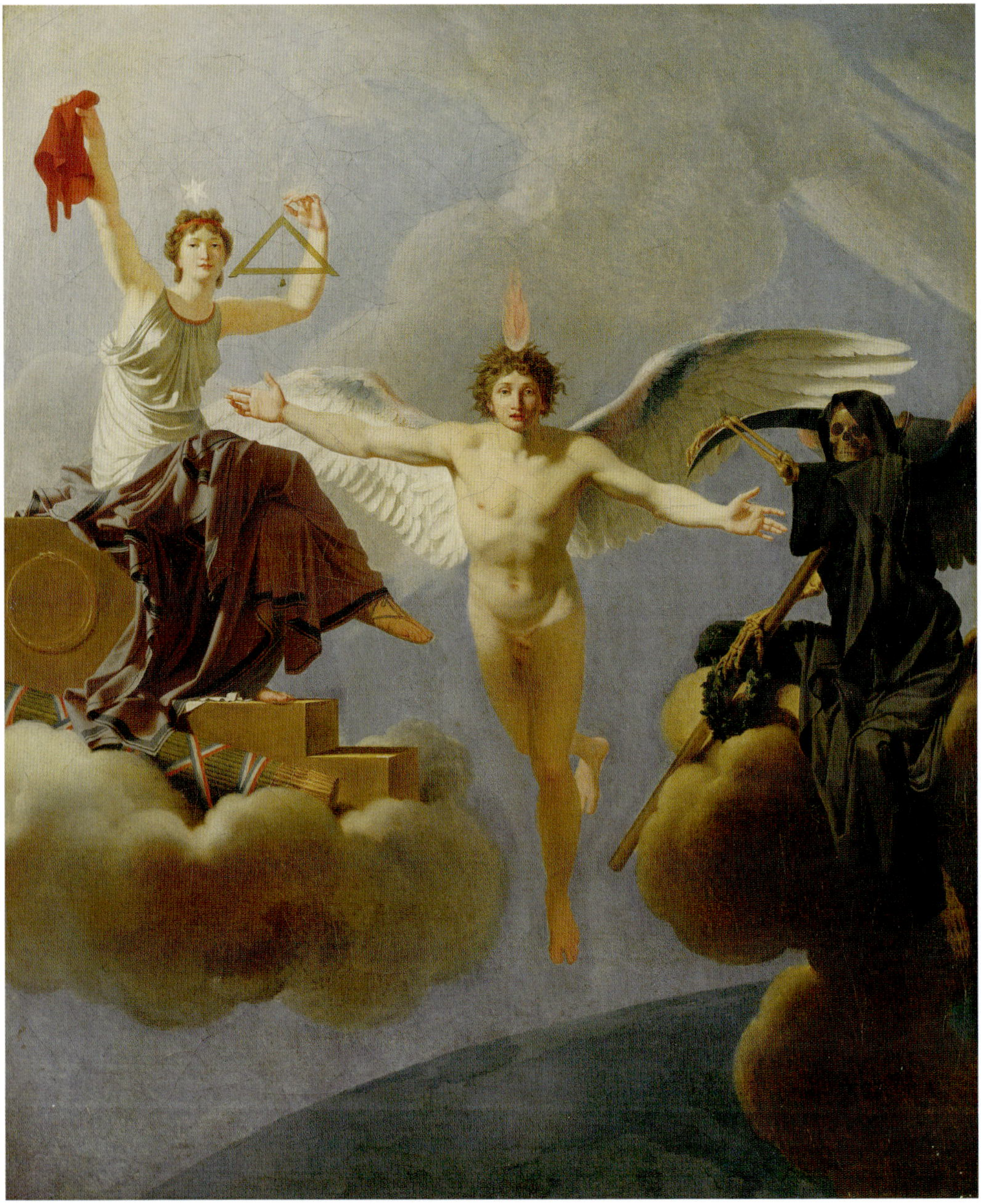

Figure 2.6
Jean-Baptiste Regnault, *Liberty or Death*, 1794–1795, oil on canvas, 60 × 49 cm. Hamburger Kunsthalle.

vignettes on official paperwork.[37] While not a representation *of* the telegraph, Regnault's painting is nonetheless one kind of "telegraphic image": reducing a complex political allegory to its simplest formulation as binary choice, a *génie* of France, its *tricolore*-tinted wings splitting the clouds, semaphores the Revolution's global message.

In the half-century of its operation, the telegraph altered skylines across France, transforming the role of buildings and the way in which landscape, urban and rural, was perceived. The visual register of the telegraph is pronounced in a work such as architect Charles Norry's *Vue de l'intérieur du Louvre d'après nature l'an 4ème* of 1799, which represents the optical telegraph on the roof of the museum, presenting the encrypted signs transmitted by the modern telegraph in proximity to the wealth of artist-made signs stored in the building below (fig. 2.7). Norry had, earlier that year, returned from Egypt, where he had been a member of the Institut d'Égypte.[38] The Louvre telegraph gave artists an unusual opportunity to represent a pressing contrast. Made in the period from 1796 to 1801 when the Louvre was closed, Norry's image focuses attention on the exterior of the building, making the telegraph the main subject of the work, hovering implacably over the ill-defined figures beneath it, transporting a sculpture across the courtyard. Yet this work, like others representing this subject from multiple angles, is not a straightforward glorification of Consular power. Rather, oscillating between dispassionate observation and allegories of representation writ large, works such as Norry's addressed the function of images across time and media, and accentuated the telegraph's modernity by highlighting its juxtaposition with an obsolete or transformed architectural milieu, which remained nonetheless necessary to its operation. The circular format of Norry's drawing lends it a telescopic air, offering the conceit that we, its viewers, are also encountering the telegraph through a lens. As is often the case in such images, smoke from a nearby chimney animates the roofline with a steady plume, its atmospheric drift drawing attention to the telegraph's jerkier interruption of aerial space.

In more rural areas, telegraphs were placed on purpose-built turrets, hurriedly constructed with practicality in mind, while within villages and towns, they frequently appeared on existing structures, such as churches, where, taking over the traditional role of church bells, which were melted down during the Revolution, they figured as secular, mobile crucifixes.[39] As late as 1840,

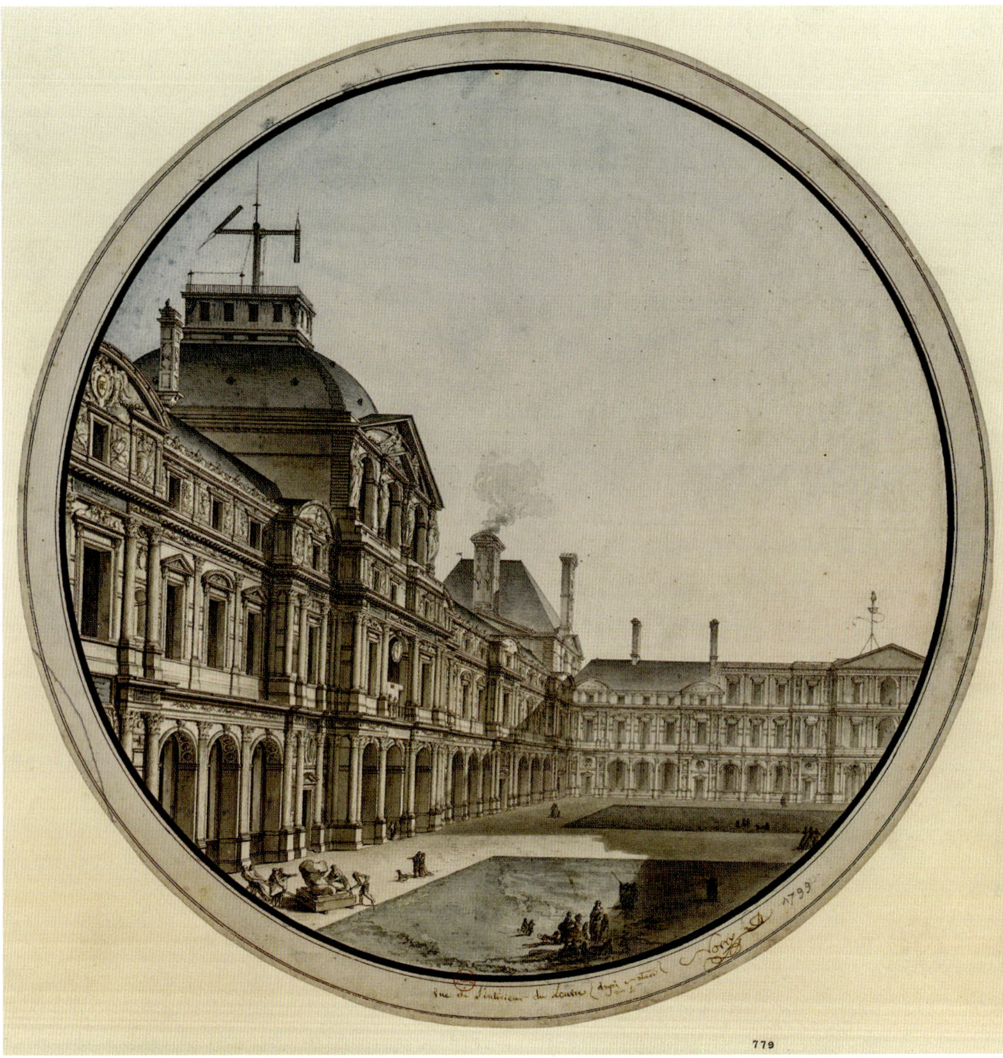

Figure 2.7
Charles Norry, *Vue de l'intérieur du Louvre d'après nature l'an 4eme* [sic], 1799, pen and India ink drawing with watercolor, 35.5 cm. diameter. Bibliothèque nationale de France, Paris.

Julien-Léopold Boilly's picturesque lithograph of the telegraph at Montpellier capitalized on the consonance between an imposing wrought-iron cross and the machine behind it (fig. 2.8). Other mechanisms of passage are shown in the image: a mail coach reloads in the foreground (coaches were frequently named after the telegraph), although a discarded sack leaking in the foreground and a few broken wheels nearby suggest it might not be entirely reliable. Boilly's lithograph replays certain elements of his father Louis-Léopold's 1803 painting of the arrival of a *diligence* in the cour des Messageries, rue Notre-Dame-des-Victoires, where a telegraph was situated from 1806, although unlike his father, he gave full voice to this disenchanting interloper. In Boilly's print it isn't clear whether the two metal devices—cross and telegraph—should be understood in relay with one another or in opposition. It is the telegraph, however, perched atop a tall building, that reaches for the heavens.

Some of the telegraph's associations were less readily locatable. As a print by Pierre-Michel Alix after Jean-Frédéric Schall shows, the telegraph invaded somewhat retro pastoral landscapes of erotic diversion, where it is deployed as both amorous semaphore and improvised missile launcher (fig. 2.9).[40] The idea of telegraphy as a means of flirtation found numerous outlets—a British fan, from 1797, is printed with advice on the art of "fanology," or how to semaphore attraction, and accurately convey ardent messages in code. While the use of fans in such a way had been well established, only now was it described as "the telegraph of cupid."[41]

The secret signals of fanology and Schall's image presented romantic liaisons as a matter of strategy in new, overtly militarized ways. As Mary Favret has proposed, "with the advent of mass media . . . and in an increasingly popular visual culture of prints, panoramas, and theatrical performances, wartime stepped forth as a mediated relationship to distant violence."[42] The Chappe telegraph, as well as its representations, should be included among the visual technologies that mediated conflict in this period—sometimes, as here, in the guise of romantic skirmish. The telegraph was, first and foremost, a machine developed for military purposes, and it mobilized landscape, sky, space, and time as parts of an extended battlefield that spread beyond France's borders. Messages that passed through the network dictated the course of conflict and shaped the temporal experience of warfare, even many miles from the front. As Favret makes clear, it is one thing to

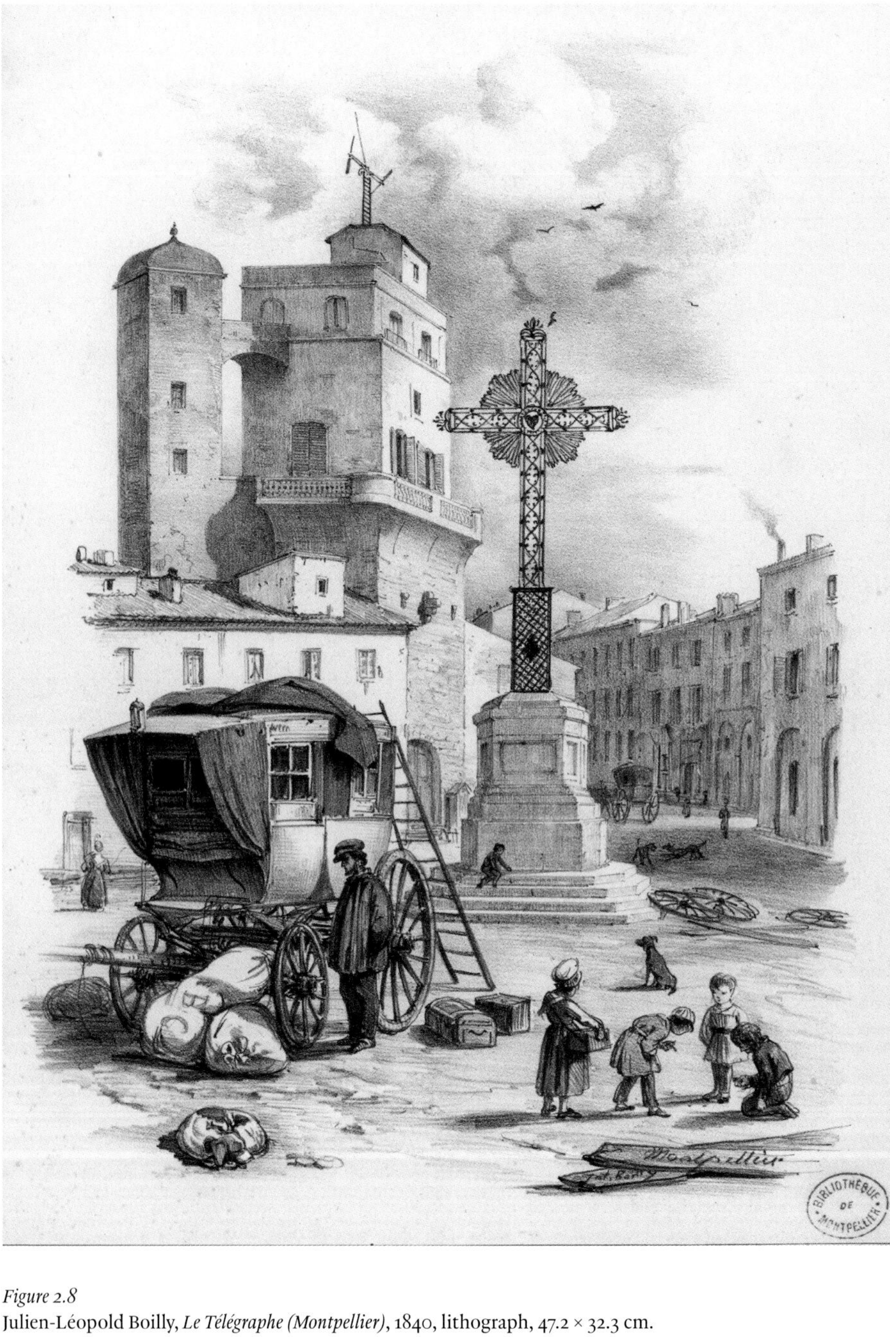

Figure 2.8
Julien-Léopold Boilly, *Le Télégraphe (Montpellier)*, 1840, lithograph, 47.2 × 32.3 cm. Bibliothèque nationale de France, Paris.

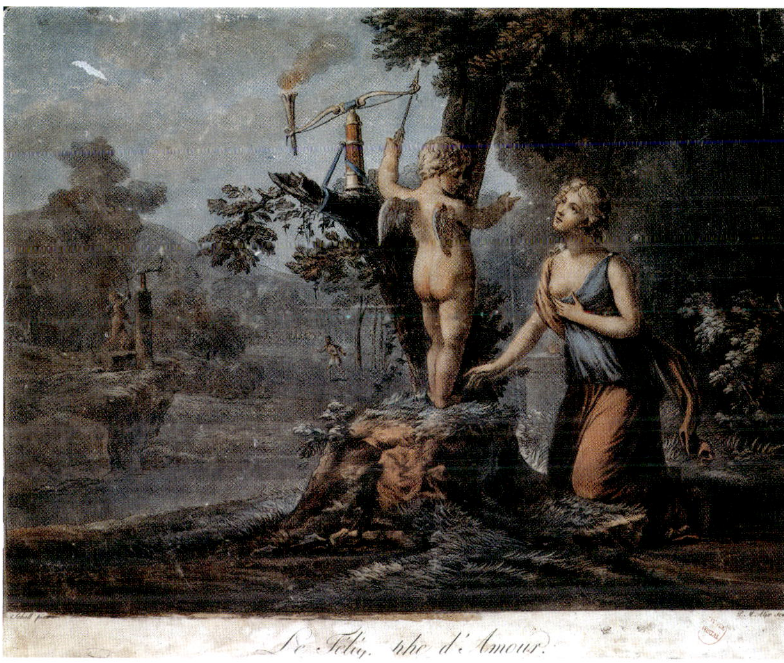

Figure 2.9
Pierre-Michel Alix after Jean-Frédéric Schall, *Le Télégraphe d'amour*, 1804–1815, color engraving, 27.5 × 33.5 cm. Musée de La Poste, Paris.

document instances of military communication, but quite another to account for the vagaries of wartime affect. Viewing a telegraph in action rendered distant conflict visible, collapsing center and periphery, forcing attention to the temporality of war and rendering it an object of potential knowledge. However, this visibility was perplexing, for it revealed none of the information being transmitted. Consequently, any knowledge produced by the telegraph, for those on the ground at least, was less about the actual events of the battlefield than it was about the visual experience of those living through a time of war. Later, military paintings of the era, even when they did not represent a telegraph, showed events that noncombatants first encountered through telegraphic news.[43] Mediations of mediations, images that invoked the telegraph provided a means to think about transmission more generally, to consider the historical dimensions of the everyday, and to trace emotional and material connections between different media.

In contrast to Schall's fantasy or the Louvre or Montmartre images where the telegraph is a principal subject of the image, in many works the apparatus appears as an indistinct silhouette on the horizon. An anonymous late eighteenth-century image of the Carthusian monastery at Vauvert, near the site of the Jardin du Luxembourg, shows the structure in ruins (fig. 2.10). Closed in 1790, this building was used as an armaments factory, before being demolished between 1796 and 1800. Whereas a representation of the telegraph on the Louvre lent the telegraph historical weight by including it in a lineage of established signifying technologies, from paintings to sculpture, and by allying it with an appropriated revolutionary and imperial space, the two telegraphs on a tall structure to the left of this image, whose irregular twin towers and placement identify it as Saint-Sulpice, are markers of temporal and symbolic difference.

The Chartreuse's shattered windows, broken roof, and air of abandonment and porosity contrast with the tightly controlled and hermetically sealed interior space occupied by the telegraph, whose telescope window can just be glimpsed at the top of one of the towers. One building looks out, back at us, while another is looked into from every direction. In this image, the telegraph, as a sign of revolutionary modernity and efficiency, is juxtaposed with the debris of medieval religious observance, although again it is not clear whether the telegraph has replaced the structures of tradition or has instead appropriated the mystical, secretive

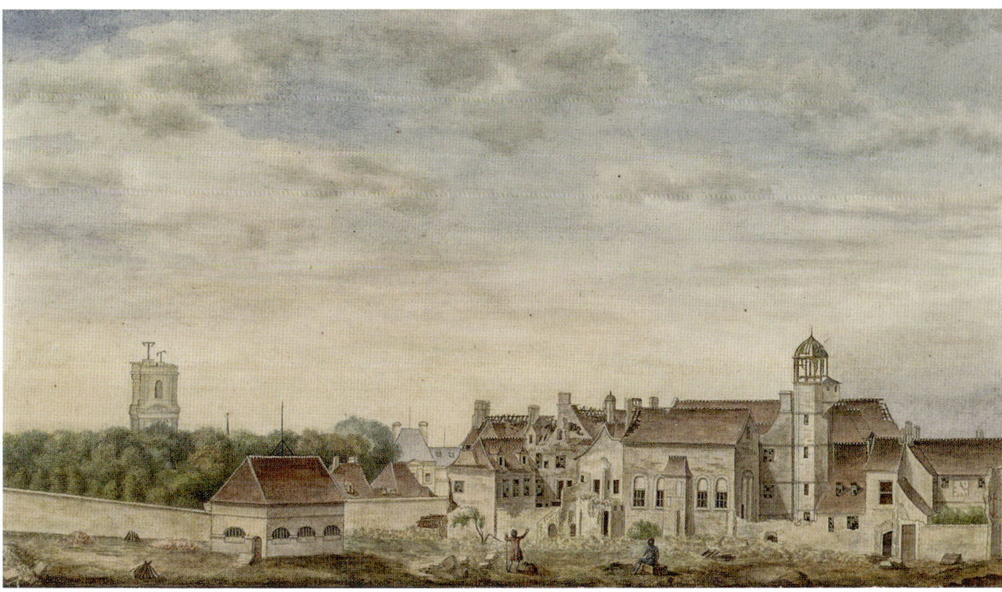

Figure 2.10
Les Chartreux [*Chartreuse de Vauvert*], n.d., pen and brown ink drawing with watercolor, 18 × 31.6 cm. Bibliothèque nationale de France, Paris.

characteristics of monastic life. Either way, the gardens in this work not only separate the city from the countryside (or at least its urban manifestation) but form a clear barrier between two historical worldviews. Overlooking the Chartreuse, the Saint-Sulpice telegraphs also moderate and police the environment they survey. This dynamic became more pronounced in the years after the Revolution, when for many the telegraph came to indicate the surveillance practices of a coercive state.

The Slave in Favor, The Master to Come

At some time in the first years of the Bourbon Restoration that followed Napoleon's defeat at Waterloo, a teenage Victor Hugo looked out of the window, scrutinizing the roofs beyond. He was annoyed by what he saw, or rather what he couldn't fail to see. Hugo later wrote: "That damned telegraph, will it finally cease / To disturb my eyes, which it begins to tire? / There, before my window! It is quite ridiculous / That someone would place a telegraph outside my room! It rises, it falls; and my distracted spirit, / In these vain movements, looks for some secret."[44] These words appeared in 1819, when Hugo was just seventeen, in *Le Télégraphe*, an anti-Napoleonic verse satire. From February 1815 until August 1818, Hugo had lived with his brother in the Cordier pension at 42 rue Sainte-Marguerite (now rue Gozlin), subsequently relocating to his mother's house nearby. He considered the Cordier pension a "prison." Both houses were a stone's throw from Saint-Sulpice, and the clanking arms of the optical telegraphs there, onto which his room faced, prompted Hugo's irritable response to the disruption of his train of thought.

Writer's block was hardly a characteristic problem for Hugo, and the lack of productivity brought about by this intrusion was momentary. Reflection on the content of the enigmatic messages passing back and forth was as much a stimulus to work as an impediment, and Hugo's exasperation with the telegraph's effects on his "sterile mind" was disingenuous. In fact, by situating himself in conjunction with the telegraph, Hugo placed himself at the center of a vast information network that efficiently allegorized his bird's-eye perspective on modern life. Yet while Hugo's satire imagines the telegraph as an analogue for writing, or rather, as a way of framing his own distraction through the act of writing, the telegraph notably figures here as a disturbance in the visual field. Although its message was not comprehensible, the telegraph produced a kind of visual "noise" that corrupted creativity and blurred lucid expression.

Yve-Alain Bois—reviving aspects of the information theory pioneered by Claude Shannon and Warren Weaver—has investigated the presence of "visual noise" in American artist Ed Ruscha's word paintings. In this context, Bois describes "what is lost, forgotten, usually unseen, unheard, taken for granted, ignored as mere noise: the size and scale of letters, the phonetic flutter, the grammatic alliterations—but also the pattern of oil droppings on parking lots, the vacant real estate opportunities in the urban landscape, the metaphors in the robotized language of mass media. In other words, the refuse, that which is destined for the gutter of oblivion."[45] The coded words of the telegraph signified a frustrating secret whose interpretation exceeded the bounds of available knowledge, even as its outward manifestation was enacted in full sight. But, in a manner not wholly distinct from Bois's account of Ruscha's declarative logos (or, for that matter, Hugo's own drawings featuring words—*Souvenir*, for instance, in which the letters hover ominously above a gloomy Paris), their presence also pointed toward something more damaged or entropic, leaving waste and decay in their wake. Telegraphy's visual acts were ephemeral; its "phonetic flutter" survived in trace forms, which themselves often had only the most fragile claim on permanence. Recalling in 1842 his youthful shock and pleasure at the installation of a telegraph in his town, writer and journalist Mathurin-Joseph Brisset found in the provisional machine—jarringly deposited on a dismantled medieval tower and so flimsy it could be blown away by an autumn wind—affinities with the Revolution itself: "Both of them with so little connection to what preceded them, both of them boldly planted on the debris, both of them resisting, like daring Titans, the storm unleashed on their lands, seemed destined to last only as long as those machines of war erected around besieged cities, which fall when they have accomplished their work of destruction."[46]

The late eighteenth century had been a turbulent time for Saint-Sulpice. During the Revolution, the church had been desacralized, operating as a Temple of Reason (simply called temple Sulpice, the "Saint" having been dropped), then of the Supreme Being, and from 1797 a Temple of Victory, before being remade as a church in 1802. The telegraph was thus only the most recent imposition in this space. The telegraph on the larger tower connected to Ménilmontant, then east to Strasbourg. The smaller tower faced in the direction of the Italian line, and from 1840 it became the head of the Paris to Bayonne line. The apparatuses can be seen in contemporary prints, where the awkward asymmetries of the telegraphs' erratically moving limbs accentuate the mismatch between Jean-François-Thérèse Chalgrin's reconstructed neoclassical tower and its smaller counterpart to the right, whose elevation to a similar height was halted by the French Revolution.

Alternatively, one might prefer to view the telegraphs at Saint-Sulpice in engravings that approached it from the Marché Saint-Germain, the machines peeking above the rooftops.[47] Or on a *faïence fine* plate, or in an inverted stereoscopic photograph of the same scene, or in a pencil sketch by J. M. W. Turner, where one of them appears in a ghostly miniature fragment to the left of the towers, which breach the top of the sheet, and from which it stands alone, as if the artist just *had* to fit it in, one way or another.[48] Not long after, Fizeau's daguerreotype, taken from his own roof nearby, shows them still in situ (fig. 2.11). In short, around the time Hugo was writing, and for some years afterward, the Saint-Sulpice telegraphs were available for consumption across a range of images and objects, each with distinctive histories of production and consumption, different clienteles and publics, and different meanings. The telegraph's "visual noise" wasn't limited to the device itself but was, rather, amplified in images that depicted it. Given his antipathy to the machine, it is no surprise that Hugo omitted it from his many drawings of decrepit towers, cityscapes, and gothic castles, but it seems to haunt them nonetheless, even in absentia. His wonderfully atmospheric image of the Mont-Saint-Michel, with the telegraph—which he loathed and was fascinated by—almost discernible at its summit, offers an eloquent graphic interpretation of what "visual noise" might look like (fig. 2.12).

Hugo returned frequently to the telegraph, that "discreet harbinger of indiscreet history."[49] Two years after Thackeray, he too inscribed telegraphic signs midtext, although he merely documented them and they passed unexplained and incomprehensible, without being manipulated to represent figures from history paintings or, indeed, anything else.[50] Following a visit to the Mont-Saint-Michel, Hugo wrote from Coutances to his wife,

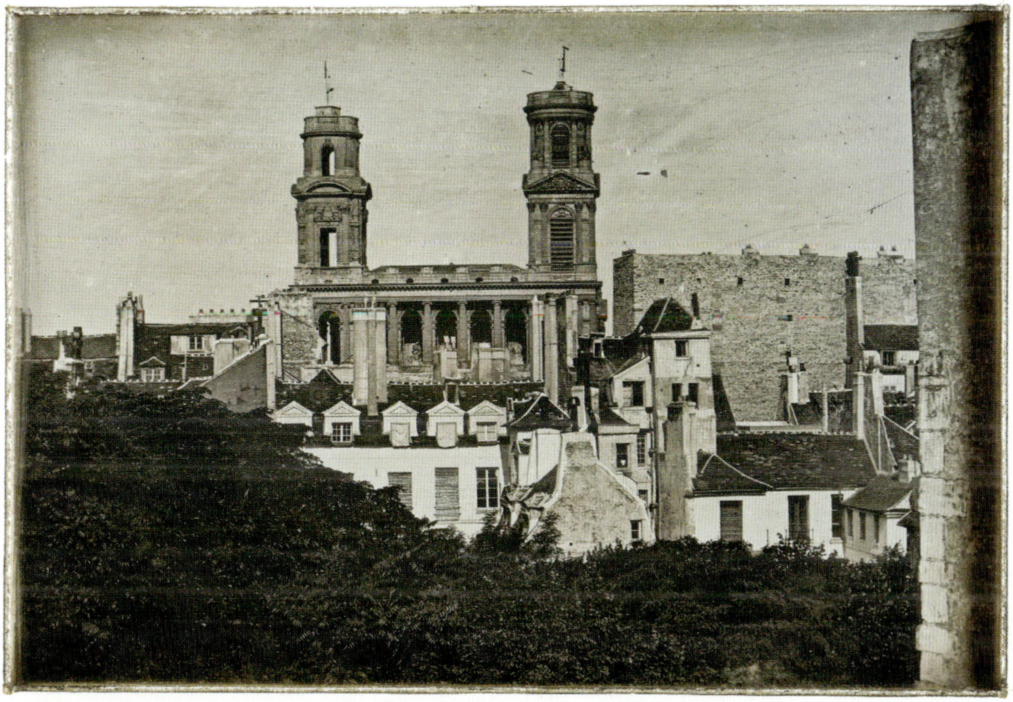

Figure 2.11
Armand-Hippolyte-Louis Fizeau, *Vue de Saint-Sulpice, prise des toits de la maison de Fizeau, 17 rue du Cherche-Midi*, c. 1840, daguerreotype, 5.5 × 8.2 cm. Musée Carnavalet, Paris.

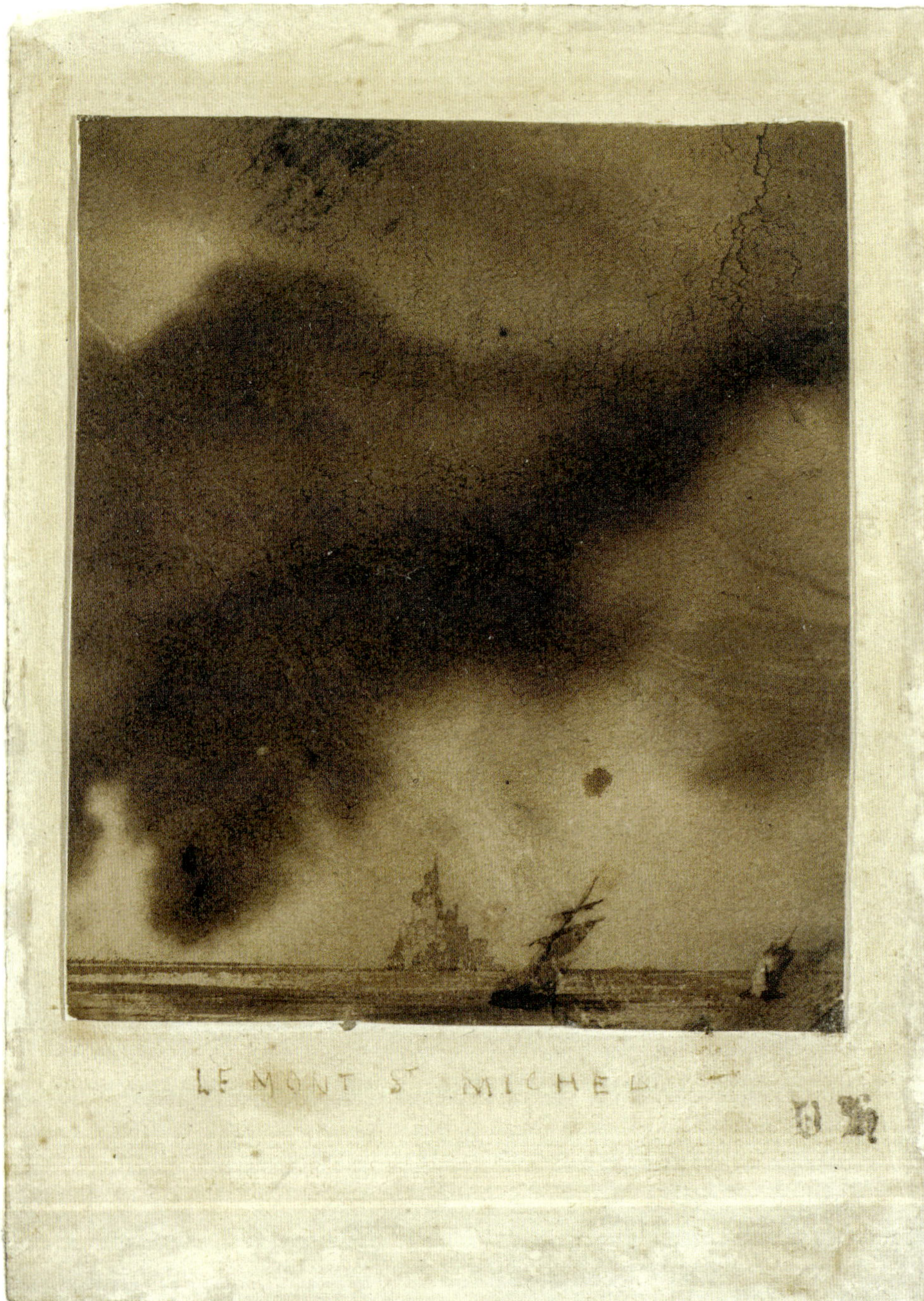

Figure 2.12
Victor Hugo, *Mont-Saint-Michel*, c. 1847, pen, brush, and ink on wove paper, 9.8 × 7 cm. Maison de Victor Hugo, Hauteville House.

Adèle Foucher, the following day, June 28, 1836: "To crown it all, at the top of the pyramid, in the place where the colossal golden statue of the archangel shone, we see four black sticks in torment. It's the telegraph. There where arose thoughts of heaven, the wretched wriggling of the affairs of the world! It's sad."[51] The Mont-Saint-Michel had had no religious function since the seventeenth century, and by the nineteenth century was mainly used as a prison. Hugo ascended to the telegraph, which was in constant use that day. He later found out, when he reached Avranches, that it was transmitting news of the latest attempted assassination plot against King Louis-Philippe. Time was out of joint, for world-historical news was encountered as it happened but was only comprehensible later, when deciphered and released.

The presence of the telegraph always promised to unsettle one with new or unexpected information, but it could destabilize in other ways too. Having been warned by the telegraph operator to avoid the antennae of the machine, so as not to be flung from the narrow precipice five hundred feet down into the sea, Hugo threw his hat into the operator's cabin and forgot the telegraph, becoming absorbed instead by the view out to the ocean from the Mont-Saint-Michel: "The sea was rising at that time. Below me, through the bars of one of these dungeons which they call lodges [*les loges*] I saw hanging the legs of a prisoner who, turned toward Brittany, sorrowfully sang a Breton song which the gust carried away across Normandy. And then there was also below me another singer who was free. It was a bird. I stood immobile above, wondering what the bars of one had to say to the wings of the other."[52] Hugo's turn away from the telegraph was not unique. Several images of the era present us with the view from the telegraph, in which the apparatus is nowhere to be seen. In works that meticulously, even disdainfully, note their production from *above* the telegraph, we are reminded again of Maupassant's enthusiasm for the restaurant in the Eiffel Tower—famously the only place in Paris where one did not have to see the tower itself.[53] However, Hugo's reverie about the relative liberty of the prisoner and the bird, each one singing a beautiful but impenetrable song in a language alien to the author, was short-lived, for "all of this was cut off by the harsh cry of the pulleys of the telegraph transmitting the dispatch of the Minister of the Interior to the prefects and sub-prefects. There are no longer any political prisoners at Mont-Saint-Michel. When will there be no more prisoners at all!"[54]

The connection that Hugo made between communication technologies and servitude or imprisonment was, in fact, firmly established. A satirical print from 1815, *La Force de l'habitude*, represents Napoleon directing a rat-operated telegraph on Saint Helena, demonstrating the extent to which the association between the emperor, authoritarian rule, and the telegraph had taken hold (fig. 2.13).[55] Mocking the folly of Napoleon's ambition, this print also associates the boredom and madness of exile with the presumed monotony and cruelty of the telegrapher's task. The rats in their traps figure in this image as early instances of what soon became a common trope—the idea that the task of telegraphy brought about a particularly enervating effect upon its caged operators, poor exhausted creatures trapped in their personal Bastilles and blind to the messages they transmitted. "Living chrysalises," Alexandre Dumas termed them in the *Count of Monte Cristo*, "poor wretch[es]," "fagged to death with cabals, factions and government intrigues."[56] Such characterizations of telegraphy's "insect-actors" had a long shelf life, continuing into the era of electric telegraphy and providing evidence to back up widespread suspicion of the telegraph's secret transmissions. Sometimes, this unsympathetic response came from those in power: an 1815 letter to Ignace Chappe from the *préfecture de police* denounced an employee of the telegraph at Saint-Sulpice for persistently making remarks injurious to the newly restored king and the royal family, in the presence of visitors to the telegraph. The letter requested that the stationnaire be taken to task for his "reprehensible conduct" and demanded that Chappe report back to the police on his character and behavior.[57]

Hugo was sensitive to the telegraph's eloquence about power and its operations, and to the way that it made the domination and reach of the modern state alternately visible and invisible. The penultimate stanza of *Le Télégraphe* warns of dangers to come and invokes the language of captivity and exploitation that became typical in talking of telegraphy: "In the slave in favor see the master to come." The machine, enslaved for now, contained the potential to rise and subjugate those who controlled it. Fear of such a role reversal accompanied—and still accompanies—industrialization and technologized intelligence of various kinds, and it was surely inflected by recent experiences of revolution in France's colonies. At the very end of *Le Télégraphe*, however, the apparatus takes a different shape. It becomes a bumbling old man, waving its arms

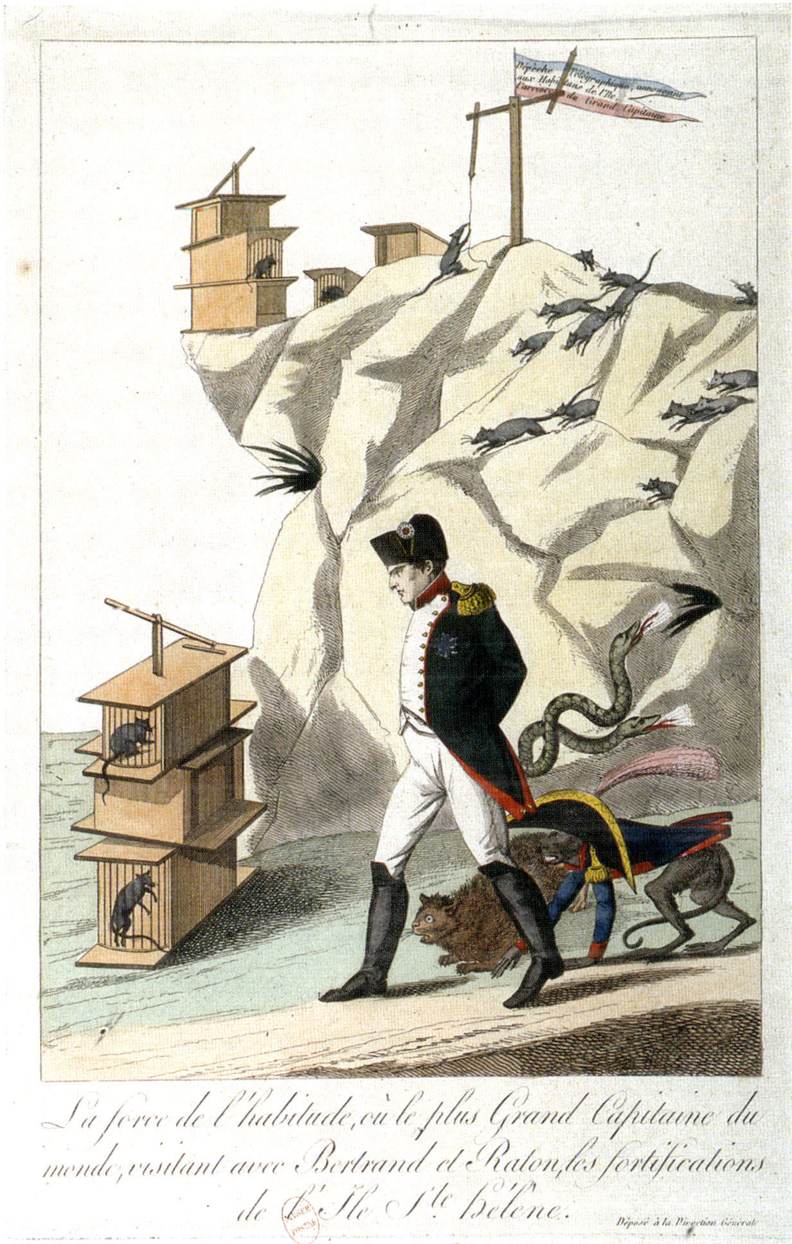

Figure 2.13
La Force de l'habitude, où le plus Grand Capitaine du monde, visitant avec Bertrand et Raton, les fortifications de l'Ile Ste Hélène, c. 1815, hand-colored etching, 43.5 × 27.5 cm. Musée de La Poste, Paris.

"Like that baron, famous for his jumble; / Who, scratching his head, with weeping eye and pale face; / Announces a great secret, which he himself does not know."[58] The "mastery" the telegraph prophesied was accompanied by blindness, incomprehension, miscommunication, and obsolescence.

In contrast to the bleak image of telegraphy presented by Hugo, contemporary utopian projects soon compensated for the telegraph's exclusivity and the lack of public access to its codes.[59] Notably, Charles Fourier and his followers were fascinated by the optical system. The telegraph's distinctive zigzag appears in designs for his ideal community, the *phalanstère*—described by Walter Benjamin as "a machine made of human beings"—showing up both in Fourier's own sketches and on a clock tower at the center of the better-known print by Victor Considerant (fig. 2.14).[60] Alongside other communication instruments, from flags and sound signals to carrier pigeons and messenger dogs, the telegraph was for Fourier an essential means to economic stimulus, enabling trade and the free communication of ideas. However, for the Fourierists, telegraphy was "above all an instrument of social harmony; it was the harmony between people that guaranteed material prosperity."[61] Telegraphy represented "the vocation of the phalanstery to communicate with the rest of the world," and it was suggested that operators should be lovers or friends, to avoid them seeing telegraphy as drudgery.[62] Chappe had imagined a future in which telegraphy would allow the government to "realize the greatest idea of power that we can conceive, by making use of telegraphs to spread directly every day, every hour, and simultaneously, its influence over the whole surface of the republic."[63] Fourier, on the other hand, saw telegraphy as a system that would exceed the limitations of a particular technical device, open for all to use and liberated from the oppressive imperatives of the state.

For the Fourierists, the technical aspect of telegraphy was secondary; it was primarily a social metaphor. From 1844, the electrical telegraph was enthusiastically received. However, as Thomas Bouchet observes, whether because of its familiarity or because electrical communication was difficult to represent with clarity, the optical device remained the most frequently depicted image of telegraphy. Its distinctive silhouette appeared in Fourierist

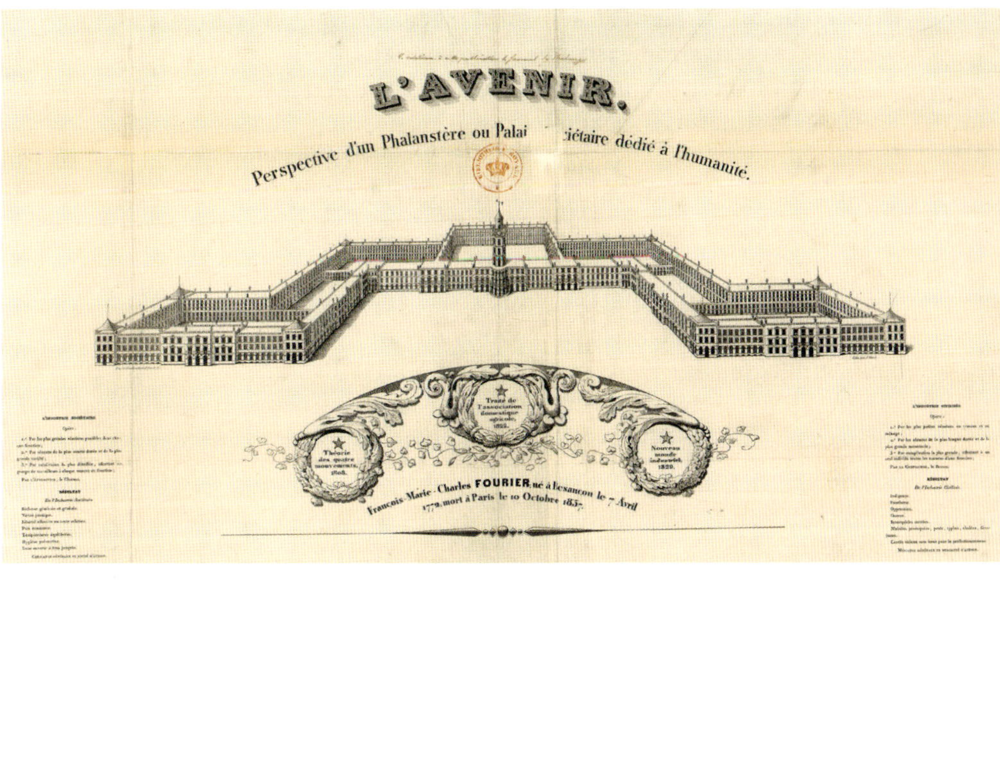

Figure 2.14
Detail of Victor Considerant, *L'Avenir. Perspective d'un phalanstère ou palais sociétaire dédié à l'humanité, d'après le plan de Ch. Fourier*, c. 1840, lithograph, 56 × 77 cm (full image). Bibliothèque nationale de France, Paris.

newspapers, in lithographs, in books published by the École Sociétaire, and of course in visual descriptions of the *phalanstère*, a motif that continued into the electrical era.[64] Some expressed anxieties about the telegraph's adoption for authoritarian uses and its "crystallization of colonial rivalries," although it retained a powerful symbolic charge.[65]

The telegraph's messianic qualities later found expression in the pasilalinic-sympathetic compass—or snail telegraph—proposed by the eccentric Fourierist sympathizer and future Communard Jules Allix in 1850.[66] Allix's telegraph, whose announcement provoked hilarity in the press, was based on the claim that mating snails produce a lifelong sympathetic bond, forming a connection that could be maintained across distances and over long periods of time. Most terrestrial gastropods are hermaphrodites, prompting speculation on their unique ability to act as both transmitters and receivers. Annexed to a simple code, this nebulous snail love—an "escargotic commotion"—could be appropriated for human communication, according to Allix. Although the results, predictably, were less than successful, the sense of telegraphy as universalizing, mystical, and socially generative carried over from more conventional Fourierist thought into Allix's snail apparatus. As a *vaudeville* performed at the Théâtre de Montansier on November 17, 1850, mockingly put it: "Here is progress. . . . Well! Everything is like that. . . . You didn't have the telegraph, and the Chappe brothers invented it. . . . After the ordinary telegraph came the electric telegraph, after the electric telegraph came the underwater telegraph, after the underwater telegraph here come the snails . . . after the snails . . . Ah! My word, after the snails, walk away . . . it's over, there's nothing more to do, the world is perfect."[67]

The use of telegraphy as a sign for the dynamics of association, over and above the detail of the technology itself, also emerged in the work of artists influenced by Fourier's ideas. Notably, the first version of Dominique Papety's *Rêve de bonheur*, painted in 1842, featured a telegraph and a steamboat in the background of the work.[68] Linda Nochlin has described the connections between Papety and Gustave Courbet, whose 1855 *The Painter's Studio* was, among other things, a Fourierist allegory.[69] Around 1848, Papety had begun work on a painting, *The Last Evening of Slavery*, for a fellow Fourierist, François Sabatier, who may have proposed its subject. Sabatier was a supporter of Courbet and a friend of his patron Alfred Bruyas (another believer). In 1854 Sabatier invited Courbet

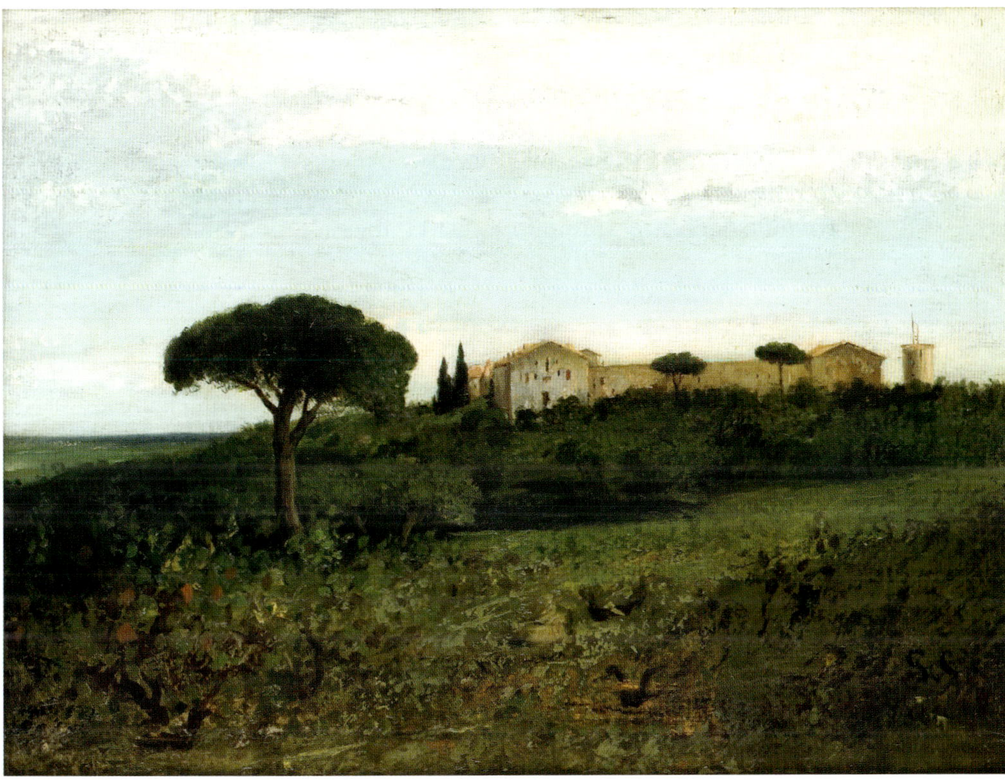

Figure 2.15
Gustave Courbet, *Vue de la Tour de Farges*, 1857, oil on board, 48 × 63 cm. Musée Fabre, Montpellier.

to visit him at his estate and vineyard at Tour de Farges, near Montpellier, where they discussed Papety's sketch for *The Last Evening of Slavery*. Nochlin suggests that this was a significant prompt for Courbet's *Studio*, a work in which he attempted to transform Papety's academic classicism into a new pictorial language.[70]

While at Tour de Farges, in addition to making a portrait of Sabatier and conversing with him on shared ideals of association and universal harmony, Courbet made a work that includes, to the right, the telegraph tower that stood on the estate (fig. 2.15). Surely, this detail is not without significance, given the context of its location at Tour de Farges, where Sabatier earnestly drew up plans for *phalanstères*? This painting can, I propose, be understood in light of telegraphy's significance to the Fourierist imaginary and Courbet's work at this time—the bucolic Tour de Farges estate appears as a kind of *phalanstère* in this image, and the telegraph would have surely prompted such connections in the minds of followers familiar with Considerant's design and with telegraphy's broader resonance within Fourierism. We might also see its incidental presence in this humble landscape as making good on the failures of Papety's *Rêve de bonheur* a decade earlier, for the telegraph in Papety's painting was not well received. Critics mocked the "poor devil" locked up in the telegraph's "dungeon," contrasting the Fourierists' high claims for telegraphy as an engine for social happiness with the labor and captivity of the stationnaire, toiling at this "most insipid profession in the world."[71] Papety replaced the steamboat with a Greek temple, and by the time the final version appeared at the 1843 Salon, he had shirked the appeal to an arcadian vision of technology and removed the telegraph from his painting too.

Telepresences

Nearly forty years after its first introduction, the optical telegraph had reappeared on the horizon of revolutionary politics in the aftermath of the 1830 Revolution. It became a focus of political satire during the July Monarchy, as if being seen again for the first time. This may account for the negativity with which critics, by 1842 trained to see the telegraph differently, responded to Papety's work. In Daumier's famous 1831 caricature, Louis-Philippe as Gargantua squats gross and immobile on his throne, his body a machine that, fed with a continuous stream of taxpayers' money, defecates sinecures and privilege in the direction of the greedy politicians who scurry around like portly dung beetles beneath him (fig. 2.16). The caricature, which sent Daumier to jail, represents the citizen-king Louis-Philippe as a technology of transformation, turning the taxed coins of the starving French people into the printed paper of state. *Gargantua*'s success as a political satire is usually attributed to Daumier's radical simplification of form, Louis-Philippe beginning his mutation into the pearlike shape that would dominate his caricatural image during the July Monarchy. In the spirit of the spatial compressions wrought by telegraphy, it is worth pausing, however, to shift the focus, to reorient this image toward a barely visible speck in the distance. In the nebulous Parisian cityscape to the right-hand side of the print—the side that seems to stand for the people, juxtaposed with the National Assembly and its corrupt politicians on the left—a telegraphic apparatus rotates atop a building whose asymmetrical towers identify it as Saint-Sulpice.[72]

Collapsing the space of the print, the presence of the telegraph invokes a form of invisible communication that troubles the insistent visibility of the lithograph. The materiality and immediacy of lithographic technique—the gestural effects of the artist drawing on stone—contrast with the intangibility and airborne inscrutability of the telegraph. At the same time, the smeary effect of the planographic process seems suited to the indeterminacy of telegraphy, for those without the code, at least. Happening above heads, in the corner of the eye, at the periphery of vision, it figures in Daumier's print as a kind of pictorial interference that structures the whole scene. Daumier recognized that while telegraphs made a substantial visual impact in the first half of the nineteenth century, even more important was what they left to the imagination. Who, or what, is this telegraph trying to contact, and what news is it hoping to convey?

Daumier himself provided something by way of response in a print made three years later, one of several lithographs he made between 1839 and 1844 that reference the telegraph directly (fig. 2.17). The print refers to a suspicious delay in the release of news, via telegraph, of Don Miguel's decisive victory over his brother Don Pedro in the Carlist war in Spain ("Nouvelles d'Espagne" can be seen written on the telegraph's regulator). This delay allowed members of the government to accrue sizable profits at the Paris

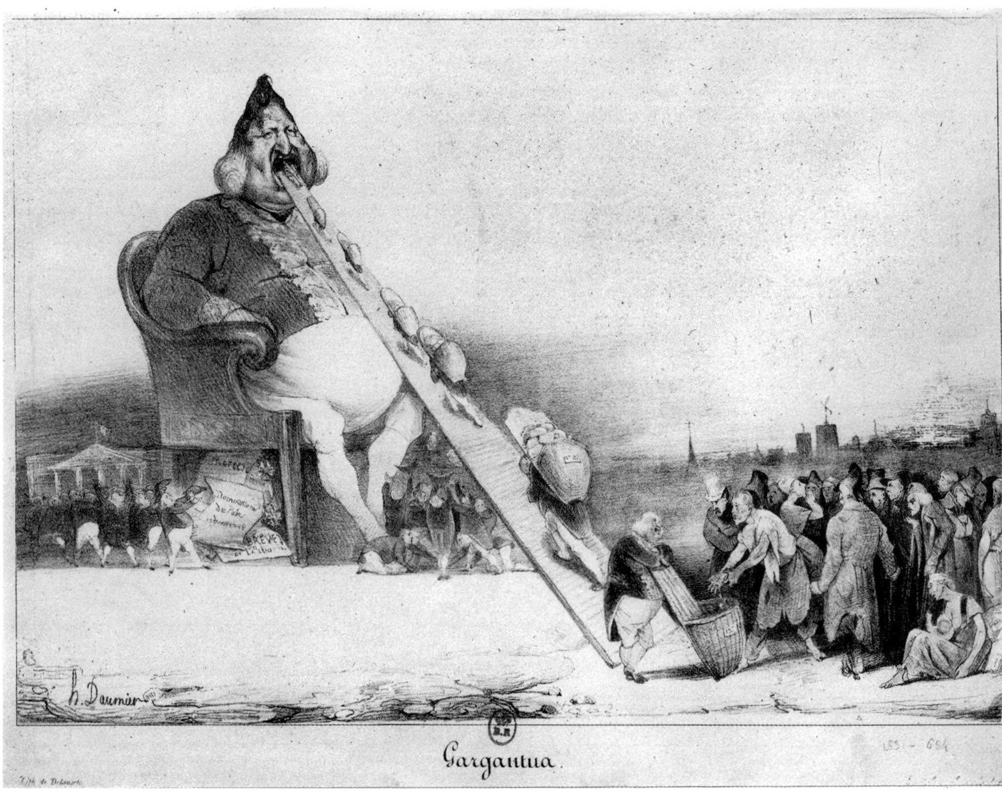

Figure 2.16
Honoré Daumier, *Gargantua*, published in *La Caricature*, December 15, 1831, lithograph, 20.4 × 30.5 cm. Bibliothèque nationale de France, Paris.

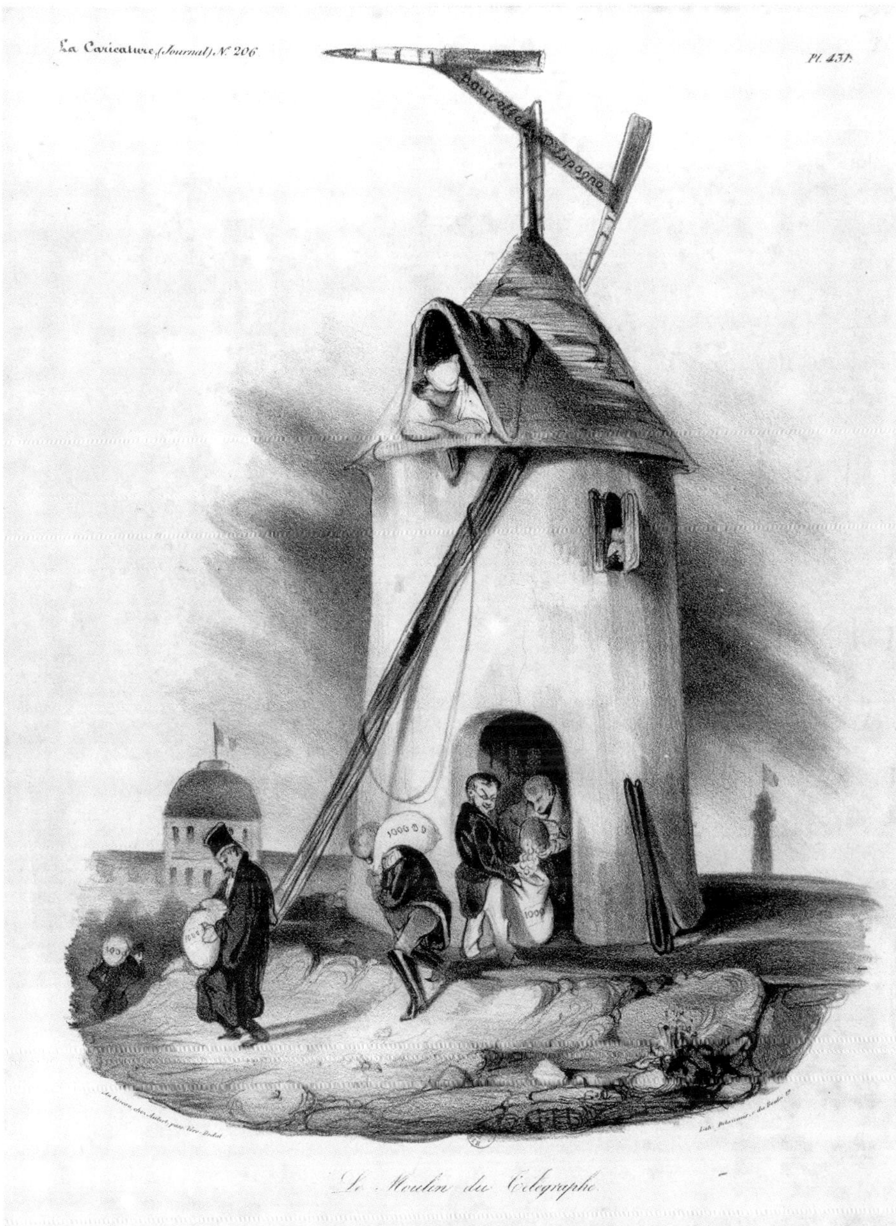

Figure 2.17
Honoré Daumier, *Le Moulin du télégraphe*, published in *La Caricature*, October 16, 1834, lithograph, 31.8 × 24.5 cm. Bibliothèque nationale de France, Paris.

Bourse. Several politicians, including Minister of the Interior Adolphe Thiers, who faced accusations of insider trading by way of telegraph, can be seen extracting sacks of cash from the mill, while Louis-Philippe looks on from the window. This print, *Le Moulin du télégraphe*, bears comparison with Daumier's design for *Gargantua*, and indeed, the two lithographs work as a pair, as if in relay. The telegraph again stands in for the body of Louis-Philippe as an engine for the production of ministerial and clerical corruption, making explicit what was hinted at in the earlier image: that *Gargantua* is, in part at least, an image about telegraphy, and that telegraphy, finally withdrawn from control of the Chappe family in 1830, was fully an agent of the state—the silent, sinister means of transmission that enabled the July Monarchy to stay one step ahead of the people.

Gargantua marks the point from which, due to his imprisonment and censorship, Daumier's attacks on *juste milieu* repression became ever more encoded. Yet telegraphy, as represented in print at this time, was increasingly presented as caricature's antithesis. Its communications were recast as blind enigmas and its cryptic secret-state modalities contrasted vividly with the oppositional potential of graphic satire. By 1831 the revolutionary potential of universal communication suggested by Lakanal and Chappe in the 1790s (and later resurrected by the Fourierists) had, in these works, become something distinctly antirevolutionary, and any possibility that it might be harnessed to speak to power, as well as for it, was extinguished with an 1837 law establishing prison sentences for transmission of "signals, from one place to another."[73] The imagined community to which the telegraph spoke was enabled, as in Benedict Anderson's classic analysis, by print capitalism, but it increasingly involved a return to arcane codes and privileged access, rather than the democratic, universally available vernacular that Anderson considered key to print's transformative role in modernity.[74]

That said, Daumier's print reanimated the telegraph as a revolutionary technology, or at least, a formerly revolutionary technology against which contemporary politics might now be judged. Part of the force of these images comes from their situation of the telegraph in a historical network and in relation to other forms of political representation. Daumier's fellow caricaturist Grandville alluded to this in a sketch for a print representing Louis-Philippe as a wizard, casting the telegraph into darkness in the middle of the day (fig. 2.18). Not only did

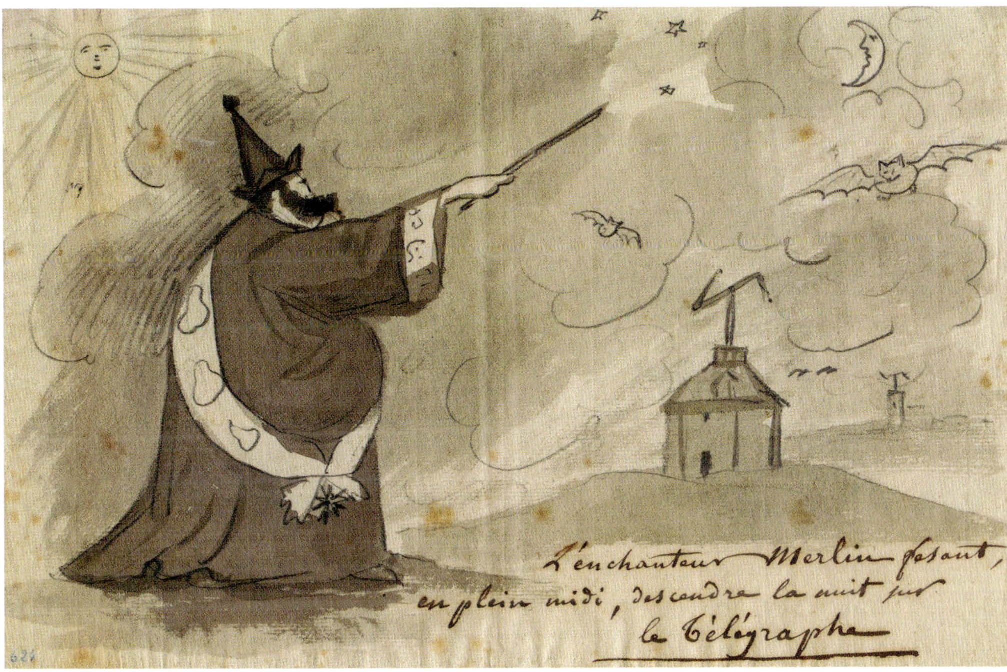

Figure 2.18
J.-J. Grandville [Jean Ignace Isidore Gérard], *L'Enchanteur Merlin fesant, en plein midi, descendre la nuit sur le télégraphe*, 1835, pen and wash on paper, 20.1 × 31 cm. Musée des Beaux-Arts, Nancy.

Louis-Philippe censor the news, but he suppressed the national lottery, causing major funding problems for the telegraph network. Embodying the practices of a particular repressive regime, the telegraph, circled by bats, also functions more abstractly here as a cipher for the mysticism and promise of communication *tout court*. The telegraph was entwined with the satirical representation of Louis-Philippe's reign (signaling to the towers in the distance, the king is, once more, a telegraph in this image, transmitting to the machines ahead of him, which sport matching pears). Yet it also signified an ideal of free communication that his censorship was shrouding in obscurity.

In these prints from the early 1830s, the telegraph thus provided a means to think of revolution beyond the bounds of its immediate manifestations, to theorize its corruptions but also its possibilities. Daumier's *Gargantua* looked back to the enlightened possibilities of telegraphy presented by its advocates in the early 1790s, but also envisioned the broader visual landscape within which it later evolved, a telegraphic world that was increasingly burdened by political unscrupulousness and constraint. Faced with the impasse presented by Daumier's image of the repressive telegraph, we might do well to return to the revolutionary origins of the telegraph, to a point in time when it seemed to presage an enlightened future of uninhibited contact between peoples, to track another hidden message.

Lisa Gitelman has observed how "even media that seem less involved with representation than with transmission, like telegraphs, offer keenly persuasive representations of text, space/time, and human presence, in the form of code, connection, and what critics today call 'telepresence,' that feeling that there's someone else out there on the other end of the line."[75] "That feeling" of the coexistence of, and perhaps even identification with, an unseen other certainly informs the meanings the telegraph was loaded with, and those it produced, in the final decade of the eighteenth century. At the inception of the telegraph in the early 1790s, the potential for communication across distance raised the possibility not only of military advantage and future commercial success, but of the transmission of revolutionary ideas. The sense of feeling someone else was on the "end of the line" could incite fear or suspicion, or it could provide concrete evidence of government surveillance, but it might equally, given the right conditions, suggest an empathetic community of like-minded citizens.

During the French Revolution, attempts to centralize authority and stabilize the boundaries of France operated in an ambivalent and at times contradictory relationship with the Revolution's global potential. While some revolutionaries imagined the Revolution as a model of emulation for other countries, others argued against the transformation of liberty into an export and suggested that France should concentrate on internal politics before attempting to revolutionize the world.[76] The outbreak of the Haitian Revolution in 1791 demonstrated powerfully that this debate could not be contained within the confines of France. Political philosophers were no longer ruminating on the problems for formal government of political relations formed between strangers within the metropole. Rather, legislators in France were forced to take the practical implications of the Rights of Man into account, considering claims for equality and citizenship from those they had believed to be less than human. As Emma Rothschild has framed it, this "new diversity of political subjects" unsettled an antique republican ideal of citizenship, whereby "all citizens should be the same sort of person."[77] Instead, communication between strangers was incorporated, in ways that were frequently unresolved and contradictory, into everyday revolutionary practice. To be free was to communicate. The utopian rhetoric of telegraphy at this moment, as enabling both communication between equal citizens *and* free trade, might be read in light of the conundrum that the Haitian Revolution presented to the French: how to justify the legitimacy of a supposedly universal and equal civil society while profiting from the sale of sugar produced by far-away slaves?[78]

The dystopian visions of exploitation and drudgery that emerged in the nineteenth century, with telegraphic workers dictated to by the whims of their machines, are obviously of a different order from the horrific brutalities of enslavement in the French colonies. However, telegraphy had, by rationalizing and collapsing abstract principles of "commerce" and "communication," already provided a language with which to speak, in a distanced way, about such matters. Although it finally came into official use six months after the abolition of slavery in French colonies in February 1794, telegraphy—more as an idea than as an apparatus—encapsulated with remarkable efficiency the twin yet contradictory desires that underpinned the initial revolutionary approach to France's colonies: the irreconcilable ambitions of universalist freedom and commercial advantage. Arguably, telegraphy's colonial associations were sublimated, almost immediately, in terms of Eurocentric resistance to "antique" or political slavery (freedom from despotism) or, later, via repeated reference to the cruel conditions of those who operated the telegraphic apparatus itself: the rats in cages on Saint Helena, the "poor wretches" described by Dumas, Hugo's mournful prisoner at Mont-Saint-Michel, and so on, whose misery bore the traces of another, more distant servitude, but worked to repress it too.[79]

Such associations were not a consequence of the parlous work conditions that accompanied industrialization but were present from the telegraph's inception. The first significant telegraphic message, regarding the recapture of Quesnoy, sent from Lille to the telegraph on the Louvre on 28 Thermidor Year II and relayed two days later by Barère to the National Convention, reported that an "Austrian garrison of 3000 slaves has put down their arms and surrendered voluntarily." "The soil of freedom is no longer trampled by the slaves of the North," Barère continued.[80] That telegraphy, optical and electrical, played a crucial role in colonial expansion through the course of the nineteenth and twentieth centuries is only part of the story. Telegraphy was imbricated, in unresolved and often unrecognized ways, in the colonial imaginary of postrevolutionary France, its mute disturbance of the visual field chiming with the silences, stoppages, and disconnections that characterize the colonial archive.[81]

Coding Commerce

Returning to this early phase of telegraphy means, however, that we must acknowledge that the "telegraphic" aspects of an image might not be coincident with the implementation of the system, and that the work in question might even predate it. An engraving by Jean-Charles Levasseur of Anicet Charles Gabriel Lemonnier's enormous painting *Le Commerce* offers one entry point for an examination of the intersection of communication and commerce at this crucial juncture (fig. 2.19). Commissioned in 1786 to hang in the Chamber of Commerce of Lemonnier's native Rouen, *Le Commerce* is a work that points in intriguing ways to the global exchange of people and commodities in the late eighteenth century, and to the manner in which reproductive media provided a conduit for the transmission of revolutionary politics across space

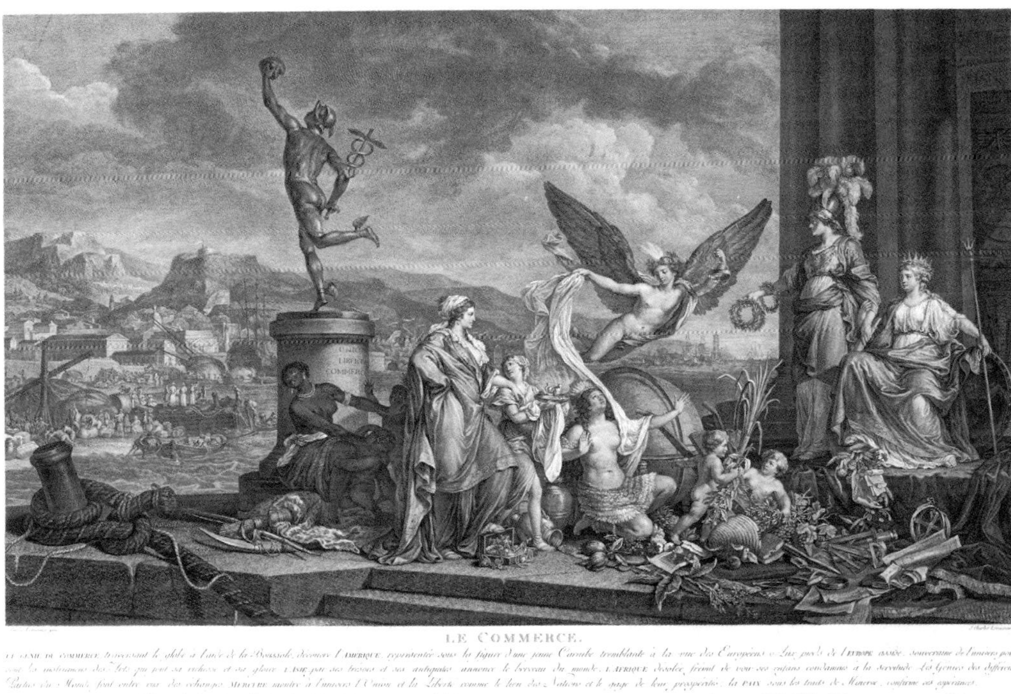

Figure 2.19
Jean-Charles Levasseur after Anicet Charles Gabriel Lemonnier, *Le Commerce*, 1791, etching and engraving, 43.2 × 71.6 cm. Musée de la Révolution française, Vizille.

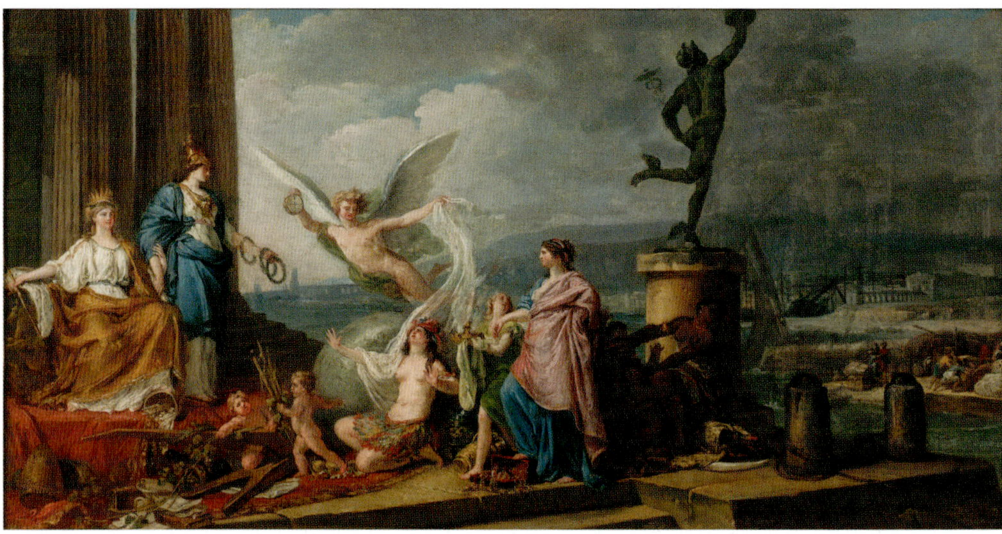

Figure 2.20
Anicet Charles Gabriel Lemonnier, *Le Génie du commerce découvrant l'Amérique aux yeux des Nations*, 1787, oil on canvas, 70 × 130 cm. Musée des Beaux-Arts, Rouen.

and time.[82] It is also an image that condenses in visual form several of the debates taking shape around telegraphy in the early 1790s.

Although a large-scale preparatory painting was exhibited at the 1787 Salon (fig. 2.20), the finished version of the work, now destroyed, was not presented to the Rouen Chamber of Commerce until 1791.[83] This was a crucial delay, for in the intervening time the artist made several changes to the canvas based upon recent revolutionary events in France as well as in Saint-Domingue, soon to be Haiti, where the first throes of the slave Revolution had radically reconfigured the insular prerogatives of revolutionary liberty. The final painting was exhibited at the Salon of 1791, while the preparatory work was shown again in 1793 and was considered a great success.[84] Levasseur's print, also issued in 1791, represents the final version of the painting, in reverse, of course. Even putting to one side the mediation and transmission of information integral to all intaglio printmaking, the meaning of Levasseur's print was notably conditioned by the images that preceded it, a system of relays that enabled it to speak to the historical character of more than one revolution.[85]

Levasseur's engraving of *Le Commerce* represents the four continents, shown as female allegories, against the background of Rouen's teeming port. On the right-hand side, "Europe," assisted by Minerva, sits in judgment over the scene. "Asia," her hand on the shoulder of a servant bearing a platter of perfume and jewels, is represented in antique garb, observing the prospect in front of her, where a plumed, bare-breasted America is unveiled by a *génie* of Commerce carrying a compass. *Le Commerce* reiterates a stock image of the birth of America that had become a common subject for print culture since the American Revolution, and it is explicit in its articulation of France's happy role in assisting American independence from British rule.[86] However, a dissonant note is struck by the presence of the figure group representing Africa, cowering at the foot of a burnished statue of a flying Mercury with liberty cap held aloft. Crucially, the "Africa" group was moved from the center of the painting after the outbreak of the Revolution in Saint-Domingue, and displaced to this more marginal space, likely to neutralize its impact, while two *putti* exchanging French and American products, including telescopes, maps, and a print of Christopher Columbus, take its place in the foreground.[87]

This image is, consequently, in keeping with the abolitionist sentiments of the merchants and Masonic lodges who had commissioned Lemonnier's painting, expressing many of the ambiguities of their liberal position. Although Rouen was no longer a slave-trading port to compete with the likes of Nantes or Lorient, the commodities that sustained one leg of the Atlantic trade in enslaved people, textiles in particular, departed from its harbor in vast quantities. The artist was overt in his own antislavery position. In Toulon in 1784, on his way back to France from his second visit to Rome, where he had spent four years, Lemonnier had met the Abbé Raynal, the *philosophe* author of the influential *Histoire philosophique et politique des établissements et du commerce des Européens dans les Deux Indes*, who was also reentering France after three years in exile. Raynal had famously predicted the emergence of a "Black Spartacus" who would free the enslaved people of France's colonies in an outburst of revolutionary violence; the realization of this prediction was often understood to be the revolutionary leader Toussaint Louverture.[88] Raynal was full of praise for Lemonnier's painting, which he saw in the artist's studio on the rue de Soubise, and which expressed visually many of the key themes of his own work (although Raynal had by this time reneged on his antislavery position in light of the events in Saint-Domingue). On viewing the painting in 1791, an anonymous critic writing in the *Mercure de France* suggested that "If Africa ever raises its oppressed head to the level of the organized nations, it will owe much to the brush of M. Lemonnier."[89] This response was not atypical in its vaunting claims for white benevolence. Although Lemonnier's painting aimed to express the Black mother's inability to face seeing her children born into servitude, this image of a dependent and despondent Africa rejecting her children and violating nature perpetuated a racist and gendered discourse of servility and ineptitude that underscored much abolitionist sentiment.[90] Indeed, this group replicates stereotypical representations of Haitian women as beautiful, virtuous, and, in Michelet's words, "in every way the most amorous," that were current in the writings of Moreau de Saint-Méry and other abolitionist authors.[91] As Robin Mitchell has described, the bodies of Black women were eroticized, mythologized, and disavowed in the wake of the Haitian Revolution, their depiction in both literature and art giving rise to "cultural discourses about Frenchness that shaped the country's postrevolutionary national identity."[92]

Despite the presence of these tensions in his work, Lemonnier, himself from a Rouennais merchant family, attempted to obscure them, stating: "Commerce, so useful, so great, in that it binds together the men of a single empire, gets an even brighter gloss from communication and reconciliation between different nations. I have considered my homeland (*patrie*) in its commerce with France and in its trade with foreigners . . . in order to deal with the history and the spirit of commerce in general."[93] Overt here were Lemonnier's location of his work in time and his sensitivity to the communication of revolutionary values, allied to a tacit understanding of how the reduction in space between people and things is embraced by capital.[94] In his turn to a world-historical perspective, Lemonnier not only reaffirmed his status as a painter of important allegorical themes in the grand manner but suggested an attempt to come to terms with broader currents of historical change: "the history and the spirit of commerce in general." Yet his ideal of commerce abstracted from it any reference to the exploitation and murder of enslaved Africans that underlay such a history. While Levasseur's print clearly reproduces the painting's message, it circulated in a revolutionary context in which communication and commerce could not be so easily reconciled.

The image is dominated by a statue of Mercury, messenger and god of commerce. It emulates Renaissance archetypes such as the statue by Giambologna, the owner of which, the duc de Brissac, was executed for his antirevolutionary sympathies in 1792. Indeed, the use of Mercury in this context was established at this point, not least by the celebrity of Pigalle's *Mercury Fastening His Sandal* (1736–1739), regarded as a sculptural ideal well into the nineteenth century, and Mercury appeared during the Revolution in a wide variety of contemporary printed images with diverse political agendas.[95] This sculptural motif finds a more cumbersome metallic echo in Lemonnier's image in the prominent bollard in the left-hand foreground, possibly a cannon captured from a British ship, a relic from the American Revolution. The cannon is not present in the painted sketch, where more conventional posts suffice; a notable addition, even in the context of the image's crazed excess of objects, inserting a militaristic motif where none had previously been present, but figuring its obsolescence as a sign of the peace conducive to commerce.

In 1782, four years before Lemonnier received his commission for *Le Commerce*, a proto-telegraphic apparatus designed by a priest, Dom Gauthey, was demonstrated in Paris (fig. 2.21). Dom Gauthey's device involved metal tubes embedded in the ground to convey

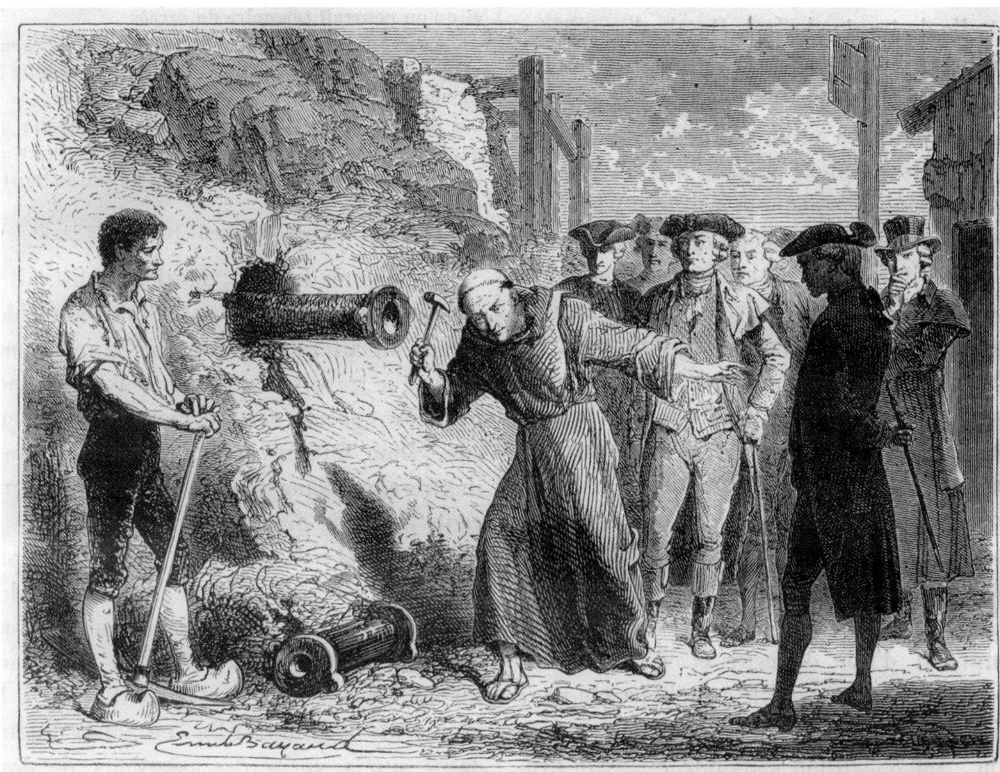

Figure 2.21
Émile Bayard, *Expérience de télégraphie acoustique faite à Paris, par Dom Gauthey, en 1782*, from Louis Figuier, *Les Merveilles de la science ou description populaire des inventions modernes*, vol. 2 (Paris: Furne, Jouvet, 1868).

sounds from one place to another, using the water pipes that ran underground from the pump at Chaillot.[96] Anticipating telegraphy's subterranean future and telephony's sensory effects—as well as the hidden wires of electric telegraphy and the underground pipes of the late nineteenth-century "télégraphe atmosphérique," which fired messages around by pneumatic pressure—Gauthey's project was nonetheless not taken up.[97] Within six months he was forgotten, and he emigrated to America. Lemonnier's painting, unsurprisingly, makes no reference to Gauthey's device, yet the introduction of the buried cannon is an image that transparently evokes themes of projection, transmission, and appropriation that were echoed elsewhere in the work, as well as in the broader scientific culture of the 1780s. Manifesting what Jonathan Sterne calls "the audible past" in the grooves of the print, the appearance of the buried cannon in Lemonnier's work in the aftermath of the revolution in Saint-Domingue resounds emphatically.[98]

Captured cannon seldom retained their material integrity at this time, for they were readily transformable into rare metal coinage. Another form of reuse was the monument made from melted-down enemy armaments. Indeed, Lemonnier was certainly aware of the contingent nature of materials and the potential for their remaking as revolutionary signs, having helped inventory Notre Dame in November 1790.[99] The presence of the buried weapon in Levasseur's engraving, in alignment with the statue of Mercury, suggests that statue and cannon may be composed (at least in the frame of representation) of the same matter, pointing to the possibility that the figure represented in the statue may be far from peaceable, and not unequivocally friendly to French interests. In this light, the hefty rope knotted loosely around the cannon (restraining it, even) serves as an analogue for the bound rods—the fasces carried by the lictors of Republican Rome—that were a symbol of revolutionary unity, justice, and in some versions, the immediacy and brutality of state-administered violence. This was a culturally central motif in 1791, following the reexhibition that year of David's 1789 *The Lictors Returning to Brutus the Bodies of His Sons*, a subject Lemonnier also represented in a sketch. Furthermore, in its binding of chromatically distinct elements, the rope—led by a white guide rope—inevitably recalls the complex coding of color in this context, what Darcy Grimaldo Grigsby has described as a kind of bodily materialization of the French *tricolore*, mapped out in racial terms.[100] Does Lemonnier represent such a system, abstracted onto the rope, as a bulwark against colonial violence or as a constraint on liberty, or both?

In the print, the statue of Mercury occupies an indeterminate space in the racial politics of an image so clearly coded in terms of difference mediated through skin color. If this difference is obscured in the print, whose monochrome tonal gradations render such distinctions more subtle, it is made explicit in Lemonnier's preparatory painting. Nonetheless, Levasseur's engraving perhaps expresses most accurately the nuance with which the racialized dominance of colonized people was understood and maintained in the late eighteenth century. In prerevolutionary Saint-Domingue, slavery was justified by the imposition of increasingly precise and intricate hierarchies of skin color between whites, Blacks, and *gens de couleur*. The fantasy of a fixed idea of "race" as a biological reality that skin color could legitimately articulate and naturalize was made present in these distinctions, whose effect, as Anne Lafont has described, was to ensure "Blacks were thus relegated to a non-shareable alterity."[101] The representation of Mercury in this manner and in this setting could not have failed to evoke the complexities of this system to contemporary viewers.

Squarely aligned with the mother and child group representing Africa and, by extension, Saint-Domingue, the figure of Mercury, read vertically, implies a positive futurity altogether lacking in the figure group beneath it (fig. 2.22). Whereas the first sketch submitted by Lemonnier had shown Mercury announcing to Neptune that thanks to the Revolution all nations could trade freely with one another, the final image represents a patinated, or even "Black," Mercury, caduceus in his right hand, revolutionary cap of liberty aloft in his left. The latter makes for a striking addition to the classical iconography. This section of the print marks a coming into revolutionary consciousness, a historical narrative of the trajectory from enslaved Africa to revolutionary Saint-Domingue, but also a recognition, and likely, a fear, of how this transmission might "feed back," for it foregrounds both the communication of Revolution to France's colonies and the returned message, as Mercury is facing inland, back toward Rouen. Is this "Black Mercury" an iteration of the "Black Spartacus" prophesied by Raynal, inflected by abolitionist anxiety about the real consequences of the attainment of liberty in the colonies? Might this Trojan horse explain, in part at least, and ambiguously so, the fact that Laocoön is invoked in the "Africa" figure group, recast to radical effect as a Black woman?

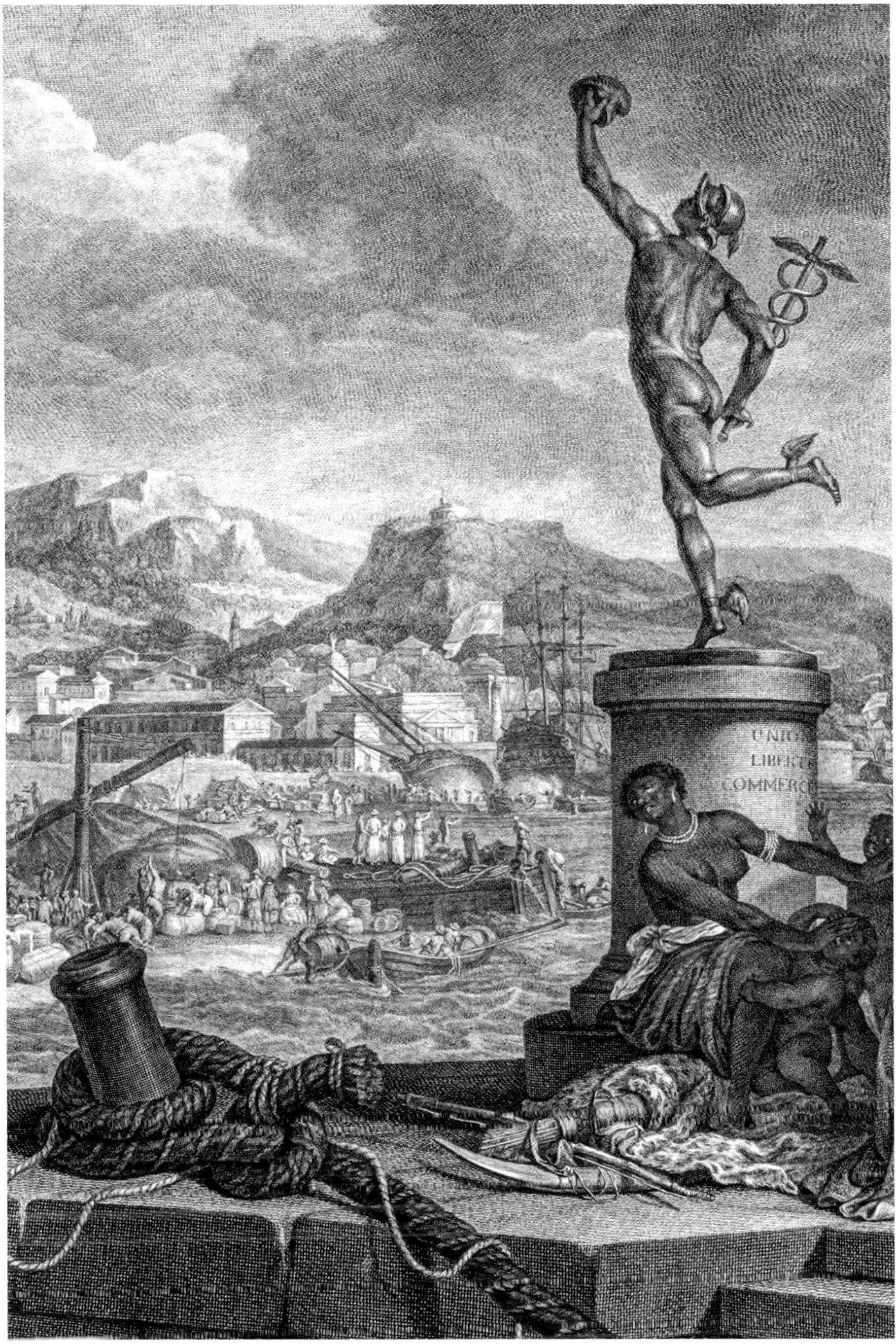

Figure 2.22
Detail of Jean-Charles Levasseur after Anicet Charles Gabriel Lemonnier, *Le Commerce*.

What is more, while the apparent visual homology between the slanted beam of the crane on the left-hand side of the image and a Chappe telegraph apparatus is coincidental, as is the collapsing of distance between the Mercury figure and the fortified outcrop in the distance which from this perspective resembles a telegraph tower, they raise the prospect of communication between Mercury and a point in space and time that is at once near and far away. In fact, this picturesque rendering of Rouen's port is not actually much like Rouen at all, or at least not *just* Rouen, for it certainly seems different from contemporary renderings of the scene (the proximity of the ruins of Richard I's Chateau Gaillard at Les Audeleys and the Barfleur lighthouse to the right of the print means that it does not make a great deal of sense as a representation of a singular place). With its mountainous landscape and stormy skies, it gestures to the topography of Saint-Domingue—can we read it, like the rope, as a binding together of disparate elements, an intervention in pictorial and geographical space that parallels the grafting and hybridization that characterized both the flora of the colonial landscape, and the population that inhabited it?[102]

The image of Mercury becomes, in this context, a kind of telegraphic apparatus, while the print anticipates and was later comprehensible in relation to the telegraph's concrete realization of abstract principles of communication. Hugo's warning about the telegraph—"in the slave in favor see the master to come"—takes on a new power here. The idea that a machine or an enslaved people (it seemed all too easy to conflate them) might rise and dominate a former master was a fantasy and a fear. Both resurfaced in the much-maligned figure of the telegraph stationnaire, so often a proxy for the telegraph itself. Philippe Toto, a Black man from Saint-Domingue recruited as part of a drive to retrain invalided military veterans as stationnaires, is known to have operated a telegraph in the Burgundy villages of Viserny and Massingy. Soldiers missing hands and legs—reminiscent of the amputee veterans who later populated the lithographs of Nicolas-Toussaint Charlet—were not always best suited to the physical demands of the machine's precarious placement (plus, according to Claude Chappe, they were often drunk).[103] The appeal was instead to a more positive model: the kind of national devotion seen in Jean-Baptiste Bizard's grisly and extreme demonstration of patriotism in 1793 (fig. 2.23). With one arm gone, Bizard's wounded revolutionary semaphores his dedication to the flag. The painting's

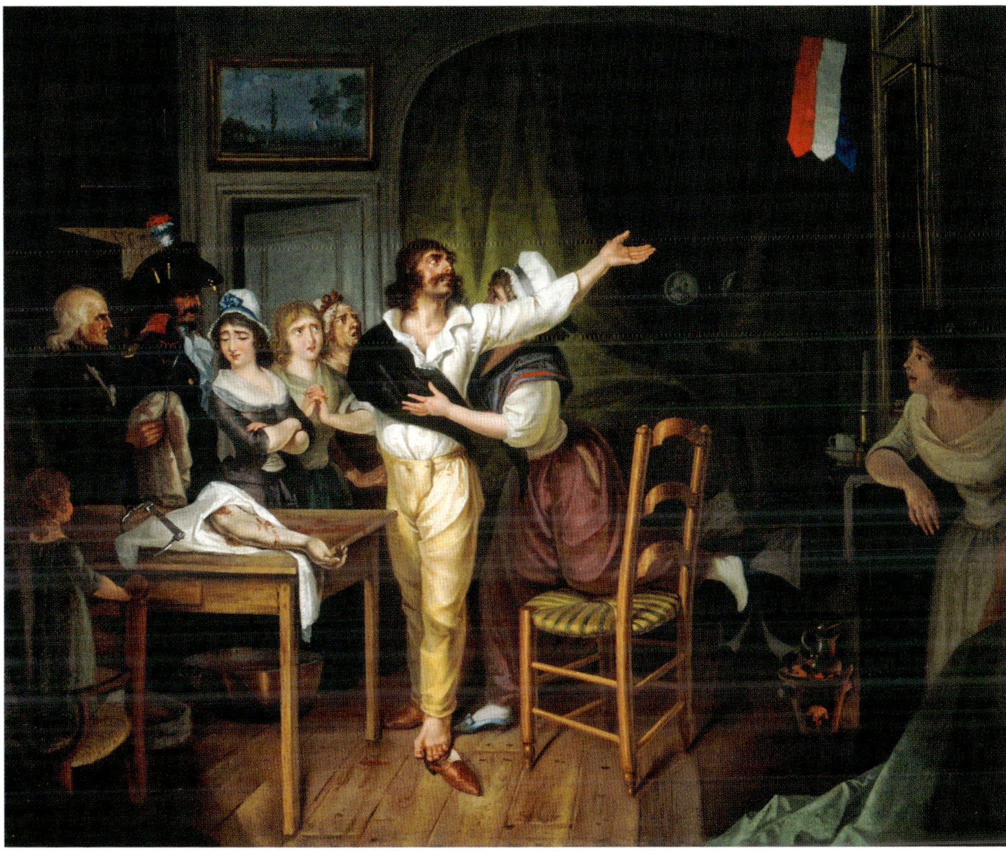

Figure 2.23
Jean-Baptiste Bizard, *Dévouement à la Patrie d'un homme qu'on vient d'amputer d'un bras*, c. 1793, oil on canvas, 53 × 64.5 cm. Musée Carnavalet, Paris.

artistic limb play is developed in the gestures of the other figures, with hands clasped, up and down, and arms reaching, leaning, and crossed. A play of gazes networks the scene—enthralled, fervent, and frankly, in the timeless sideways glance of the folded-arms figure, grossed out. Ultimately, disabled veterans only comprised a limited number of employees—around a fifth at most—and their presence in the system waned after the end of the Napoleonic wars. Nonetheless, their bodies, broken by war, and in Toto's case by violent conflict in the Caribbean, where amputations and disfigurements were among the many punishments imposed on slaves, were for a time integral to the functioning of the prosthetic telegraphic arms they operated. The horrors inflicted on enslaved people's bodies in Saint-Domingue are, of course, invisible in the ideal body of Lemonnier's telegraphic Mercury, which alludes to but at the same time represses its traumatic context.

In 1791, the year that Lemonnier's finished painting made its debut at the Salon and Levasseur's print appeared on the market, Claude Chappe also made his first live forays into telegraphy. Soon after, the figure of Mercury traversing the seas was reimagined as a symbol of communication rather than commerce, although the commercial possibilities of telegraphy were not disregarded, but were built into early premonitions of the future course the technology would take. An image of Mercury flying from port to port was, from 1799, the natural choice for the letterhead used by the Chappe brothers for the voluminous correspondence relating to the administration of their network (fig. 2.24). By 1815, the terrestrial context was removed, with Mercury shown flying through the clouds, and in 1823 the letterhead changed again, featuring an angel on a monument, attended by a quotation from Virgil: "I fix no limits in space and time." The July Monarchy retained the Mercury image, however, which had become central to the optical telegraph's visual identity.[104] A print from 1848 shows Louis-Philippe, as the "patron saint of thieves," in the guise of Mercury, flying improbably high above the seas, unimpeded by his ever-expanding girth (fig. 2.25).

In Lemonnier's image, telegraphy was evoked *avant la lettre* in an allegorized fantasy of communication benefiting commerce that was, at the time, more a concept than a fully realized

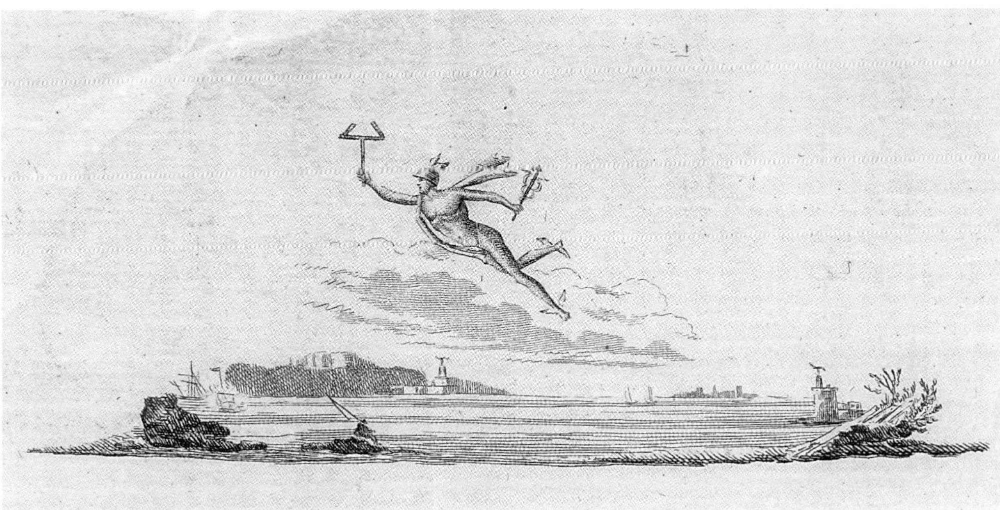

Figure 2.24
Letterhead for Chappe telegraphic network, 1809, engraving. Dibner Library of the History of Science and Technology, Smithsonian Institution, Washington, D.C.

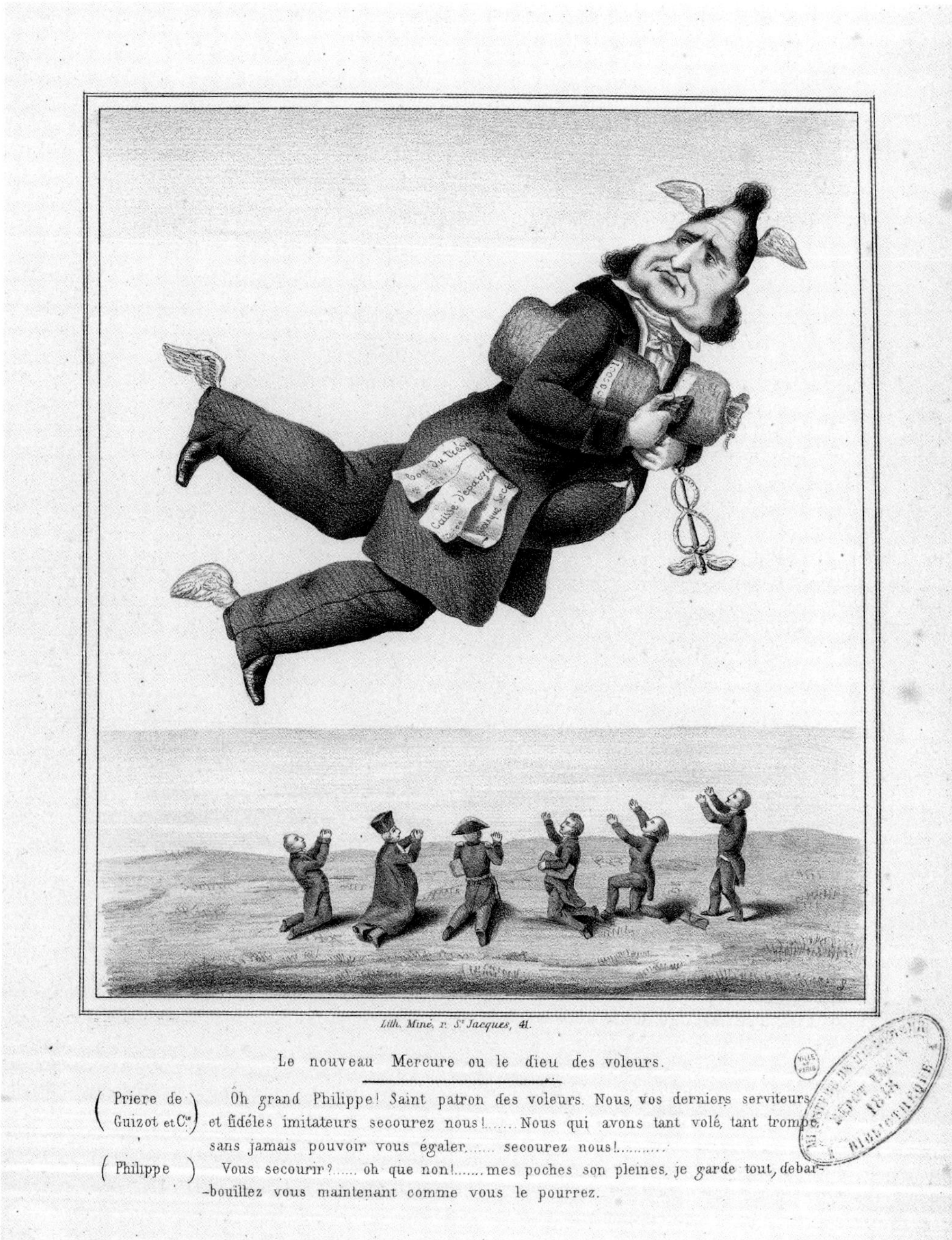

Figure 2.25
Jacques-Denis Miné [lithographer], *Le Nouveau Mercure, ou le dieu des voleurs*, 1848, lithograph, 42.3 × 30.5 cm. Musée Carnavalet, Paris.

technology. By foregrounding relations of space and time, telegraphy provided a powerful means to think about history, yet these possibilities did not emerge from a vacuum and were not, crucially, produced by the technology itself. When Ignace Chappe published his history of telegraphy in 1824, focusing on the role he and his brothers had played in the introduction of the optical system in France, he stressed the historically determined nature of telegraphy, which, after decades of similar proposals, only came to fruition in congruence with a unique set of historical circumstances.[105] Telegraphy was, in a sense, always already there.

In its early days, the Chappe telegraph was considered inseparable from the new commercial opportunities it might offer, and it was impossible to talk of commerce in the 1790s without invoking the Atlantic trade, even if the words themselves were not spoken. However, telegraphy, by opening a space for thinking about communication between equal citizens in (almost) real time, also provided a model for imagining resistance to a dominant message. "Our great discoveries will perhaps not make us better," Ignace writes, "but they help make us as strong as those who would oppress us; and oppression is the worst of all the evils to which the human race can be prey."[106] Telegraphy, for Chappe, was a specifically antidespotic technology, and as such, for a brief period in its utopian early 1790s mode, it raised the possibility of confronting the real, not merely conceptual or ideal, conditions of enslavement. However, an Enlightenment object par excellence, the telegraph was correspondingly paradoxical, and its use was accompanied by anxieties about the possible effects of the challenge it posed—universalism tempered by a fear of outsiders gaining access to its messages or by the realities of colonial Revolution. Such ambivalence was carried far into the future, into the era of electromagnetic systems. One of Puvis de Chavannes's mural designs for the Boston Public Library, made in his Neuilly studio and shipped across the Atlantic, where it was installed in 1896, shows good and evil tidings flying through space down a telegraph line. Ostensibly an allegory of Physics, the sketch for this image in the Louvre is titled "Télégraphe."[107] Although in this drawing good news runs just barely ahead, its negative counterpart, covering its face, is but a heartbeat behind. The wish-image of telegraphy, Levasseur's print of Lemonnier's painting is similarly double-faced, representing overlapping understandings of empire and global trade marked as much by continuities from previous political regimes, and versions of the image, as by ruptures in this stable order. In this sense, too, it was a telegraphic image, bound to and materializing complex overlapping histories. Purporting to speak in favor of Liberty, *Le Commerce* did so, like the telegraph, at the edge of vision, its message hidden in plain sight.

Stoppages

In 1838 Samuel Morse, himself an artist, presented his new system in France, although it wasn't introduced definitively until twelve years later. The first electric telegraph branch in France was established in 1845 with a line from Paris to Rouen. Until that point, Lemonnier's painting notwithstanding, Rouen had not hosted an optical telegraph charged with transmitting official news. The city had, however, been the site of a private optical line—an exception to the usual prohibition against commercial operation—approved in 1833 but abandoned two years later.[108] A second electrical line was introduced along a more traditional route from Paris to Lille in 1846 (the device used was not Morse's machine but one made by Breguet, whose grandfather had helped design the optical telegraph). The 1848 Revolution produced a decisive shift, the election of Louis-Napoléon as president in 1849 accelerating policies that foregrounded commercial success and technological advantage. The optical telegraph's days were numbered, although its messages continued to filter through.

As a youth, Claude Chappe attended the same establishment in Rouen, the Lycée Pierre Corneille, as two artists whose interventions shaped the meaning of European art at either end of the long nineteenth century: Eugène Delacroix and Marcel Duchamp. They would both, surely, have been aware of their illustrious inventor forebear, albeit with different senses of proximity and urgency. In 1914, Duchamp's *Network of Stoppages* mapped a diagrammatic landscape of chance, making a network of the threads seen previously in his *3 Standard Stoppages* (1913–1914), which was in turn based on the meter, that most revolutionary of measurements. He also made work that referenced telegraphy directly, notably *Du Tignet* (1959), a drawing of telegraph poles and lines traversing a landscape that has been connected to both *A Network of Stoppages* and the machines of *The Bride Stripped Bare by Her Bachelors, Even (The Large Glass)*.[109]

More powerfully, however, twenty-eight years after *Le Commerce* was first shown, and the same year that Hugo penned his account of telegraphy's invasion of his mental world, a work by another Rouennais artist, Théodore Géricault's *Le Radeau de la Méduse*, offered a striking, negative postscript to Lemonnier's "telegraphic image" of Rouen's port (fig. 2.26).[110] This is, surely, the nineteenth century's most insistently telegraphic work. William Kentridge, for one, has recognized this aspect of the painting. The silhouette of the Chappe telegraph features several times, moving from position to position, in his 2012 five-channel video installation *The Refusal of Time*, made in collaboration with Peter Galison (Kentridge made a large-scale wheeled wooden replica of the telegraph to achieve this effect), while in a preparatory drawing Kentridge depicts Géricault's raft with a giant megaphone attachment and a Chappe telegraph as a mast, shown midsignal, inching its way through the seas, "off the edge of the world—into the black hole."[111] Gesture and speech coincide again, but head nowhere. Crary has compared the sail in Géricault's raft to the curved canvas of a panorama, and we might profitably view it too as another iteration of the parabolic screens and sound amplifiers prophesied in the 1789 *Mémoire*.[112] However, Géricault's image trades in speechlessness and confusion, as well as in contemporary news and its transmission. Kentridge's Chappe telegraph sail makes this point more forcibly. In Géricault's silent scream of a painting, and the studies that preceded it, a tangle of limbs substitutes for the clarion legibility of the telegraph's arms, and the frantic semaphore of the *Méduse*'s multiracial crew—these people of no state—exposes the frank horror of the colonial encounter that is repressed in Lemonnier's image (fig. 2.27). Displacing the machine onto the bodies of the survivors, *Le Radeau* figures telegraphic communication, once again, as a means of historical comprehension, but one that is failed, illusory, and exploitative, in which, this time, the return message is left unanswered.

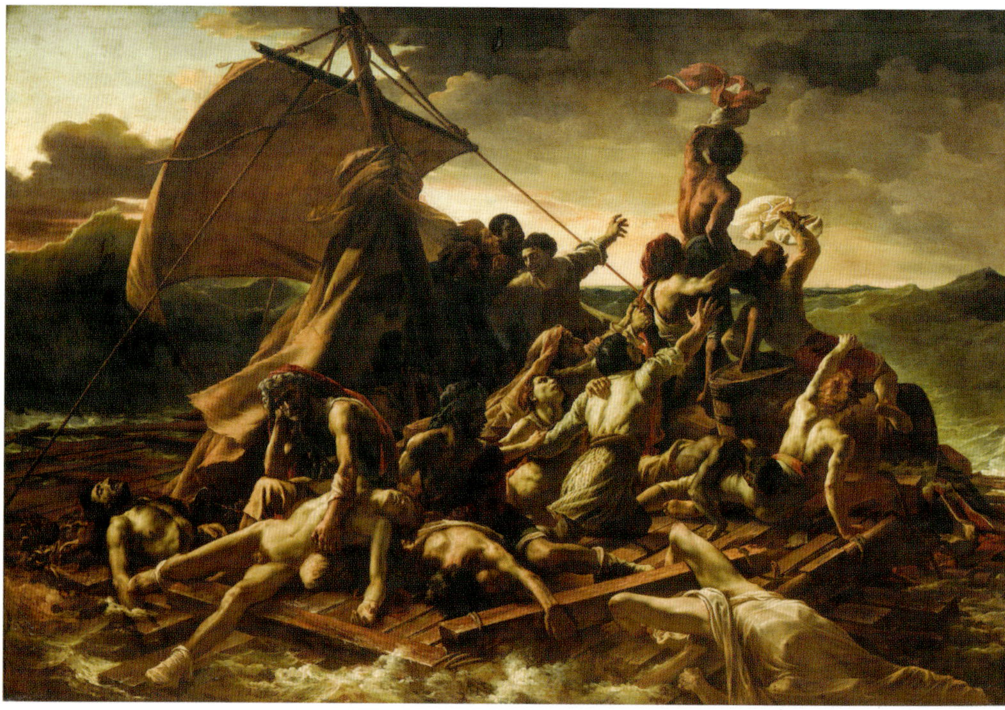

Figure 2.26
Théodore Géricault, *Le Radeau de la Méduse*, 1819, oil on canvas, 4.91 × 7.16 m. Musée du Louvre, Paris.

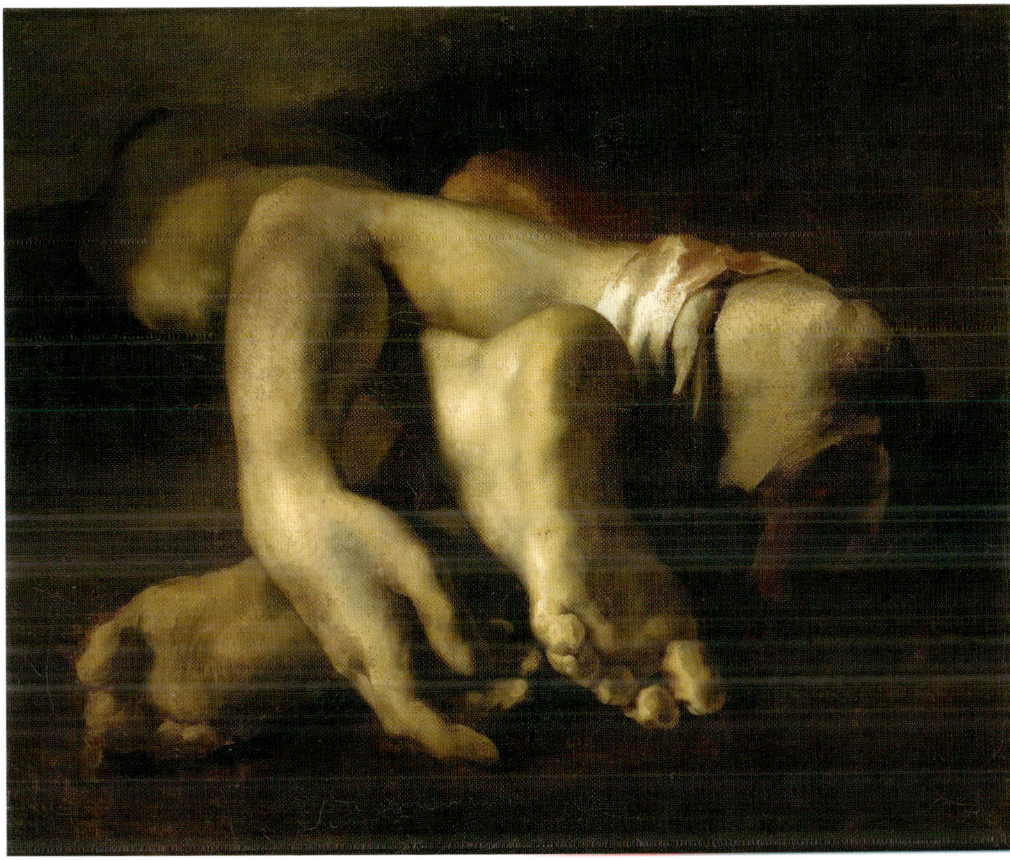

Figure 2.27
Théodore Géricault, *Étude de pieds et de main*, c. 1817–1819, oil on canvas, 52 × 64 cm. Musée Fabre, Montpellier.

3 *Lines in Sand*

"The most beautiful pages that the human soul has dictated to human genius."[1] Not one to run shy of hyperbole, this was how photographer and inventor Félix Nadar described a text he considered foundational to the modern era, chemist Antoine-François Fourcroy's 1795 report to the National Convention, *Sur les arts qui ont servi à la défense de la République, et sur le nouveau procédé de tannage découvert par le citoyen Armand Seguin.*[2] Somewhat incongruously, these words by Nadar appeared at the beginning of a chapter of his eccentric 1900 memoir *Quand j'étais photographe*, which he dedicated to the Belgian photographic pioneer Doctor van Monckhoven, who had died eighteen years previously.[3] "Whatever opinion one might have," Nadar wrote, noting with affection the pompous tone of French Revolutionary official language, and to some extent replicating it himself, "it is impossible not to feel a profound emotion before this document of the admirable effort of a people determined not to perish or to retreat under the universe conspiring against them."[4]

What was it that Nadar found so compelling in Fourcroy's report? As Stephen Bann, Eduardo Cadava, and others have shown, Nadar was particularly attuned in his writing of this period to the telescoping of historical time, and to the ways in which past media might reappear to new effect in the present.[5] Often this took the form of reminiscence on the graphic satire of the 1830s, inflected by his early career in lithography, a form of printmaking whose urgent temporality enabled later reflection on the historical possibilities of the still new medium of photography.[6] But Fourcroy's account, written over a century earlier, was seemingly disconnected from such visual concerns. Operating in the present tense of revolutionary conflict, its tone was briskly practical, permeated with a pugnacious sense of patriotic militarism. Indeed, it was preoccupied, for the most part, with leather; especially shoes, those most earthbound of objects. Seguin's new tanning procedure, which made him very wealthy, increased the speed at which boots could be produced for the military, a matter of urgency in wartime.[7] Recounting this innovation, Fourcroy's text is suffused with muted pride in an ad hoc, makeshift approach to the production of new materials for new objects in a time of crisis, under enormous pressure for immediate supply. As he explained, "the most important materials necessary to their fabrication didn't exist . . . resources had to be taken almost entirely from the soil of the Republic."[8] Or, as Nadar put it more emphatically: "We will have it."[9] This responsiveness to the contingencies of the historical moment provided Nadar with a powerful analogy for the early history of photography, and he found in this heroic generation of revolutionary scientists—the likes of Condorcet, Lavoisier, Monge, Chaptal, Lalande, Vauquelin, and Fourcroy—antecedents for the tactical genius of figures such as Doctor van Monckhoven and, presumably, himself. During the French Revolution, ropes were made of plants, tar was extracted from peat, and potash derived from soda; in due course, photography too would be improvised, seemingly from nothing.

Despite the deliberate, rather mannered contemporaneity of Fourcroy's report, Nadar discerned in the chemist's approach, hovering behind the words themselves, a broader sensitivity to the significance of the achievements documented. In fact, in his technological imaginary, Fourcroy demonstrated a keen awareness of his historical situation, introducing his account of Seguin's contribution with a careful summary of the accomplishments of France's revolutionary inventors, who had saved the Republic by way of their "simple machines, found everywhere."[10] The manufacture of saltpeter and weapons and the development of ballooning as a military technology are discussed at length in this "patriotic breviary."[11] Ballooning was of particular interest to Nadar, who had been preoccupied for many years with his own experiments in human-powered flight.[12]

Fourcroy also praised Chappe's optical telegraph, that "industrious machine" whose lines had multiplied across the nation in the previous year, transmitting semaphoric code through the air across fields and rooftops. The cryptic silhouettes of the telegraph were to be a persistent object of fascination for Nadar, who can be seen in a youthful photomontage striking a telegraphic pose (fig. 3.1).[13] The photographer's artificially elongated limbs drawing crazed signs in the air, the ambiguously phrased legend—"Au grand homme le daguerreotype instantané reconnaissant"—echoes that inscribed on the portico of the Panthéon. Implying that Nadar is atop that building—even, perhaps, a candidate for inclusion inside it, as the displaced "grand homme" of the work's title (although who needs that when one has photography?)—it recalls a lithographic "panthéon" representing the great men of French letters that Nadar had produced a couple of years previously. In the photomontage, the *Panthéon Nadar*'s serpentine line of celebrities is revived in the form of Nadar's zigzagging limbs.

Almost exactly coincident with the replacement of Chappe signals by electrical transmissions, Nadar's photomontage made a declining technology central to the aesthetic and historical possibilities of a new one.[14] Further down the line, Henri Théophile Hildibrand's wood-engraved illustration after Émile Bayard for Jules Verne's 1870 *Autour de la lune* represents the French explorer Michel Ardan in a strikingly similar pose, paying mischievous homage to Nadar's burlesquing of the telegraph by cavorting in space on the nose of a rocket (fig. 3.2). Bayard, a former assistant to Nadar, likely saw the montage in Nadar's studio, while Verne's character of Ardan was based, none too subtly, on Nadar: his name was an anagram for that of the photographer, and Bayard's image of Ardan clearly sports what Jillian Lerner has described as Nadar's "physical signature."[15] A series of coded messages, arranged in relays that traverse time and space, connect these images, separated by at least a decade, the ghost of optical telegraphy's simple mechanism haunting this projection into a futuristic alternative present. Telegraphic communication inevitably invoked duration, the distance between the timing of a signal and its reception. As George Kubler put it, "the nature of a signal is that its message is neither here nor now, but there and then."[16] Sometimes, though,

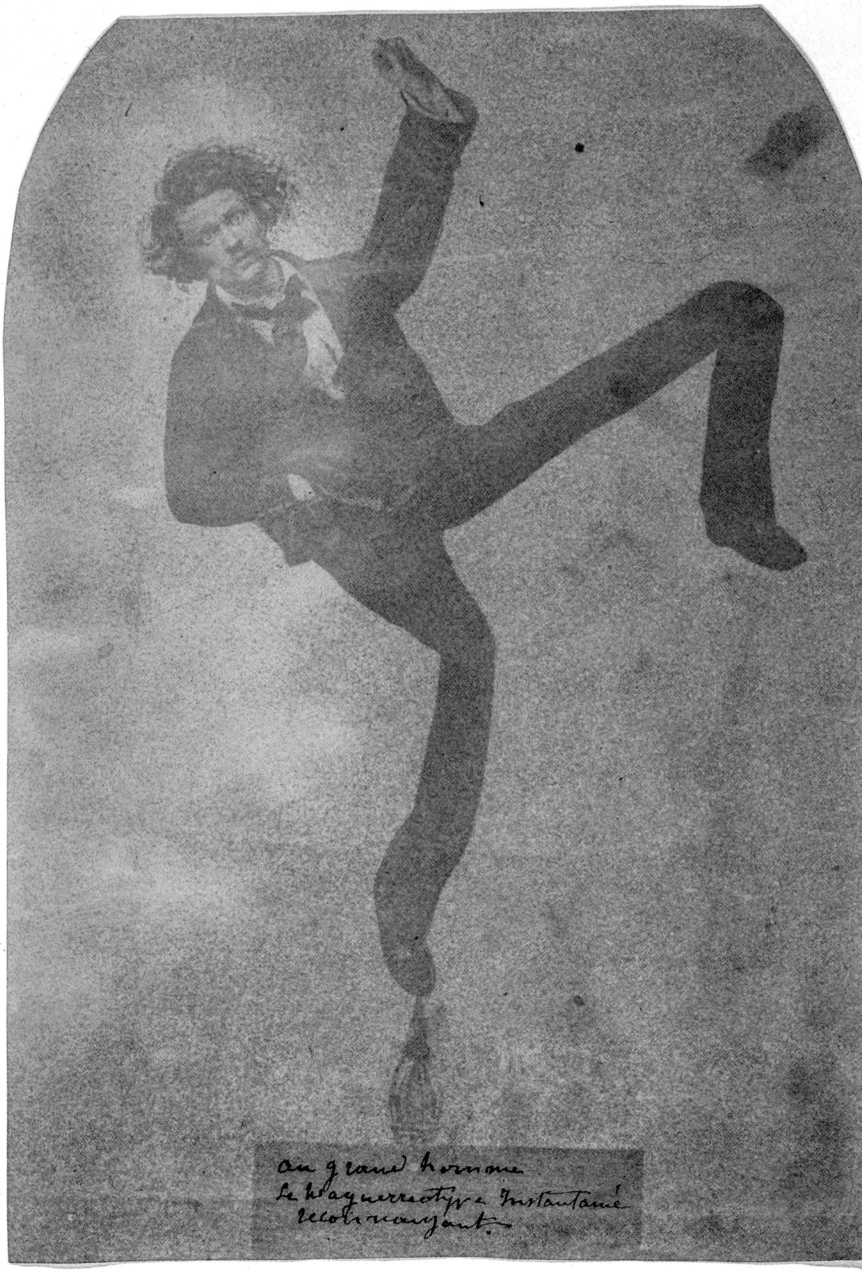

Figure 3.1
Nadar [Gaspard-Félix Tournachon], *Photomontage of Félix Nadar*, 1855–1860, salted paper print, 19.8 × 13.9 cm. The J. Paul Getty Museum, Los Angeles.

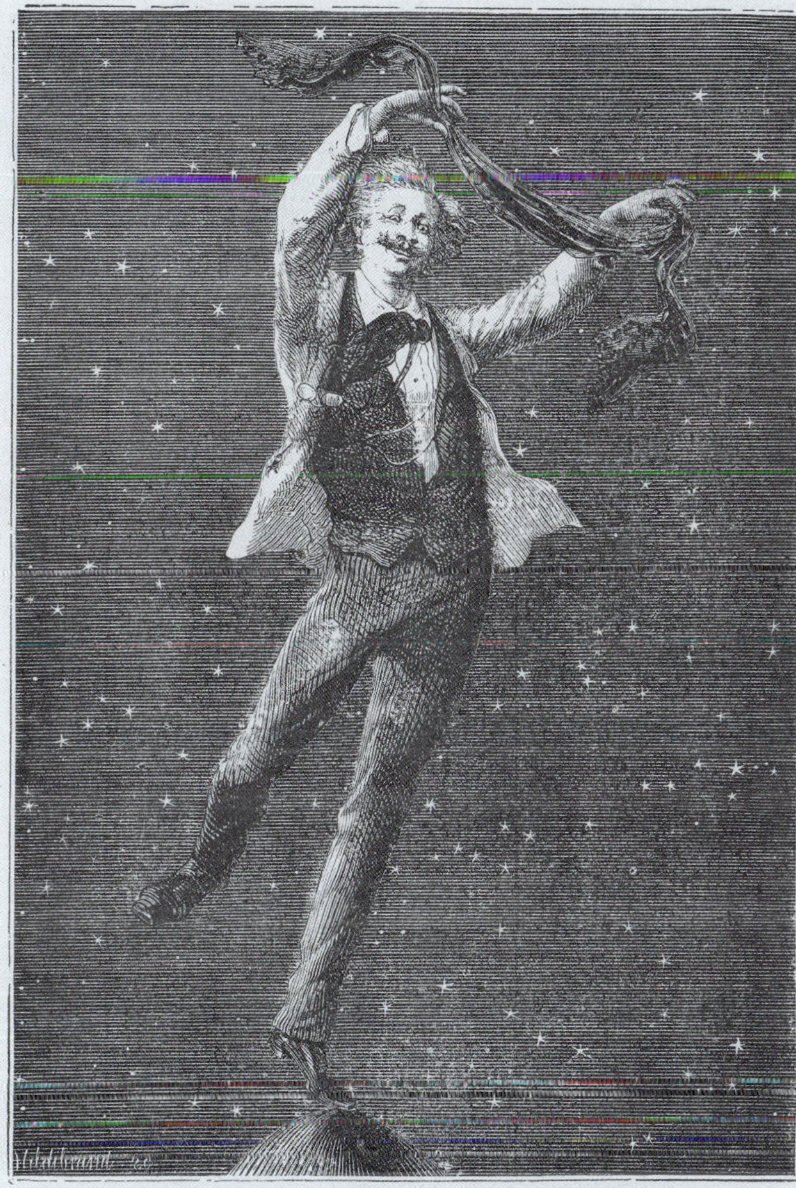

« J'aurais pris des attributs de chimère. » (Page 60.)

Figure 3.2
Henri-Théophile Hildibrand, "J'aurais pris des attributs de chimère," engraving after Emile Bayard, illustration from Jules Verne, *Autour de la lune* (Paris: J. Hetzel, 1893). Bibliothèque nationale de France, Paris.

the "pastness" of the optical telegraph made it suitable for thinking about signals to come.

Consonant with this recuperation of past media to describe the contours of those with which they later overlapped, the pinnacle of Nadar's "Panthéon" in his self-portrait photomontage is, in fact, more than just an architectural feature. For Nadar appears to be pirouetting on the tip of a lithographic crayon, or a pencil—to be precise, a *porte-crayon* of the type commonly seen in self-portraits of artists from Jean-Baptiste Greuze to Anne-Louis Girodet, Élisabeth Louise Vigée Le Brun to Gustave Courbet (and visible in the central figure group of Samuel Morse's 1833 *Gallery of the Louvre*), a metal drawing instrument with a grip at one end to hold a piece of chalk, lead, wax, or charcoal.[17] He seems to hold another piece of crayon in his hand. What was telegraphy, after all, if not drawing in the air? These details return us to Fourcroy's text as later reconstituted by Nadar. Both Fourcroy and Nadar finally alight on a category of object that might, at first sight, appear insignificant, but which, as Fourcroy put it, "becomes of great importance for the men who take care of the needs of a great people," and which bore an immediately discernible symbolic weight.[18] Faced with the naval blockade that prevented the import of graphite from England, Nadar writes: "Conté, on his first encounter with our mountains, extracts the iron carbide from which he will form the pencils that our students still use today and which England in turn will borrow from us."[19]

This was not the only occasion on which Nadar celebrated the work of the polymathic artist and scientist Nicolas-Jacques Conté—probably best known today, in art historical circles at least, for the smudgy black crayons to which he lent his name. Those with different investments in the period might recognize him as the inventor of modern processes for manufacturing pencils, for his chemical or aeronautical research, or for his involvement in the Napoleonic expedition to Egypt. Thirty years earlier, writing during the Siege of Paris in 1870, Nadar returned to the example of 1794, and to more recent American uses of military balloons in 1862, to chide the lack of seriousness with which airborne warfare was currently taken in France, at a time when Bismarck was attacking French *aérostiers*.[20] Conté was, here, his primary point of reference. Nadar, who owned a copy of Edme-François Jomard's early biography of Conté, found in him a fellow inventor, balloon enthusiast, and pioneer of inscriptive image technologies: "This marvelous Conté . . .

who made everything, from electric telegraphs, machines to strike money, windmills, barometers, balloons, etc.—right up to the pencils with which our children today learn to draw, which Conté was also very well known for having learned like all the rest, all alone and without a master."[21] Conté functioned, for Nadar, as an origin myth, as an allegory or shorthand for the production of objects from thin air, or which traversed thin air, and as a model of historical comprehension. An aspirational aesthetic, we might call it—air, animation, and influence all bound together. Chappe's telegraph was, from the start of the century, embedded as the primary means of transmitting state information, but it did not have a monopoly on the telegraphic imagination.

How might the Revolution's own turn to originary motifs—the regenerative effects of a revivified antiquity, recourse to natural time, or the production of new objects and technologies from the soil of the Republic—fit with a longer posthistory of the Revolution? What was it, at heart, that Nadar found so significant in the work of both Chappe and Conté, and what might we deduce from his fascination, in 1900, with the scientific world on the threshold of the previous century, the incredibly productive convergence of art and science in the wake of the French Revolution? Drawing, certainly, was one line that connected them, and it is perhaps unsurprising that Conté was so significant for the photographer, given his novel mediation of graphic marks via increasingly technical means. In the immediate postrevolutionary context, drawing, as mythical origin point for representation itself, brought artistic and scientific invention into conjunction with political renewal, an outlining of a new society on a clean surface. It provided a model for thinking through ideas to do with individual subjectivity, as well as neutrality or objectivity, especially as they evolved in Romanticism, and as an artistic practice it was well suited to the more democratic artistic climate created by the Revolution. In due course it offered, too, an immediate metaphorical language to both telegraphy and photography. Drawing with light was, to some degree, another iteration of drawing at a distance.

But what happens, in light of Nadar's enthusiasm, when such origin stories—the emergence of drawing as an activity with particular postrevolutionary implications, annexed to Conté's specific role in its material practice—appear at the end, as it were, just a few short years before photography's self-animating "pencil of nature" and other technologies of mechanical reproduction reconfigured forever the valence of the hand-drawn line? Perhaps that was partly the point: that the simple practice of drawing, like photography, was made anew via new apparatuses in postrevolutionary France, the contingency of these machines' production swiftly acquiring an aura of historical inevitability. Chappe's optical telegraph was an archetype for such a machine, but it was not alone. In the final decade of the eighteenth century, Conté's "simple machines," produced "all alone and without a master," point to both a beginning and an end. They also offer a dynamic case study for thinking about inscriptive media as a means of negotiating spatial and temporal distance, for articulating a fraught concept of imperial power, for balancing clarity of description with artistic expressivity, and as a route to the representation of an atmosphere (political, social, technological, or meteorological). While foregrounding Conté's contribution to the production of images, I am not concerned here with tracking straightforward patterns of influence, mapping the causal effects of technical invention on artistic practice. Rather, as Nadar's reading of Fourcroy revealed, the responsive character of Conté's inventions and their often-arbitrary genesis exposes the unanticipated, historically mutable nature of media, complicating the linear, sequential narratives that describe them, and laying bare the extent to which they might produce effects distinct from those they were intended to achieve. This story also points to telegraphy's central role in the exercise of Napoleonic power, and to the ways in which that technology operated as a guiding concept that intersected with several other methods for framing time and space, for drawing lines and making patterns in the air.

Against Origins

Conté's career is not easy to summarize, given his varied contributions to histories of art, technology, and administration, as well as his rich mosaic of research in chemistry, physics, and hydrodynamics.[22] Later in life, he was a significant figure in the formation and teaching program of the Conservatoire National des Arts et Métiers, which had its own role to play in associating technical drawing with Republican pedagogy.[23] "Conté," his collaborator Gaspard Monge wrote of him, "had all the sciences in his head, and all the arts in his hand"—a conventional partition of knowledge that, in Conté's case, might just as easily be reversed.[24] Monge too had telegraphic ambitions: a few years earlier, he had worked with Claude Chappe on a seven-armed version of the optical telegraph, which increased substantially the informational capacity of the line, although financial constraints prevented it ever being completed, and both Monge and Conté were intrigued by the possibilities of long-distance communication. Although Conté's visual practice has been somewhat overwhelmed by the sheer abundance of his technological vision, before he achieved renown as a scientist, Conté was trained as an artist in the studios of Greuze and, later, the miniaturist Pierre-Adolphe Hall, after he moved to Paris from Normandy for the first time in 1776.[25] Working as a portrait painter under the sponsorship of the duc d'Orléans, Conté began to participate in a dizzying range of scientific experiments, most notably in the field of ballooning; in 1784, he launched one of the first Montgolfiers from the terrace of the unfinished tower of the Cathedral of Sées.[26]

It was the Revolution, however, that transformed Conté's career. Clients for his artistic work disappeared, but, guided by prominent figures such as Jacques-Alexandre-César Charles, Louis-Nicolas Vauquelin, and Jean-Baptiste Le Roy, his scientific research gathered pace.[27] As he adapted Antoine Lavoisier's experiments of ten years earlier and developed a new method to synthesize hydrogen in large quantities without using sulfuric acid, the direction of Conté's scientific career chimed with the prerogatives of revolutionary science. His research facilitated the use of balloons as war machines, and Conté subsequently became director of the Société des Aérostiers at Meudon, establishing himself as an inventor of remarkable facility and range, a detail-oriented bricoleur for whom the production, from unexpected or found materials, of the tools to make the thing to make the thing became a specialty.[28]

A watercolor by Conté from this period shows a tethered balloon used by Jean-Marie-Joseph Coutelle for aerostatic reconnaissance during the Siege of Mainz; it was the second time, after the Battle of Fleurus in June 1794, that balloons had been used for this purpose in the field of conflict (fig. 3.3). At Fleurus, flag semaphore was used to convey the balloon's observations to the front. In Conté's painting, piercing views to an unseen adversary are implied

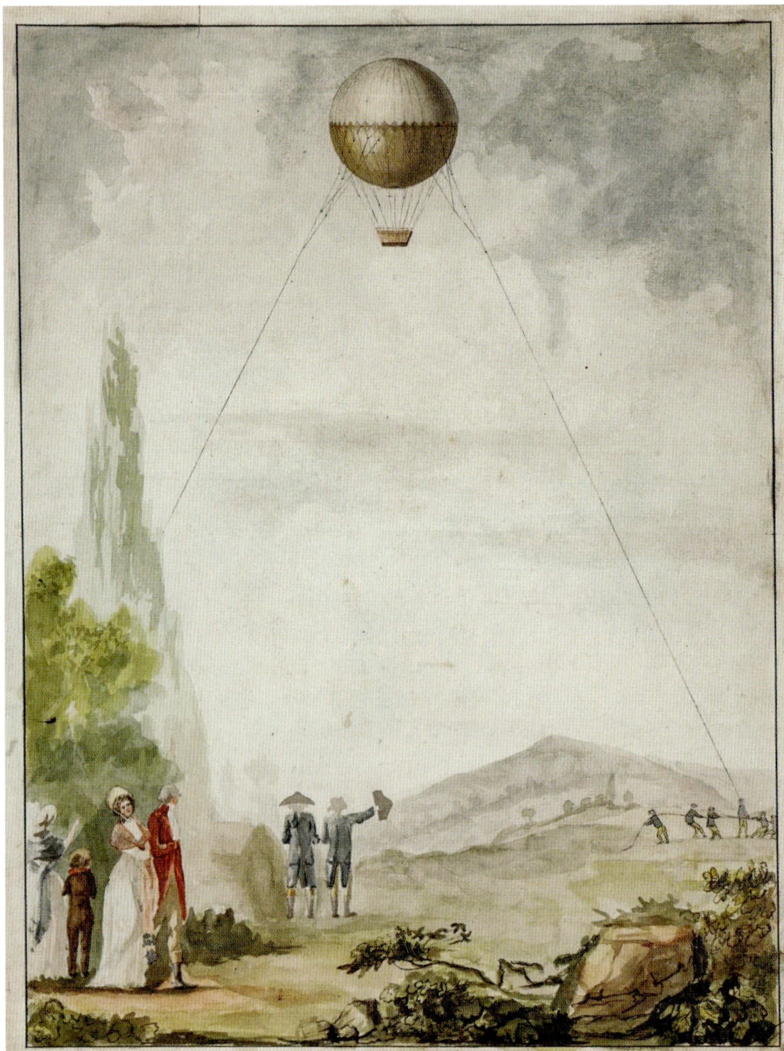

Figure 3.3
Nicolas-Jacques Conté, *Aerostatic Reconnaissance before Mayence, Besieged by Lefebre*, 1794, watercolor, 66 × 53 cm. Science Museum Group Collection, Penn-Gaskell Collection.

but not revealed. The hawsers restraining the balloon lead our eye to this strange object in the sky while a small group of men to the right heaving on a rope expose the labor, otherwise obscured in this calm image, of holding it in place. A small group of passersby has come to gawp. We look at their looking, and as if to make sure we don't miss that this image is all about lines of sight, about information thwarted and made visible, Conté includes a figural *repoussoir* in the form of a bourgeois woman who meets our gaze as her partner looks the other way, turning his eyes to the sky. As a footnote to Lakanal's 1794 report to the National Convention introducing Chappe's telegraph had put it, "it is therefore demonstrated today that an army that enters into battle without an observation balloon is like a man who fights a duel blindfolded."[29]

Being blindfolded, however, did not limit Conté's enterprise. It is fitting that an aesthetic of the line, evident across so much of Conté's practice, governed even his own representation, his face partly obscured by a black ribbon eye patch in one of the few contemporary portraits we have of him, an image reproduced in the *Description de l'Égypte* (fig. 3.4). Conté lost an eye following a laboratory explosion, an accident attributed to a servant leaving a door open, replaying the sloppy domestics in the genre scenes by Greuze that he had copied in his youth. The intense gaze of Conté's remaining eye produces a telescopic effect, making it a compelling and instantly recognizable portrait, while the dramatic sash that cuts a diagonal across his face accentuates his large cranium.[30] It is unsurprising that few other portraits followed it; again, in Conté's visual world, every beginning was also an ending, a chance to do something new. Like the eyes of surveillance that populated passports, banknotes, and other documents of state during the 1790s, Conté's portrait interpellates its viewers, subjecting them to a scrutiny that is at once scientific and mystical, a quality enhanced by the Egyptian pyramids that occupy the background in some versions of the print.

In examining Conté's contribution to the visual culture of postrevolutionary France, it is important not to reproduce uncritically mythic narratives of the heroic male inventor-genius and the deterministic narratives of media change that accompany them, hagiographies rendered ever more problematic by their colonial context. However, these accounts are worth attending to closely, for not only do they comprise the majority of contemporary responses—reinventing "invention" as a peculiarly

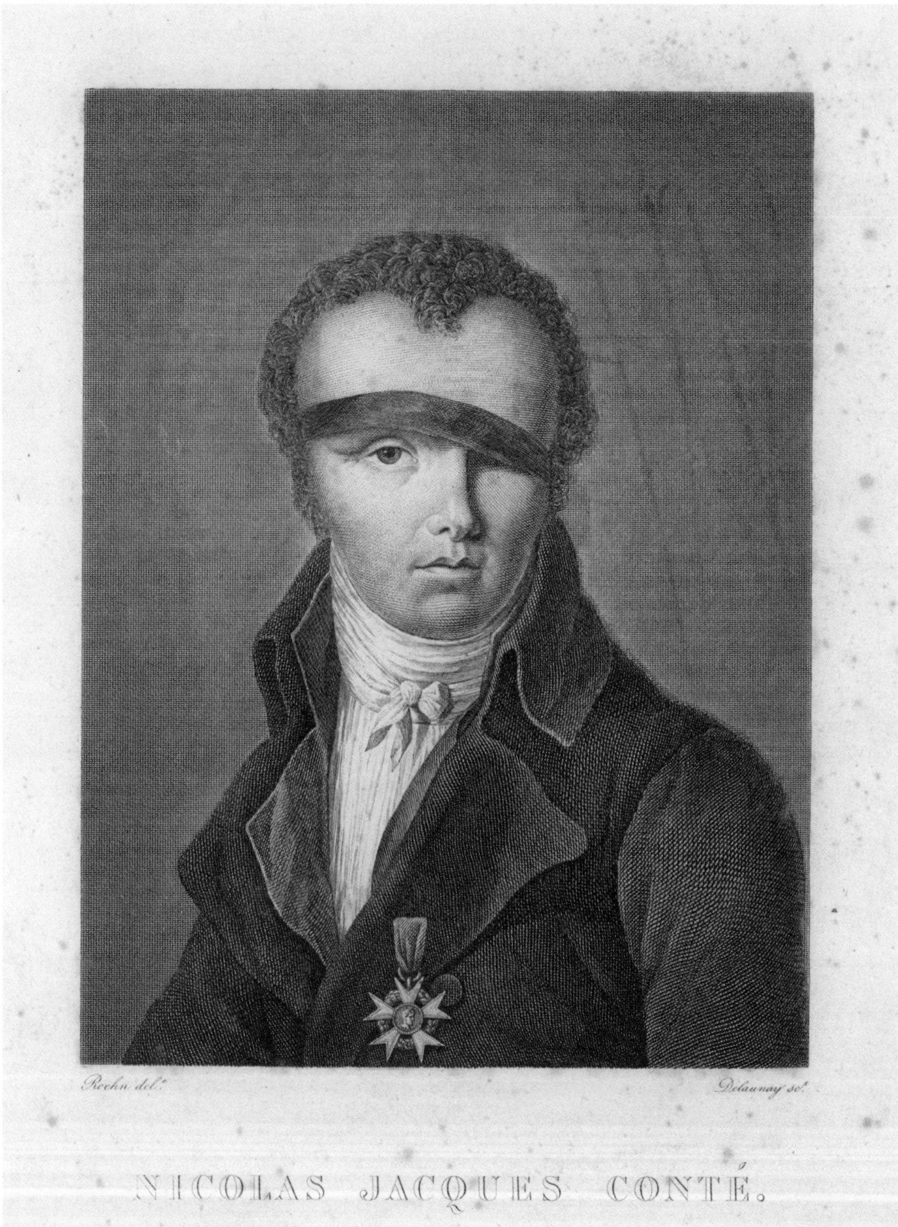

Figure 3.4
Adolphe Eugène Gabriel Roehn [draftsman] and Nicolas Delaunay [engraver], *Nicolas-Jacques Conté*, c. 1800, engraving, 70 × 50 cm. Tissandier Collection, Library of Congress Prints and Photographs Division, Washington, D.C.

postrevolutionary psychosocial phenomenon—but they also foreground the way Conté functions as both a central figure and a problem for the autonomy of drawing as it took shape in proto-Romantic scientific culture. In fact, Conté himself is strikingly absent from many of these narratives, evacuated by dint of his refusal to register copyrights on his inventions and his unwillingness to write much regarding his discoveries. As his friend and first biographer, Jomard, described, "everything was well written in his head," and he shared with fellow scientist Montgolfier a "repugnance for fixing his ideas on paper."[31] This characteristic—provisionality as ethical desideratum—was also discerned in far more widely published contemporaries such as Monge; in 1819, Charles Dupin bemoaned the latter's unwillingness "to entrust to paper the fruit of his meditations."[32] Yet Conté's aversion to fixity is particularly interesting, given that the means by which such attachment might be brought about loomed large in his work and came to dominate his reputation (Conté's inventions, Deanna Petherbridge observes, had "as radical an effect on writing as on drawing").[33] Conté, as an 1806 eulogy claimed, emerged fully formed, his early work carried out in secret, his inner life a blank page.[34] As I have already intimated, this combination of erasure and inscription situates Conté within a revolutionary context whose Ur-motifs might be the *tabula rasa* and the freshly engraved tablet of law. I want to propose a similar, temporary, erasure here, of drawing itself—or rather, of drawings—to ask how far we can think drawing through its materials, and to question at what point drawing stops being drawing and starts being something else. If, as Erika Naginski has suggested, "mark-making can be understood to function not as premeditated origin at all, but as accidental *consequence*," Conté prompts the question: "a consequence of what," or of "what things"?[35]

The Inventions of the Pencil

Conté's name is today connected most firmly with the invention whose expedient, improvised nature was, as Nadar noted, typical of his work: in 1795, responding to a dearth of imports of plumbago, so called "lead," from the mines in Cumberland, England, as a result of the naval blockade, Conté developed the modern pencil. The company formed in the aftermath of this innovation still produces the distinctive artists' materials that bear Conté's name, the flat, dark crayons associated indelibly with the development of several artistic careers and styles—most notably the work of Georges Seurat, who exploited the fuzzy effects of the crayons' disintegration on a rough matrix of laid Michallet paper.[36] Displayed at the first Exposition Industrielle in 1798, the pencils were swiftly reproduced elsewhere upon the lifting of the blockade, for instance by established German manufactories such as Faber, active since the 1760s, or by the 1830s, Staedler.[37] Conté's technique consisted of mixing potter's clay with finely powdered graphite enriched with lampblack. This mixture was fused at high temperature and poured into a mold, yielding stems that were encased mechanically in grooved wooden halves. A nineteenth-century demonstration plaque issued by the manufacturer Desvernay et Compagnie, which took over the production of Conté's crayons, shows the successive stages of this process, beneath a printed reproduction of Jules Droz's statue of Conté, erected in Sées in 1852 (fig. 3.5).[38] From the beginning, this was imagined as a project of Enlightened transfer, the assumption being that Conté's works would serve not only the nation from which they originated but "European commerce" more generally, as Lazare Carnot put it, in Floréal Year IV (April 1796), urging approval for the pencil's fabrication.[39] Although it had the pleasing effect of wresting control of the manufacture of pencils from the English, as with François Arago's announcement of the daguerreotype in 1839, this product was developed in France, but offered as a gift to the world.

The pencils were divided into two general types, a split evident in the H and B division (hard and black) used today to categorize pencils. The first group was geared toward the production of clear lines; these pencils were considered especially suited to technical drawing and writing of various kinds.[40] They varied in density, numbered from one to four. The softest, numbered one, was understood to be particularly effective for drawing and writing. The second was expected to have bureaucratic applications. The third was harder, and considered the best for drawing plans, machines, maps, and architecture. The hardest was appropriate for the most delicate lines. The second category of pencils, again numbered from one to four in order of hardness, was aimed explicitly toward drawing (*dessin*). The softest, which mimicked the properties of *mine de plomb*, was thought to be best for drawing landscapes and figures. Number two was suited to highlights and for delicate

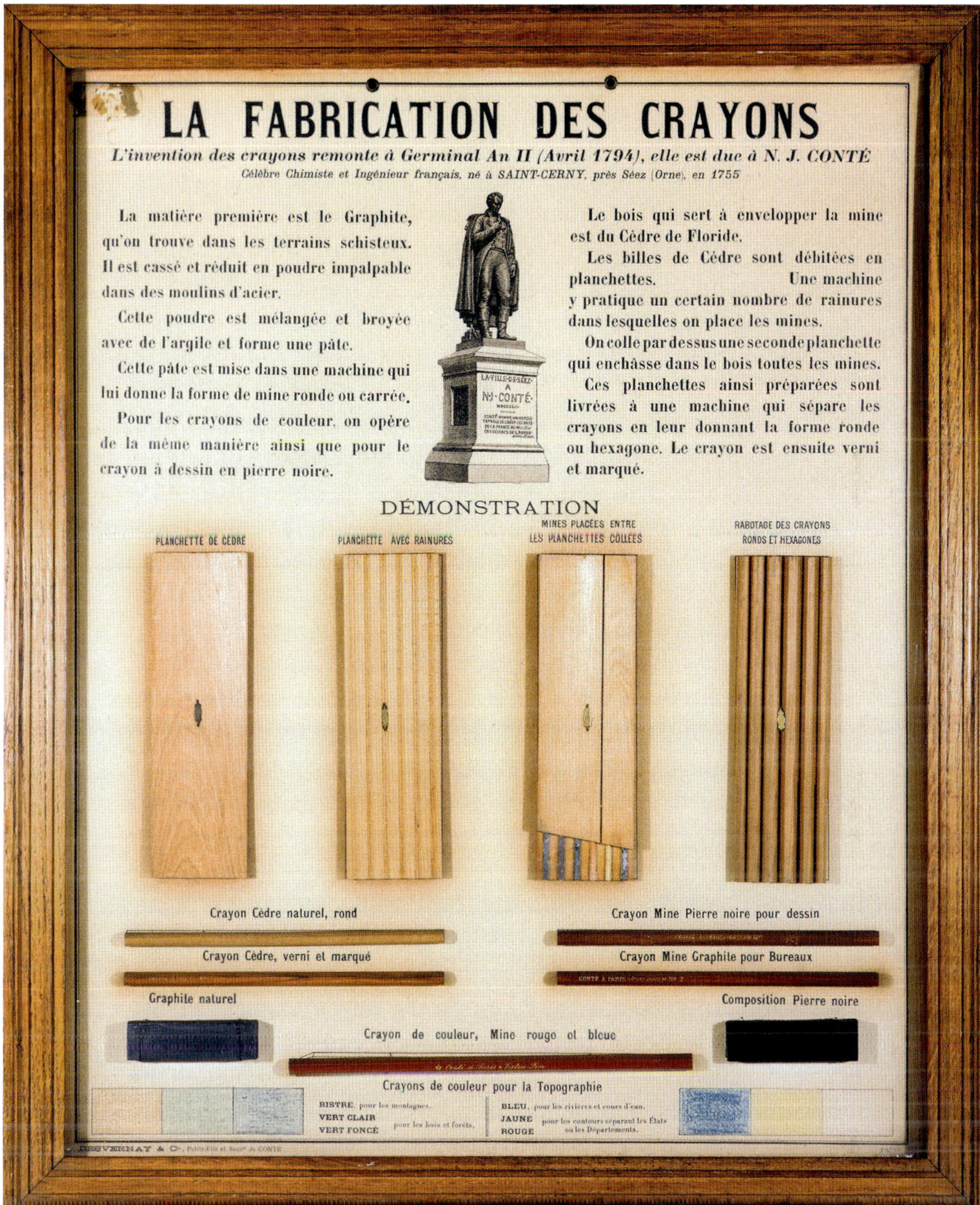

Figure 3.5
Desvernay et Compagnie, *Tableau des opérations de fabrication des crayons Conté*, nineteenth century, mixed media, 67 × 55 cm. Musée des Arts et Métiers, Paris.

drawings. The darker number three was closer in type to *pierre d'Italie* but produced a more "vigorous" line. Number four was a thick matt black. The division between Conté's two categories was not just of use, but of effect, for the first set—the "technical" pencils—suggested a gradation from softness to hardness, whereas the second, "artistic" set tracked a tonal spectrum characterized by the relative darkness of the marks made. The distinction, then, was between impression and texture. In addition to these pencils, Conté was also able to supply "velvety black pencils with different degrees of firmness and color," which the artist Jean-Baptiste Isabey used for his *manière noire* drawings, shown at the Salon.[41]

Drawings *à la manière noire* were popular from the early 1790s, enjoying a vogue at the democratic Salons initiated in the Revolution. These were images, known at first as *à la manière anglaise*, that simulated the effect of English mezzotint engraving. Unlike mezzotinters, who worked from dark to light, artists drawing *à la manière anglaise* achieved their effect via stippling with Conté's crayons, their pitchy surfaces punctuated by bright white highlights so they would stand out when displayed. Isabey's drawings, though clearly amenable to mechanical reproduction, especially given their duplication of its effects, were subsequently reproduced in stipple engraving because English techniques for mezzotinting were not practiced in France at this time. As Tony Halliday has observed, the stipple engravings failed to capture the full complexity of Isabey's tones.[42] Drawing, then, was situated, in this new world of artistic materials, posterior to engraving rather than as a preparatory method as one might expect, and it suited the openness and mobility of a reformed artistic display culture. Legitimized by a popular taste for imported engravings, and presenting a range of fresh connoisseurial possibilities distinguished by Conté's original materials and the novel uses to which they were put, this method found favor among the new audiences for art produced by the Revolution. However, criticisms arose, in particular from artist and journalist Charles Paul Landon, for whom the newfound aesthetic autonomy of drawing (as well as miniature painting, pastel, and anything made by a woman artist) threatened the authority of history painting.[43] Moreover, drawings *à la manière noire* were suspect because of their foreign inspiration. As Halliday points out: "It was not merely mezzotinting, however, which came to be identified with the alien and corrupting influence of English taste during this period, but all tonal and quasi-tonal engraving methods. . . . Line engraving was soon being proclaimed as the only truly French method—a view which received official endorsement in 1803 with the appointment to the Institut of the engraver Bervic, stern votary of the clear cut line."[44]

Conté's work in this area was roughly coincident with Alois Senefelder's invention of lithography, the calligraphic possibilities of which exceeded those previously available to printmakers, allowing them to produce tonal effects not dependent on line. Yet Conté's pencils (if not his famous crayons) represent a converse tendency, imparting to drawing something of the objective allure of engraving. A similar causal connection has been argued between the introduction of commercially available steel pens, which enabled harder, brighter lines, and the development of artistic practice after 1800. Seymour Howard has suggested that the steel pen lent itself to an objective visual language appropriate to industrial design, and, complicating more straightforwardly determinist accounts, that the enthusiasm for this mode of representation had been prefigured in artistic practice, especially by neoclassicists such as d'Hancarville, David, Tischbein, Flaxman, and Ingres, "who practiced and promoted a revolutionary use of pure outline illustration."[45]

If Conté's invention pointed toward an industrialization of drawing, it did not do so—as the case of Isabey's *manière noire* drawings suggests—exclusively by way of hard lines. Furthermore, before the introduction of lithography in France, Conté independently developed a similar technique to reproduce the hieroglyphics of the Rosetta Stone discovered in 1799 during Napoleon's campaign (its code remaining undeciphered until Champollion's breakthrough in 1822).[46] Significantly, the tonal effects of Conté's pencils were apparent from the beginning, and were crucial to the artistic dimension of at least half of them.[47] In addition to a scientific report that commented on the "technical" pencils, prepared by Fourcroy and Pierre Bayen and presented by Fourcroy to the Institut National des Sciences et des Arts, the four grades of pencil geared toward artistic drawing were examined and unanimously approved by a committee at the Louvre, headed by the history painter François-André Vincent, and comprising additionally the flower painter Gerard Van Spaendonck, the architects Charles De Wailly and Antoine-Joseph Peyre, and the sculptors Pierre Julien and Jean-Antoine Houdon.[48] As Vincent relayed, the committee appreciated not only the quality of Conté's materials but their versatility, in terms that undercut somewhat the rhetoric of figures such as Landon. For, while the pencils were superior in both range and finesse to their English counterparts, the unencased crayons could be used, with a *porte-crayon*, to draw on a large scale, suggesting that they could also be appropriate to the preparation of ambitious academic paintings.

Conté's invention—or perhaps we should call it, more properly, an adaptation—sits somewhere between eighteenth-century encyclopedic knowledge, based on ordering the world into its constituent parts, and the emerging positivism of the nineteenth century, in which phenomena demanded more pragmatic forms of resolution. One element singled out for praise in the reports—technical and artistic—was the consistency of Conté's pencils, which resisted deterioration or mutation over time.[49] Concurrently, Conté was conducting his long-running experiments with "a new genre of color" for painting in oil, pastel, miniature, watercolor, and enamels.[50] Although never completed, these experiments were taken to an advanced stage and were mentioned in Fourcroy's report to the National Convention on the progress of the arts during the Revolution, and right up until his death Conté was preoccupied by the fading of colors (a suitable concern, perhaps, for one's last days). Jomard recalled discussing this "inalterable palette" with Conté shortly before he died, when the inventor expressed his desire to rescue the great painters from the damage to their reputations caused by the deterioration of their works' hues.[51] Conté's colors were distinguished by their solidity and durability—the white especially—which purportedly resisted the effects of light and atmosphere for twenty-five years.[52] Meanwhile, the marks of his pencils could not be shifted with the usual crumb of bread. In other words, Conté's pencils stopped time. Their marks were indelible; they seemed as permanent as an engraving, and as such they made a particular claim on the production of history and, in a sense, worked against drawing's vaunted mobility and provisionality, its translation of the mind's erratic workings into graphic traces. Furthermore, and crucially, Conté's crayons were not derived from nature. This was artificial lead, and the status of the pencil as a "simulacrum" was never in doubt.[53] So drawing's fidelity to the real was produced through something artificial, powered by human reason, yet severed from the past; at the moment of its introduction as a "modern" technology that lay claim to the future, the pencil disavowed origins.

Conté's second major graphic innovation was a more complex machine, although it should be understood in close dialogue with his earlier work with the pencil. It was also connected with contemporaneous means of transmitting information, such as Chappe's network, for this device and its application firmly appended the *tele* to the *graphic*. In 1798, Conté traveled to Egypt as part of the Napoleonic expedition. He wrote back to his colleagues from Alexandria, describing the images he was making there of local customs, trades, and costumes, as well as of the ruins of Memphis, his letter a series of graphic marks whose sand-stained support bears the traces of its transmission.[54] Conté's work in Egypt, as one of the famous group of *savants* accompanying Napoleon, was incredibly diverse: he engaged in money-making, weaving, glassmaking, and bridge-building; the manufacture of syringes, keys, weapons, cutlery, barrels, cardboard, and clothing; the improvisation of barometers, graphometers, windmills, and brassworks; and the production of designs for pumps, bellows, saws, machines to elevate water, and a device for measuring time in milliseconds.[55] He also documented the arts, trades, and techniques of modern Egypt. "Conté . . . a universal man . . . capable of creating the arts of France in the middle of the Arabian desert" was how Napoleon described him, the only member of the Institut d'Égypte to be mentioned in the former emperor's memoirs.[56]

One achievement stands out: Conté developed an engraving machine that enabled the production of the huge elephant folio sheets that would compose the *Description de l'Égypte*, the monumental multivolume document of the expedition.[57] He also supplied a number of designs for its nine-hundred-plus plates (containing over three thousand images), in a restrained style that recalled his genre training; his image of a weaver's workshop is typical among Conté's drawings for the *Description* in its representation of harmonious labor in an interior scene (fig. 3.6). Conté did not live to see the first volume of the *Description* published in 1809. He died in Paris in December 1805—eleven months after Chappe—as the city reverberated to news of the French victory at Austerlitz. However, his presence can be felt throughout the project, and he appears himself on at least two occasions, his distinctive eye patch making him easy to identify.[58] The engraving

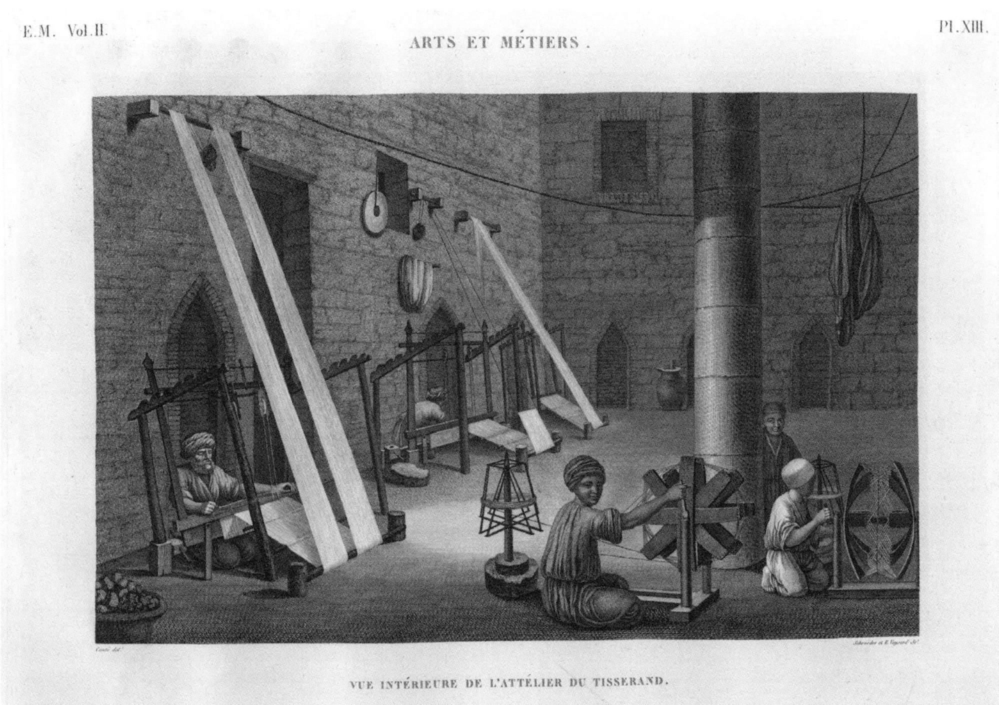

Figure 3.6
Nicolas-Jacques Conté [draftsman], Friedrich Schroëder and Étienne Voysard [engraver], "Arts et métiers. Vue intérieure de l'attélier du tisserand," from *Description de l'Égypte: ou, Recueil des observations et des recherches qui ont été faites en Égypte pendant l'expédition de l'armée française*, État moderne, plates: volume 2, 1809–1828, engraving, 50 × 70 cm. Rare Book Division, The New York Public Library.

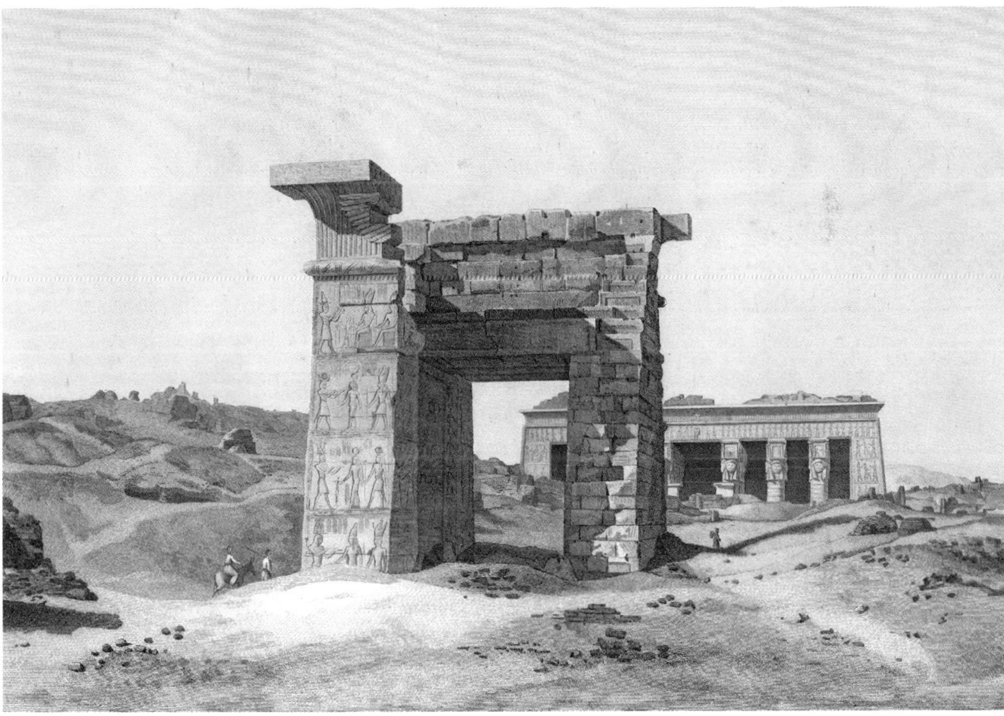

Figure 3.7
André Dutertre [draftsman] and F. B. Lorieux [engraver], "Denderah (Tentyris). Vue de la porte du nord," from *Description de l'Égypte: ou, Recueil des observations et des recherches qui ont été faites en Égypte pendant l'expédition de l'armée française*, Antiquités, plates: volume 4, 1809–1828, engraving, 50 × 70 cm. Rare Book Division, The New York Public Library.

machine made it possible to complete the gradations of the backgrounds more quickly than would otherwise be possible.[59] Eight months of work could be carried out in two to three days. The machine needed to be versatile enough to produce different backgrounds, so as to provide aerial perspective for diverse scenes, while technical illustrations, architectural plans, reliefs, and elevations had to be produced in a uniform tone (fig. 3.7).

If the pencil had subterranean origins—the task falling to Conté following extensive pressure on the Agence des Mines to find a suitable alternative—the engraving machine was all about the air. While most of the plates in the natural history volumes of the *Description* isolated creatures and objects against plain white paper, the backgrounds of the sections dealing with ancient and modern Egypt were often monotonous desert scenes or skies pitched against acute horizon lines, a difficult task for an engraver to accomplish on such a scale by conventional means. Indeed, engravers traditionally filled such backgrounds with stormy, irregular clouds to cover their inevitable mistakes. With Conté's machine, developed with mechanic Antoine-Clerc Gallet, such correction was no longer necessary.[60] Having set the parameters of the machine in advance, regulating the distances between lines and determining the kind of inscription one wanted to make, it could be operated even by someone with little knowledge of engraving. The machine comprised a rectangular wooden base set atop a table, across which a fixed burin could traverse on both axes (fig. 3.8). While the operator still had to move the carriage containing the burin back and forth across the plate by hand, the physical effort of engraving was vastly reduced. Later, this device was adapted by the Maison Oberkampf for the production of printed textiles and was also reconfigured for engraving on steel.[61]

The engraving machine was not unique in offering a regular impression of lines on paper. For example, in November 1791, Rohberg-Herr de Vausenville obtained a ten-year *brevet d'invention* for a new technique first announced in the 1760s.[62] This "art gammo-graphique"—likely a contact-printing method of some kind—promised to produce lined paper far more quickly than was possible with conventional printmaking techniques, and was recommended for printing musical scores, administrative documents, and all manner of tables and registers. Arguably, it was during the Empire that this kind of paperwork fully came into its own, the printed form functioning as a powerful technique of the

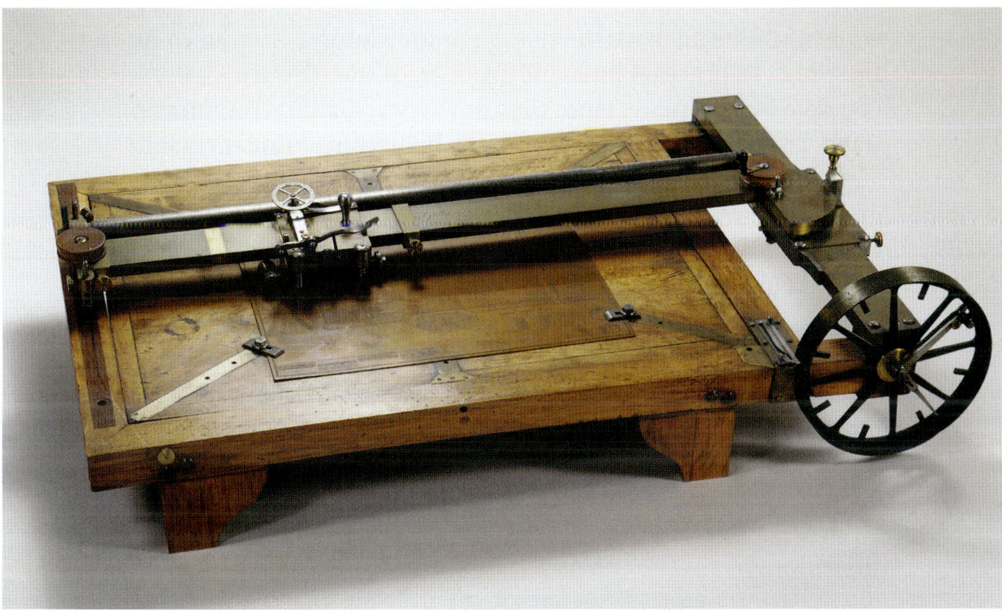

Figure 3.8
Nicolas-Jacques Conté and Antoine Clerc Gallet, *Machine à graver les fonds et les ciels*, 1803, steel, wood, and brass, 35 × 87 × 140 cm. Musée des Arts et Métiers, Paris.

modern state, determining the possibilities and limits of citizenship. The machine-engraved plates of the *Description* expanded and abstracted this phenomenon, its lines figuring as an index of the empire's universal ambitions.

Conté's machine, as a remarkable demonstration plate from the *Description* shows, was able to produce parallel lines, lines that converged on a point, differences in intensity on the same line, bendy lines as well as straight ones, and, with a few minor alterations to the setup, all the other effects of the burin, whether drypoint or etching, or acid on copper (fig. 3.9).[63] Due to the scale of the plates from the *Description*, they do not reproduce well, for a reduction in size drains them of much of their visual intensity, while also diminishing the notable bodily address produced by the project's physical bulk and the task of maneuvering its vast pages. With this plate, more than any other, digital reproduction struggles to cope with the effect of the lines, scrambling them into a rippled pattern of interference or flattening them into a static grey (fig. 3.10). Attention to this phenomenon reveals aspects of the medium that were, perhaps, there all along. A certain derangement troubles the regularity of Conté's design, exacerbated when this plate, made to aid reproduction—or to represent its possibilities—is itself later remediated by way of a different technology.

The *Description de l'Égypte* has been invoked, most famously by Edward Said, as one of the key documents in the construction of an orientalist worldview, produced to "render it [Egypt] completely open, to make it totally accessible to European scrutiny."[64] As Said observes, "the Orient was reconstructed, reassembled, crafted, in short, *born* out of the Orientalists' efforts. The *Description* became the master type of all further efforts to bring the Orient closer to Europe, thereafter to absorb it entirely and—centrally important—to cancel, or at least subdue and reduce, its strangeness," a process in which photography would later play a defining role.[65] "Reconstructed, reassembled, crafted, in short, *born*"—this could almost be a description of the how both Conté and his machines figured in the postrevolutionary French imaginary. For Said, the *Description* was an attempt, via both text and image, to impose order on disorder, to organize and control, and to represent the Orient as both unchanging and in need of regulation.[66] The rhetoric surrounding the expedition furthered this agenda. Conté was not only described, in Joseph-Marie Degérando's account, as a man thrown by a storm on a desert island, "a messenger that European

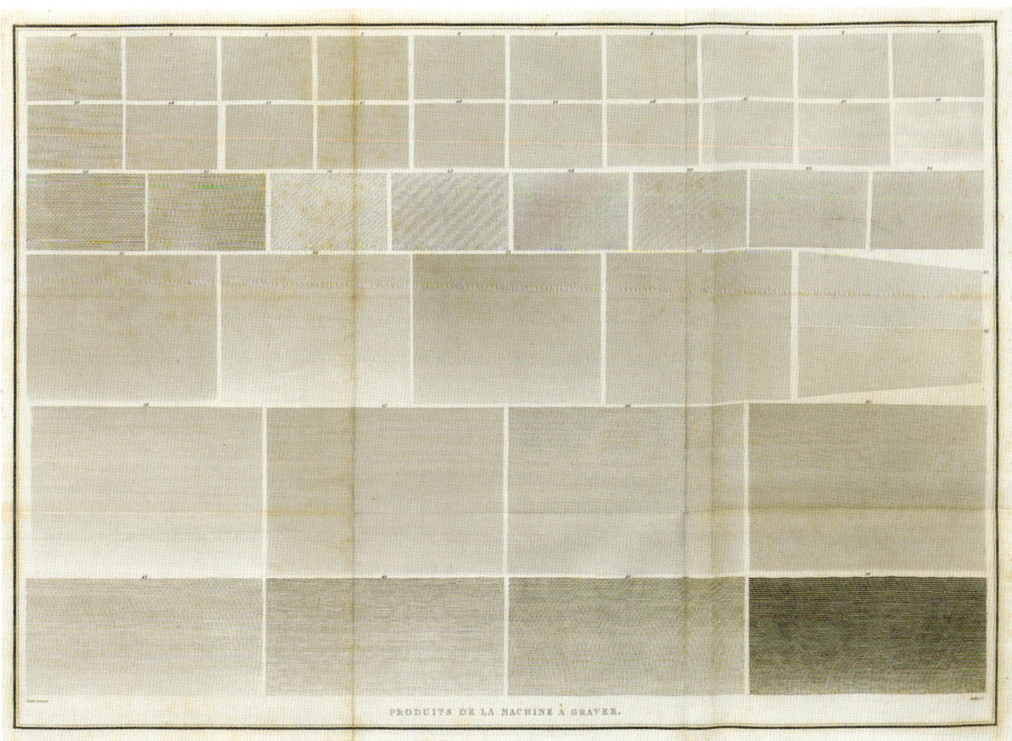

Figure 3.9
Nicolas-Jacques Conté [designer] and Antoine Clerc Gallet [engraver], "Produits de la machine à graver," from *Description de l'Egypte: ou, Recueil des observations et des recherches qui ont été faites en Egypte pendant l'expédition de l'armée française*, État Moderne, text: volume 2, 1809–1812, engraving. Heidelberg University Library.

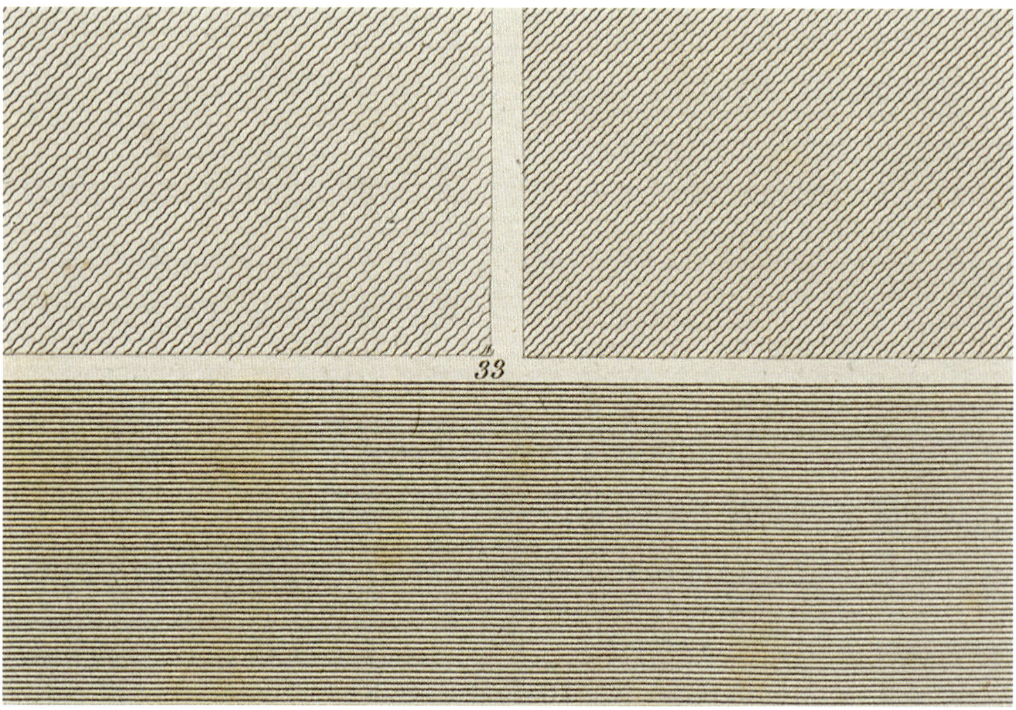

Figure 3.10
Detail of Nicolas-Jacques Conté and Antoine Clerc Gallet, "Produits de la machine à graver," from *Description de l'Egypte*.

industry sends to this other world," but as ultimately disappointed with "the stupid indifference of the natives," who had refused to adapt and develop his inventions. In a melancholic afterthought, Degérando added that "the genius of the arts has crossed this land and has not been able to leave any traces there."[67] The engraving machine seems at one level an attempt to make good on this failure, traversing the expanse of the plate with the sole aim of leaving legible marks behind.

Yet, while Said notes the effect of scale in the plates of the *Description* as an index of its grandiose ambitions, the mechanism by which the plates were produced, and the ways in which atmosphere itself, and not just people and things, was organized and displayed, is not a feature of his analysis; he pays little attention to how the *Description* achieved its aims visually. More recently, Darcy Grimaldo Grigsby has described how nineteenth-century colossal engineering—the Eiffel Tower, Suez Canal, and Panama Canal in particular—bracketed a colonial imaginary to the production and measurement of lines, cut through continents.[68] Showing how the monumental structures of ancient Egypt continued to haunt French engineers and artists into the twentieth century, Grigsby argues that mass, so often evacuated in these projects, returns as the repressed, bearing with it the ghosts of the laborers whose lives were sacrificed for the fantasy of scale. Conté's engraving machine, a device anchored to the production of the gigantic, can surely be considered in similar terms, as the sublimation of a desire to mastery, an exercise of power camouflaged as a labor-saving device. Yet, as Grigsby observes regarding one of André Dutertre's plates for the *Description*, to convey a sense of monumentality, "the image must counterpose contingency and order, randomness and coherence, formlessness and form, accident and intention."[69] For Dutertre, and for the many other artists who contributed to this exhaustive atlas, line was a disciplinary measure but also a means to expose the mutability of form. Arguably, such a tension was also discernible in the backgrounds themselves, which give the impression of total uniformity, even though they are subtly anything but. On a few occasions, even, clouds return, while elsewhere the ratio of line to blank paper varies, and in some cases, such as when a sky meets irregular foliage, empty space can be seen around the ends of the lines, a minor fault or error in registration rendered visible by the large scale of the plates.

As a report signed by Conté, Montgolfier, and Molard in June 1804 stressed, "to speak to the eyes is the surest means to reach the intellect."[70] While evidently sharing something with the empiricism of Locke and Condillac, this appeal to optical reason was, however, distinct from the absorptive model of the previous century. The Revolution's embrace of the power of signs was accompanied by an equal fear of their consequences and of a collapse into treacherous, *ancien régime* dissimulation.[71] This was the line that telegraphy also trod, between clarity and obscurity. Under the Consulate and Empire, a turn to spectacular effects sought to ameliorate this impasse. And this is where the limitations of a reading of the *Description* solely in terms of its ordering agenda are exposed. For, as much as the *Description*'s regulated lines controlled space, they also contained the potential to produce new visual effects in the minds of their (mainly elite, French) viewers. If the purpose of the *Description* was, as Said suggested, to subdue Egypt's strangeness, the facture of the plates, and the uncannily serene visual effect they created, generated a strangeness all of their own. As Jomard wrote of the uses to which the engraving machine might be put: "A sky being destined to serve as background to a perspective view of any object that traces its outline on the horizon, one must render the imperceptible deterioration of the light reflected by the atmosphere, which, from little density at the zenith, increases in brightness up to the same horizon."[72] Inasmuch as the *Description* focused a forensic lens on Egypt's ruins and its present-day trades, inhabitants, flora, and fauna, the representation of an atmosphere was a matter of perception.

The striated backgrounds produced via Conté's machine are indeed oddly of a piece with contemporaneous scientific projects undertaken in Egypt, in particular Monge's discovery of the cause of mirages in the desert, to which the shimmering skies of oscillating lines in Conté's prints speak, despite their alleged renunciation of illusionism.[73] As has been argued recently, the mirage has long figured as a sign for Oriental despotism, a corruption of vision that counters the transparency of the colonial gaze.[74] Despite their ultimately discombobulating effect, Conté's backgrounds—particularly those rendered on a large scale—attempt to stabilize the images in front of them, marking a distinction between the

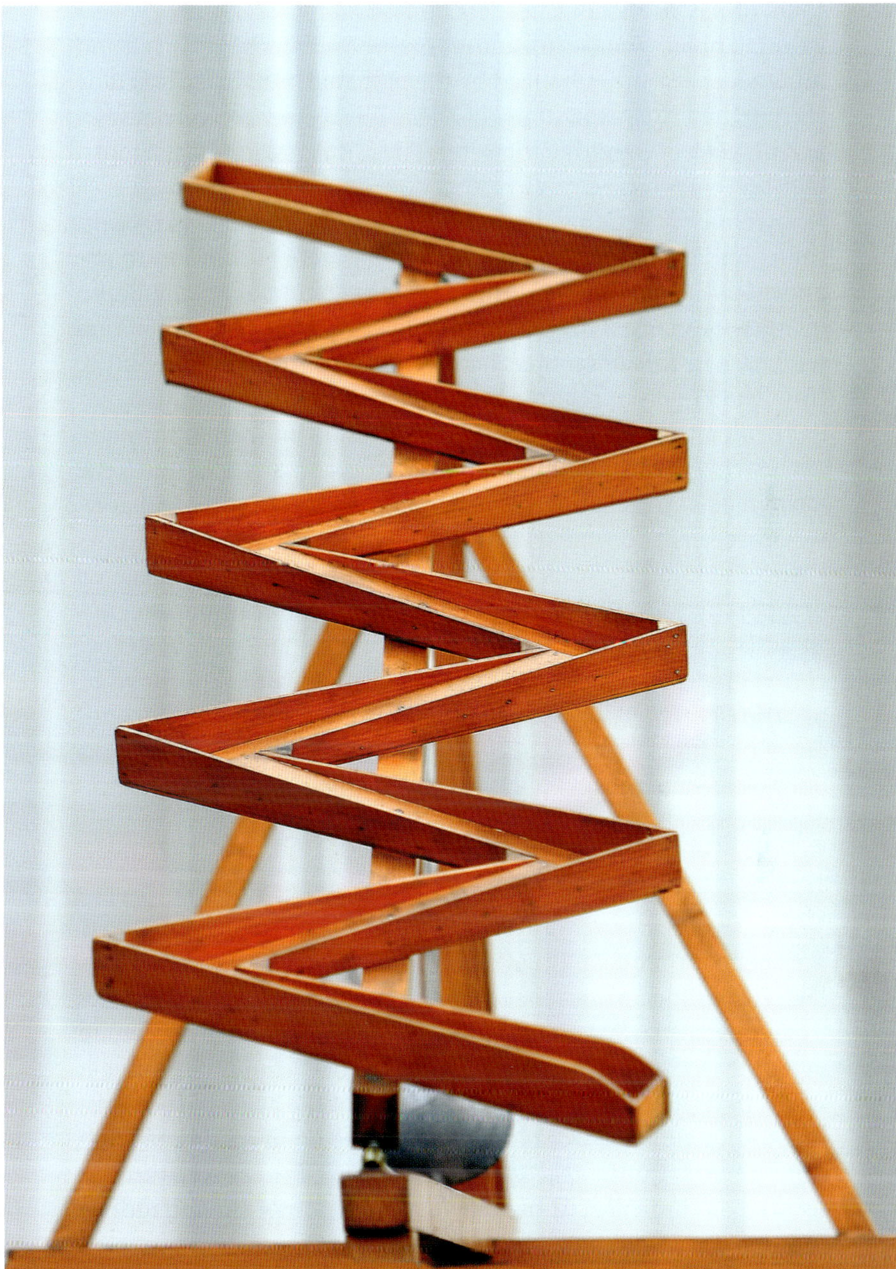

Figure 3.11
Nicolas-Jacques Conté, *Modèle: Auge oscillante en zigzag pour élever l'eau*, c. 1800, wood, ferrous alloys, paper, 49.5 × 84 × 38 cm (in profile). Musée des Arts et Métiers, Paris.

hard lines of European analytic inquiry and the crumbling contours of past ruins. But in their precise *horror vacui*, they evoke a world of objects whose meanings might as easily be reversed and refracted, a feature enhanced by the palpable heat experienced by the parched figures in many of these images, who expire, sweat, and beseech one another for water. These images, more than most, register temperature.

Around 1800, Conté designed a device in Egypt that plays with these ambiguities, and which offered a practical solution to the aridity of the environment (fig. 3.11). A machine to elevate water, in line with many of his Egyptian projects, it was made to function with ease in this unfamiliar habitat, and Conté drew inspiration from the Egyptian techniques he surveyed. A pendulum, hanging from a triangular frame, drives water up an inclined trough. Each swing of the pendulum makes the water advance up to the next angle of the channel. The machine makes distance and gravity subservient to time, syncopated by its oscillating weight. As Patrice Bret notes, the elemental geometrical forms of this jagged orange structure are "a harmonious blend of poetic invention, purist aesthetics, and minimalist technical rationality."[75] An object entirely focused on a precise and necessary practical purpose, it, like the *Description*, exceeds its function at every turn. We encounter it as a big toy or a minimalist sculpture, as Conté's straight lines were, here most explicitly, accompanied by formal and conceptual zigzags.

Another of Conté's Egyptian inventions is noteworthy in this context. He outlined plans for a telegraph connecting Cairo with news from Alexandria. Seventeen wooden apparatuses were developed in 1801, based on his own system rather than Chappe's, although it seems that they were never used because of the arrival of the British at Aboukir a few weeks later.[76] Building on his earlier balloon experiments, Conté also imagined a relay of moored balloons, secured at thirty feet above the ground. Each of these balloons contained two cylinders of black taffeta or waxed canvas, which could be extended along strings attached to the ground at varying levels corresponding to an alphanumeric code. This would be viewed by an operator in the next balloon, who would pass the message along the chain. In two or three minutes, this new telegraph could carry information ten leagues.[77] We have no images of the project, which was not ultimately implemented in Egypt. While the Chappe telegraph gradually became integrated into the landscape, Conté's system would surely have been spectacular, and impossible to miss.

Conté described his plans for an aerial telegraph in a letter to his wife, outlining how it would be "very simple, cheap, and . . . of greatest service in a country where the post is ignored and where couriers are exposed to being murdered every day."[78] Conté's Egyptian balloon telegraph—he called it an *aérographe*—emerged from a very similar project he had developed in France. On May 21, 1796, he had submitted a report to Pierre-Louis Prieur, deputy of the Council of Five Hundred, describing this plan and suggesting that his simple machine would be able to communicate with people on the ground as well as those in other balloons.[79] Starting in August of that year, he experimented with the *aérographe* at Meudon, transmitting messages by this method from *l'Aigle*, an aerostat stationed there, to *Télégraphe*, a fortuitously named balloon situated at Pantin, and subsequently at Livry and at Dammartin-en-Goële, forty kilometers northeast of Meudon. Poor meteorological conditions and atmospheric pollution above Paris prevented him from continuing the experiment to Soissons, and the *aérographe* was not taken up in France.[80]

In 1826, Claude-Joseph Ferry, a former professor at the artillery school and examiner at the École Polytechnique, as well as a deputy of the National Convention during the Revolution and one of the designers of the Republican calendar, described the historical context for aerographic telegraphy using metaphors of breeze and breath: "the storms of the revolution were not appeased; but *savants* were no longer dragged to the scaffold. We were beginning to breathe."[81] Political storms were not the only issue. The lightness of the *aérographe* made it both portable and credible for suspension from a balloon, but the fragility of its cords and slight fabrics meant that it risked being silenced by strong wind. As numerous dispatches via Chappe's optical telegraph record, the weather was frequently an impediment to successful telegraphic communication and a consistent topic of interest.[82] In his 1793 report to the Convention, Romme had praised the telegraph's signals for being "as simple as the straight line of which they are composed" but had warned, at its inception, that "a thick mist, such as occurs in the north, in watery regions, and in winter" might pose an obstacle.[83]

In urban France, this concern was exacerbated by human contamination of the atmosphere, particularly by the burning of fossil fuels, which threatened to thwart a clear view of optical signals. More familiar semaphoric systems were later deployed in Algeria (fig. 3.12); in Egypt, for the first time, the stiller, clearer atmosphere offered an opportunity to realize fully the fantasy of direct communication, uncluttered by the clouds, fog, and smoke that had plagued the implementation of the telegraph in France.

Yet fog or pollution had not been the only obstacle to the transmission of Chappe's optical telegraph messages. As one account noted, "the undulations of the atmosphere in hot weather often made signals very difficult to see. The position of the posts had a great influence on these undulations; they were particularly troublesome when the visual ray between two posts skimmed over wet ground."[84] If the distorting mirage effect produced by hot weather was a hazard for telegraphy in France, it was all the more so in Egypt, bending even the straightest line. Subsequent forms of optical telegraphy also aimed to meet the challenge of local conditions, such as Jules Leseurre's *héliographe*, developed for use in southern Algeria in the mid-1850s. The *héliographe* used mirrors to reflect the sun's rays to transmit flashes of Morse code, and was later adopted for military purposes as an alternative to electrical field telegraphs (fig. 3.13). The *aérographe*, however—influenced undoubtedly by the success of the Chappe network, as well as by its competitors in France and imitators across Europe—replaced the Chappe system's semaphoring arms with spectacular airborne barcode. Resurrected in Egypt, its invisible lines in the air correlated with Conté's contemporaneous attempts to develop new ways to fill the skies of the *Description* with information.

Such a comparison works the other way too. For the engraving machine produced what we might term a "telegraphic effect." If the "far-writing" of telegraphy collapsed time and space, the precoded backgrounds made by the engraving machine also covered distance and did so at unheard-of velocities. Like the telegraph and Conté's pencils, the skies of the *Description* also generated an unexpected surplus in the visual field. Yet as John Durham Peters has observed, media—among which the sky itself should be counted—are powerful "civilizational ordering devices."[85] Conté's engraved skies and his *aérographe* embraced this aspect of the environment they represented or inhabited, repressing the ways in which it might not comply with their version of order or of civilization.

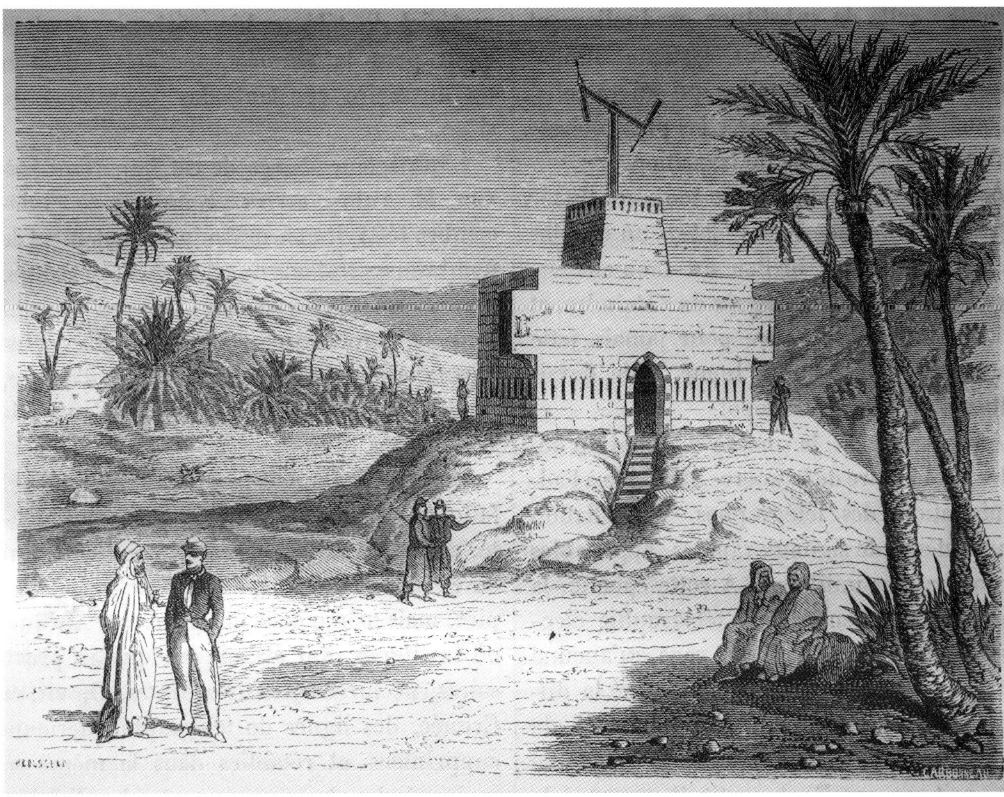

Figure 3.12
H. Rousseau [designer] and Jean-Baptiste-Charles Carbonneau [engraver], "Poste télégraphique française en Algérie," from Louis Figuier, *Les Merveilles de la science, ou description populaire des inventions modernes*, vol. 2 (Paris: Furne, Jouvet, et Cie, 1868).

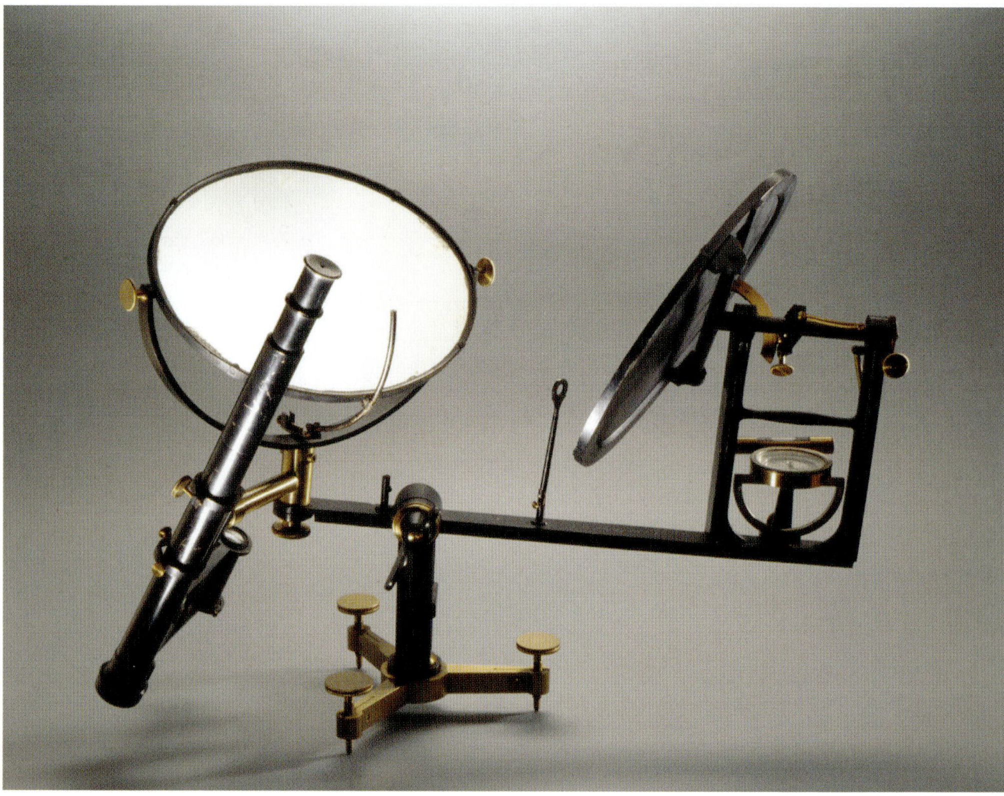

Figure 3.13
Jules Émile Leseurre, *Héliographe à deux miroirs*, 1855, metal and glass, 40 × 40 × 70 cm. Musée des Arts et Métiers, Paris.

Informational lines in the sky and practices of inscription continued to be closely associated long after the end of the Empire, although to quite different ends, in the hands of other artists. Often, they recalled Conté's earlier inventions, interleaving perceptively his graphic practice with that of his contemporary Chappe. For example, Swiss caricaturist Rodolphe Töpffer's *Monsieur Pencil*, drawn in 1831 in the aftermath of the July Revolution but not published until 1840, shows the chaotic consequences of a dog owner's attempts to retrieve his enthusiastic pet from the wing of a Chappe telegraph relay (fig. 3.14). In several scenes from Töpffer's proto-comic strip we witness the ease with which a chain of random accidents might initiate an international crisis, as the dog's antics scramble the telegraph's code so that cholera is mistakenly believed to be advancing on Paris—only the telegraph could outpace the disease. Preparations for the arrival of cholera in the capital had begun in 1830, when it first appeared in Europe; the epidemic of 1832, which killed 100,000 people in France, haunted Parisians for years afterward. Cholera was an agent of chaos, politically fraught, and accompanied by strange new bodily dramas and even forms of virtuosity. For Nina Athanassoglou-Kallmyer, it found its paradigmatic figure in the composer Niccolò Paganini—her description of him is inescapably telegraphic: "Tall, lean, and bony . . . his body jerked, swung forward and back, and twisted awkwardly while his legs kicked in all directions and his arms flew up, as if out of joint."[86] In Töpffer's image, which makes light of a climate of mounting fear, we have drawing/telegraphy as disorder, a connection intensified by the free, energetic style of *Monsieur Pencil*.[87]

This theme was developed at the end of the century in a pair of drawings by Georges Quesnel, representing two inhabitants of vaguely defined African spaces, an Arab horseman and a giraffe, attending to but also disrupting the electric telegraph's carefully laid lines (figs. 3.15 and 3.16). Here, indigenous people and animals corrupt the network, either by revealing the line hidden in the sand or by tangling their long necks in the wires. Vainly protected by barbed wire on the edge of a jungle that threatens to overwhelm, the telegraph functions in these drawings as a marker of colonial violence, of the brutality of this imposition in the landscape. Generating a depressingly familiar Conradian account of Africa

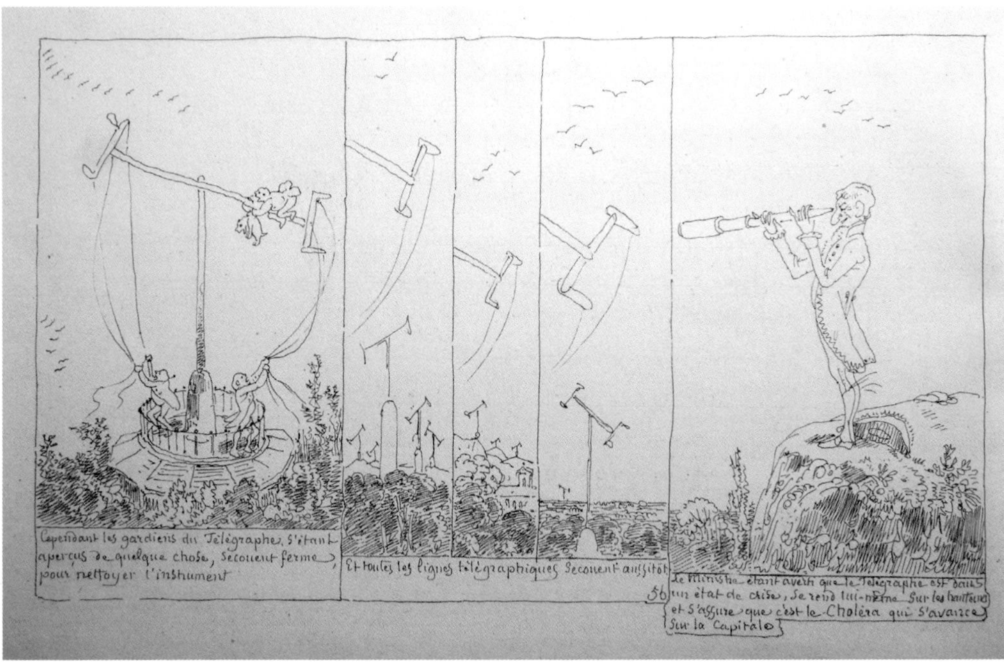

Figure 3.14
Rodolphe Töpffer, *Monsieur Pencil* (Paris: Cherbuliez, 1840), lithograph, 16 × 26 cm. Private Collection.

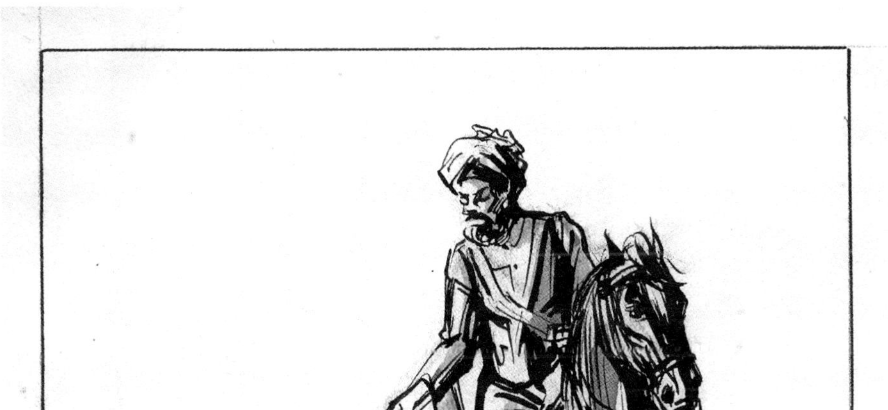

Figure 3.15
Georges Quesnel, *Le Télégraphe au désert*, late nineteenth century–first quarter of twentieth century, ink, watercolor, and gouache on paper, 16.5 × 14.5 cm. Musée de La Poste, Paris.

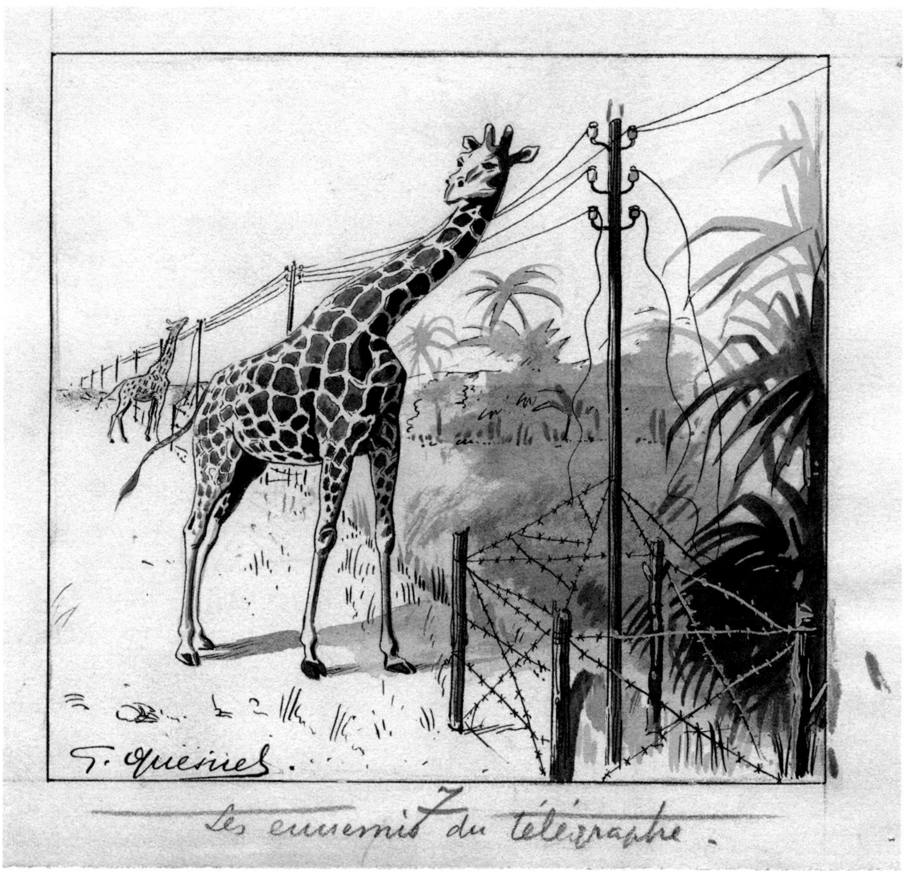

Figure 3.16
Georges Quesnel, *Les Ennemis du télégraphe*, late nineteenth century–first quarter of twentieth century, ink, watercolor, and gouache on paper, 16.5 × 14.5 cm. Musée de La Poste, Paris.

as untamable, these images also reveal a pervasive anxiety about African resistance through the disruption of colonial infrastructure. Despite their different geographic context, Quesnel's works are close in spirit to Claude Lévi-Strauss's description, in *Tristes Tropiques*, of his travels along the Rondon telegraph line in Brazil's Mato Grosso, where isolated telegraph operators might expect to find themselves decapitated, their telegraph keys placed neatly atop their severed heads. Completed in 1922, almost coincident with the invention of radiotelegraphy, the line was almost immediately obsolete, and "was no sooner finished than it became an archaeological relic of a previous technological age."[88] Lévi-Strauss noted how "The contorted silhouettes of the poles and the arcs of wire linking them one to another seem like incongruous objects floating in space, such as can be seen in Yves Tanguy's paintings. Being evidence of man's former presence and of the futility of his efforts, they mark the extreme limit he has tried to exceed. . . . The erratic nature of his enterprise and the failure by which it has been punished confirm the authenticity of the surrounding wilderness."[89] Anticipating Lévi-Strauss, in the 1830s artist Jean-Baptiste Debret had documented, in his *Voyage pittoresque et historique au Brésil*, the use of semaphore telegraphs in Brazil, where as soon as a station had been decommissioned, "the debris of its frail construction" disappeared entirely from view.[90] With these accounts in mind, the shifting sands and encroaching jungle of Quesnel's drawings tell a history of technological entropy and ungovernable "wilderness" that imparts a dubious "authenticity" to otherwise nebulous African topographies. More specifically, Caran d'Ache's 1893 cartoon *Le Télégraphe au désert* relocates the action back in Egypt, where giraffes, tethered at their feet and throttled at their necks by telegraph cables, provide the infrastructure for an infinite line. In contrast to the chaos and disruption of Töpffer's and Quesnel's images, this is drawing/telegraphy as order (fig. 3.17).[91]

Caran d'Ache's masterpiece, his 1886 *l'Epopée*, comprised a series of zinc cut-outs maneuvered in front of a lantern, projecting silhouettes onto a screen.[92] Shown at Rodolphe Salis's *Chat Noir* cabaret, *l'Épopée* narrated a hagiographic history of Napoleon's Grande Armée. Evoking the mythic origins of drawing, a key scene of the jingoistic spectacle featured Napoleon standing above a cheering crowd, his outline instantly recognizable (fig. 3.18). This scene was mechanized from behind, so that Napoleon could rise above the crowd. At the crucial moment, a tiny lever allowed the

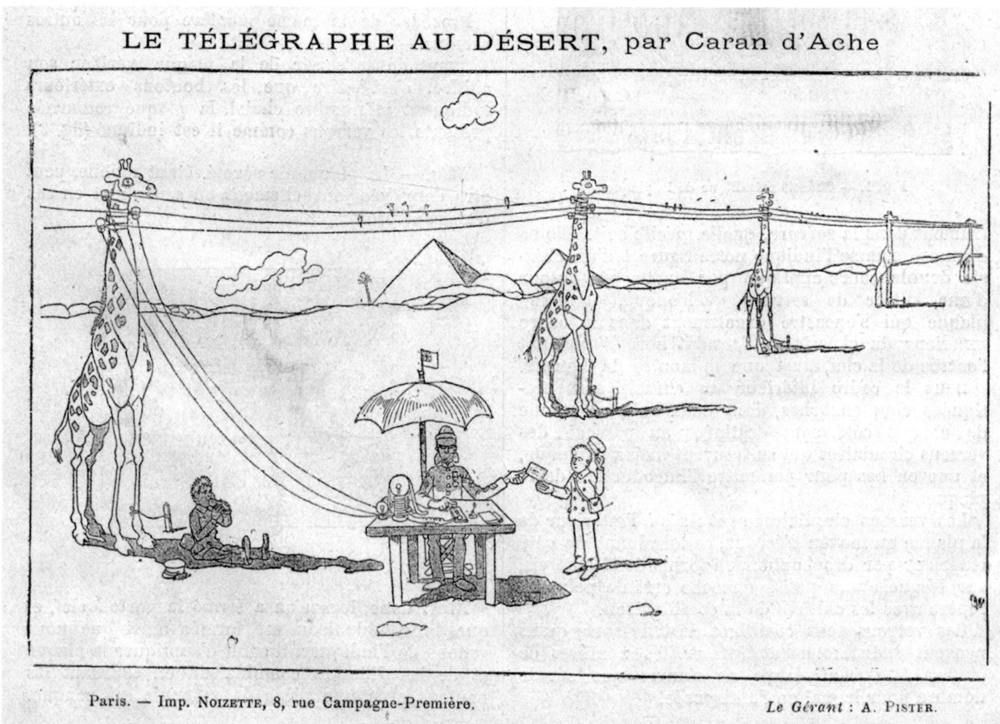

Figure 3.17
Caran d'Ache [Emmanuel Poiré], *Le Télégraphe au désert*, published in *Journal des connaissances utiles*, February 1893, 11 × 16 cm. Private Collection.

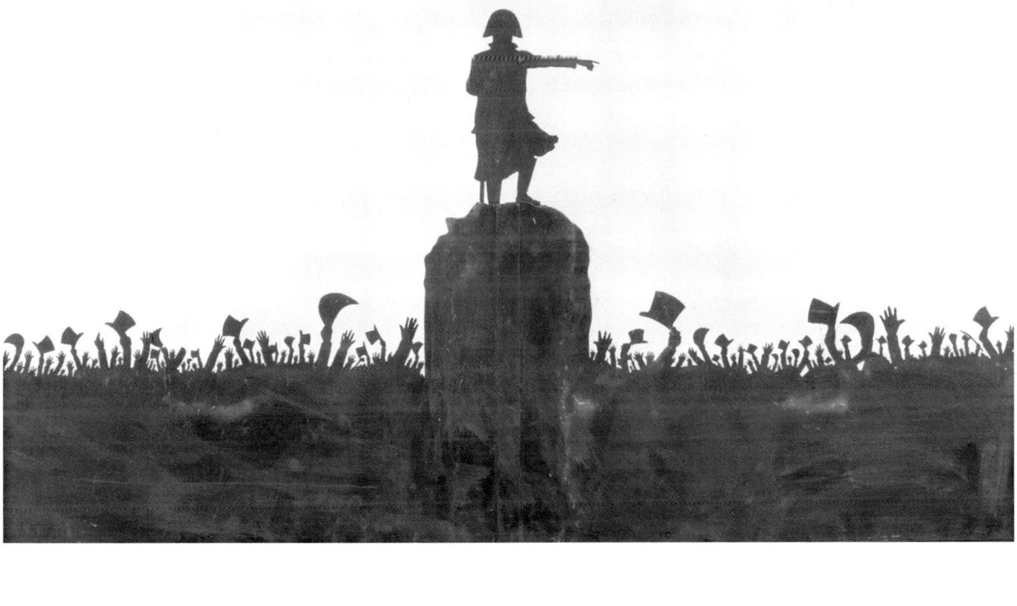

Figure 3.18
Caran d'Ache [Emmanuel Poiré], *Napoléon est acclamé par la foule*. Plate from tableau 26 of act I (*Apothéose*) of *l'Épopée*, performed at the Chat Noir cabaret, Paris, 1886, painted zinc cutout, 50 × 146 cm. Musée de l'Armée, Paris.

operator to elevate and lower the emperor's hinged right hand, "disposing of human destiny by a sort of infralanguage consisting of pure demonstration," as Barthes put it.[93] Again and again we see the emperor "surrounded by motionless air, raising a hand huge with every signification at once, designating everything and nothing, creating with a terrible languor a future of unknown acts." Power, externalized, is "a gesture arrested at the most fragile point of its fatigue."[94] The numinous gesture Barthes identified in Napoleon's staged comportment was replicated and repeated in miniaturized form in Caran d'Ache's zinc cutout. It also, of course, found emblematic expression in the optical telegraph's moving limbs, which in the absence of legible meaning, stood for power itself. In Wolfgang Schivelbusch's memorable description, the steam engine operated as "a kind of technological Napoleon" in histories of nineteenth-century industrialization, but it was not the only technology to prompt such comparisons.[95] Moreover, it is worth noting here that Caran d'Ache, the pseudonym for Emmanuel Poiré, comes from the Russian *karandash*, meaning "pencil." The pseudonym was later appropriated for the Swiss manufacturer of writing implements. Drawing, telegraphy, and the control of colonial space were frequent bedfellows, anchored, as images by Töpffer, Quesnel, and Caran d'Ache demonstrate, by Conté's invention of the pencil, which traversed and rendered interchangeable fictions of both "expressive" semaphoric and "regulated" electrical lines.

The distinction mapped here, between expressiveness and regulation (or between resistance and restraint) had temporal consequences, the impact of which was felt for decades. As On Barak has described, Egyptian time—understood as always slow or lagging behind—was coded in the nineteenth century as inferior to punctual European clock time.[96] Grounded in the colonial history of Egypt, this purported distinction was later advanced by Egypt's increasing significance to the British Empire, which marginalized Egypt politically but made it a vital strategic hub in attempts to streamline connections to India. Technologies of transport and communication—coach, steamer, canal, railway, and telegraph—were vital to this hypostatization of time in Egypt as a matter of peculiar colonial morality, with Western efficiency and precision set in opposition to Egyptian sluggishness and indolence (a juxtaposition the *Description de l'Égypte* had, arguably, already set in process). In 1839, as part of the drive toward modernization, a semaphore telegraph, based on the Chappe model, was installed between Suez, Cairo, and Alexandria.[97] Writing of Algeria in this period, Arthur Asseraf has noted the discrepancy between the speed of communication by sea or by land prior to submarine electrical signals. Asseraf gives the example of news of the landing at Sidi-Ferruch in June 1830, which took nearly four days to travel 750 kilometers between Algiers and Toulon by sea, but only twelve hours to cover the remaining 700 kilometers between Toulon and Paris, once the information was relayed via optical telegraph.[98] The 1839 Suez-Alexandria semaphore rectified the possibility of a comparable shortcoming by exploiting telegraphy's superior capacities to rush ahead, the telegraph giving Alexandria twelve hours' advance notice of when the Indian mail arrived at Suez. As Barak puts it, "it was as if the steamers arriving at Suez and Alexandria were actually 'seen' from Cairo."[99] The imposition of this technological acceleration was justified by accounts of Egypt's sluggishness. However, while the *Description de l'Égypte* had contributed to this image of Egypt as languidly stuck in the past, in contrast with the speed with which Conté's process covered its surfaces, it too was subject to temporal lag. By the time the final volume of the first edition was published, in 1822, the French involvement in Egypt was long over, and the *Description*'s protagonists would have appeared inhabitants of a very different time.

Repentance

Conté's engraving machine had presented an intriguing component missing in histories of printmaking, and in its relationship to drawing, but it threatened to destabilize them both. There were parallels here with the relationship between speech and its mechanized form in telegraphic communication, as it was being worked out at the same time, which similarly clouded the legitimacy of natural versus artificial signs. Prints have always, of course, been made by machines of one kind or another. Yet the prints of the *Description* transformed the embodied knowledge that up to that point defined the engraver's task. While the finished results appeared to be traditional copper engravings in every way, they were, like the marks made by Conté's pencil, products of celebrated artifice. By reducing the labor of the engraver and ensuring that lines, spaced apart from one another to the minutest degree, in precisely calibrated distances, never overlap, break, or converge, the machine seemed to forge an irrevocable schism between marks made by pencil and those made by burin. But Conté's pencils had, in truth, already blurred the boundary between the "artistic" and the "technical," between expressive marks and hard lines, and the future of drawing was, in any case, as the century progressed, increasingly mechanical.

Nonetheless, the infinite calligraphic possibilities of Conté's pencils certainly seem to be compromised by the effective yet limited range of his engraving machine (fig. 3.19). There seems to be no doubt here, no risk, no opportunity to make a mistake, no *repentance*, as Françoise Viatte has evocatively described what she views as one of drawing's defining characteristics.[100] Perhaps no space for apologies, either. It is notable that a number of the plates in the *Description* feature a miniature draftsman, dwarfed by the surrounding monuments and suffering stoically through the desert heat, sweltering under an umbrella in a premonition of the British telegraph operator in Caran d'Ache's later cartoon (fig. 3.20). The faceless, draped figure detailed in this plate of the ruins at Karnak, recumbent like something from a de Chirico painting, imparts an uncanny air to the vignette, suggesting Egyptian languor and amplifying the fortitude of the industrious French draftsman. These heat-wracked figures also evidence an apparent desire to return to the project a sense of human presence, of witnessing and direct inscription, that the shimmering machine-engraved skies work against.[101] But does the apparently "unrepentant" mechanized nature of the *Description*'s backgrounds wholly obviate their status as drawing, understood in its broadest sense? And can we, in an unexpected return of subjective experience, discern a transference of drawing's ad hoc, risky pleasures onto the early biographers' documentation of Conté's career—his miraculous invention, his ability to conjure something from nothing, his embrace of the provisional or unexpected, his pristine appearance on the revolutionary scene, and, too, his self-erasure—fulfilling the desires we ask for in the act of drawing, even as the things he made himself paradoxically threatened those characteristics?

It is, finally, tempting to focus on that which is lost in Conté's hybrid technologies, operating as they do as a hinge between the haptic and the optical, between the natural and the artificial,

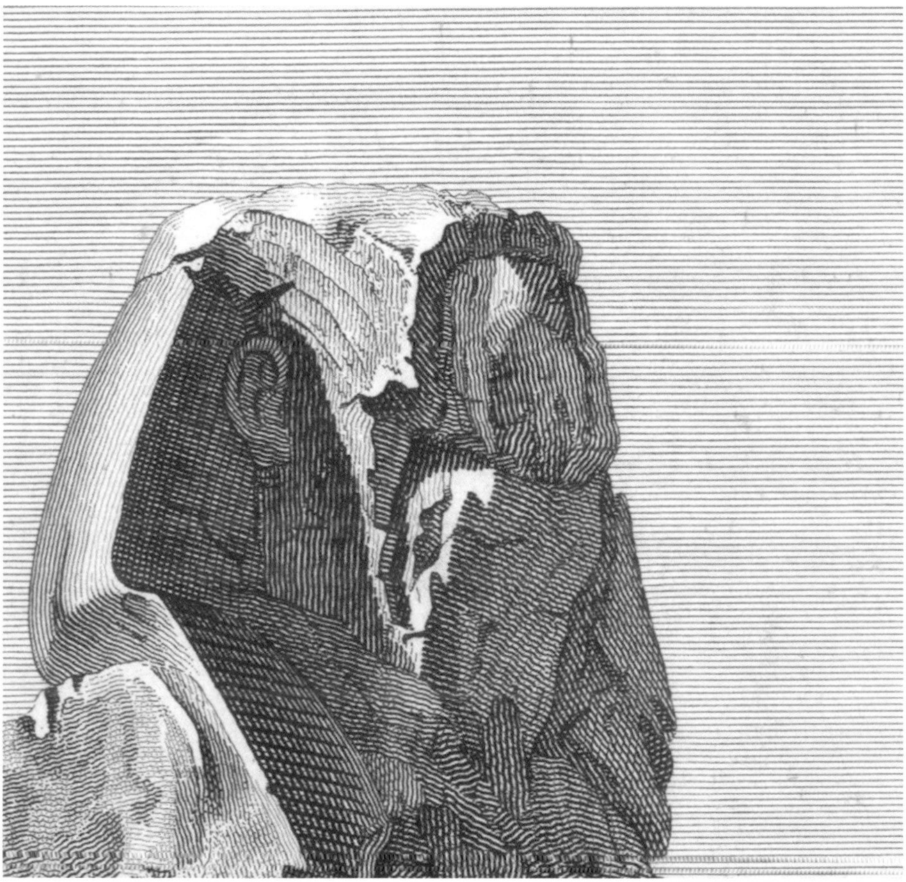

Figure 3.19
Detail of André Dutertre [draftsman] and Louis-Pierre Baltard [engraver], "Thèbes. Memnonium. Vue des deux colosses," from *Description de l'Égypte: ou, Recueil des observations et des recherches qui ont été faites en Égypte pendant l'expédition de l'armée française*, Antiquités, plates: volume 2, 1809–1828, engraving, 50 × 70 cm. Rare Book Division, The New York Public Library.

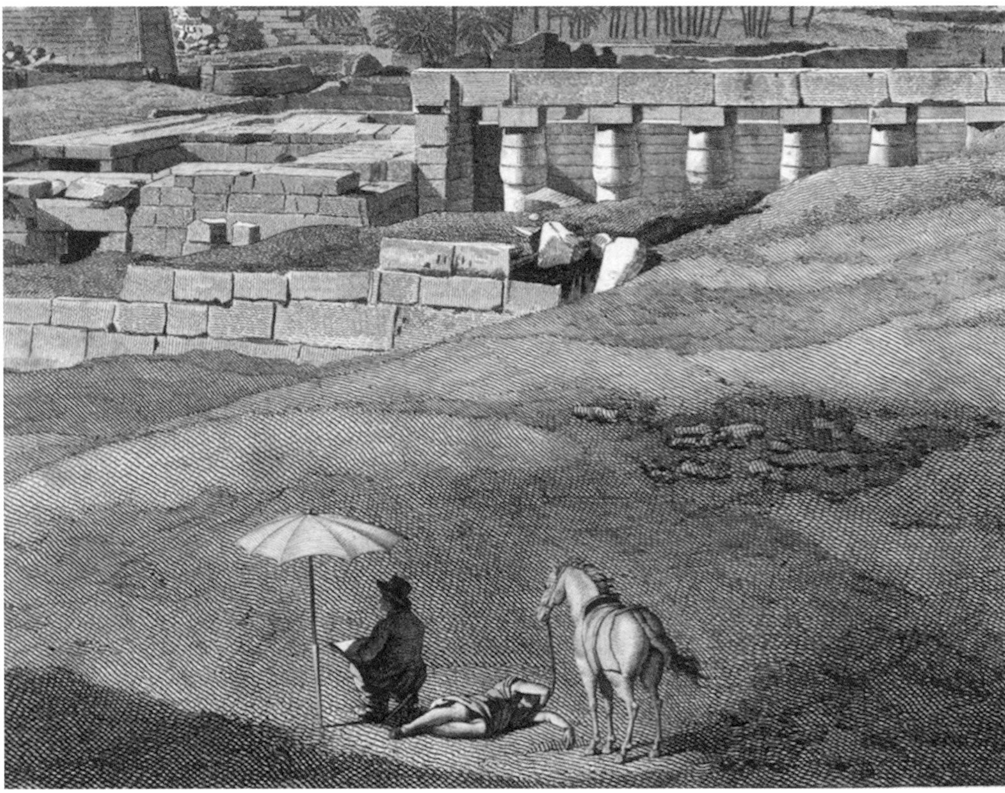

Figure 3.20
Detail of André Dutertre [draftsman], F. B. Lorieux [engraver], Desmaisons [etcher], "Thèbes. Karnak. Vue générale des ruines du palais, prise du nord-ouest," from *Description de l'Égypte: ou, Recueil des observations et des recherches qui ont été faites en Égypte pendant l'expédition de l'armée française*, Antiquités, plates: volume 3, 1809–1828, engraving, 70 × 105 cm. Rare Book Division, The New York Public Library.

between lead and air, and between the old-yet-new world of the pencil and the "pencil of nature"—the intersection so crucial to Nadar's investment in Conté's practice. Conté's images for the *Description* are, for a start, haunted by absence—of the inventor himself, who did not live to see the finished plates—and by the failure, despite the series' grand ambition, of the Egyptian campaign. Yet straight lines, as Tim Ingold has observed, have succeeded in Western culture "as a virtual icon of modernity," and Conté's systematically banded atmospheres—which, tellingly, Nadar did not discuss—speak to a kind of future contained most powerfully in that inviolable structure of modernism, the grid.[102] Grids, while declaring the autonomy of art and grounded in an unshakable materialism, appear "even where we do not expect to find them," as Rosalind E. Krauss writes, for "behind every twentieth-century grid there lies—like a trauma that must be repressed—a symbolist window parading in the guise of a treatise on optics."[103] I introduce this comparison not to claim a meaning for the plates of the *Description* that resides primarily in their foreshadowing of certain transcendent modernist principles, but to point to some of the ways in which the regulated lines of its skies produced effects that disordered their apparent objectivity, effects that were also palimpsestic, citational, and historical in nature. This is not to diminish a political interpretation of the plates of the *Description*—far from it—but to acknowledge that the ordering effect of the plates, their excess of representation and eerie side effects, were accompanied by failure, by erasure, by invisibility, by repressed traumas more powerful than any symbolist window, and that these were also characteristics of the French colonial project in Egypt.

In a well-known essay on the American artist Agnes Martin, Krauss highlights the insufficiency of a concept of the "abstract sublime" in the interpretation of Martin's works, many of which were painted and drawn over a number of years in the New Mexico desert (fig. 3.21). Rather than diagnosing natural allusion in Martin's lines and grids, so that abstract work might always be "decoded by its romantic double: Rothko read out through Friedrich; Pollock by Turner's storms; Martin by Turner's skies," Krauss cites Martin herself, who unexpectedly sees her work as drawing on a classical tradition that is "Coptic, Egyptian, Greek, Chinese."[104] For Martin, "Classicists are people that look out with their back to the world / It represents something that isn't possible in the world /

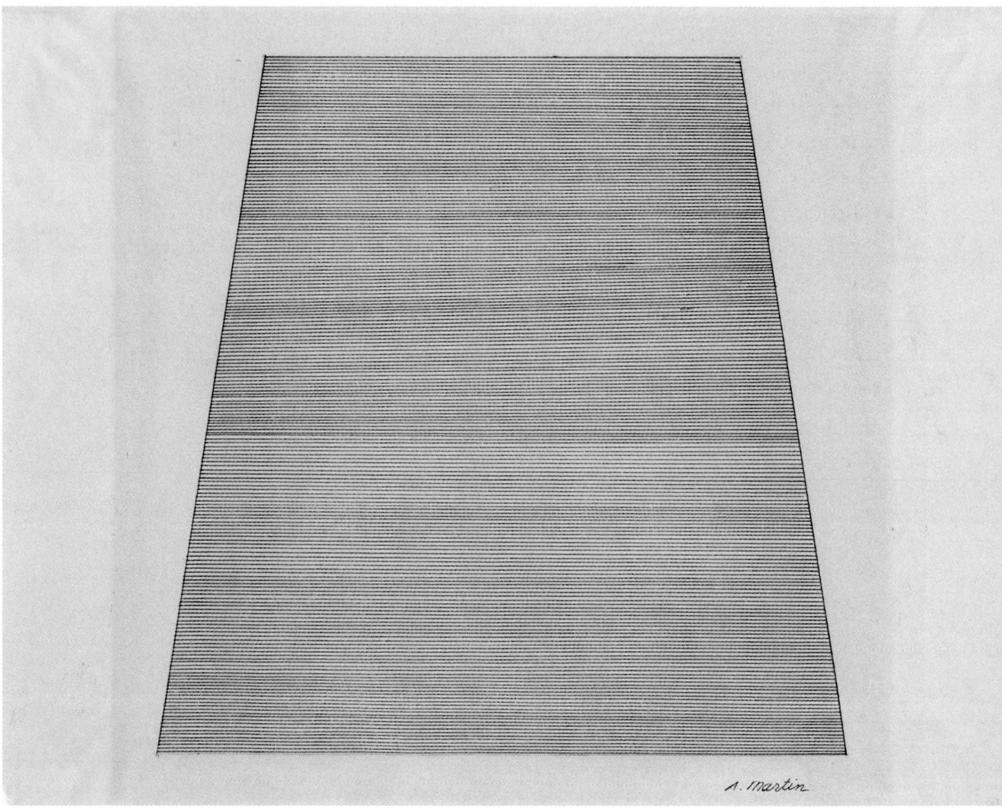

Figure 3.21
Agnes Martin, *Mountain*, 1960, ink and pencil on paper, 24 × 30.2 cm. Museum of Modern Art, New York.

More perfection than is possible in the world / It's as unsubjective as possible."[105] Drawing heavily on Hubert Damisch's semiotic theory of /cloud/, Krauss argues that we should take Martin's paradoxical classicism seriously, and that "atmosphere" is not a natural given of the paintings, to be discerned as if a neutral component of the work, the paintings' signified or ostensible "content," but rather a system of shifting signifiers, "the open member of a differential series."[106] Science, rationality, perspective—all are accompanied and undone by their opposite, the unformed, the /cloud/.

That shift between the objective and the *informe* might be seen to characterize Conté's skies too. Roughly coincident with Jean-Baptiste Lamarck's pioneering attempt to provide an objective taxonomy of clouds in the 1802 edition of his *Annuaire météorologique*, Conté's machine set out to strip the printed image of the clouds that obscured the *repentance* of the engraver, and the subjectively experienced "fleeting states of mind, feeling, and atmosphere"—of thought itself—described by Mary Jacobus with regard to Constable's cloud studies, and also found, as the next chapter will describe, in the work of Georges Michel.[107] However, the *moiré* effects of the mirage of serried lines in Conté's plates return us closer to /cloud/ in the terms outlined by Damisch, as a dialectical play of signifiers that oscillates between the ephemeral and the material, between subjectivity and objectivity, rather than indexing an actual cloud as such. The violence of their rigid order was not absolute, but turned on itself. When, at the end of the century, Nadar wrote of Fourcroy and of Conté's simple machines, he was expressing admiration not simply for a forerunner, or for Conté's technical acumen, but for the ways in which a search for new expressions of reason, and, it must be added, new territories and people to exploit, were subtended by happenstance discoveries and accidents that ultimately impinged on the ways in which they might be encountered over time. To recognize the history of the graphic in the photographic, and to see how telegraphy overlay both, it was necessary to look between the lines. Conté's machine-tooled skies, and his standardized pencils too, for that matter, hold in suspension the spur-of-the-moment, improvisational qualities discerned in them by Nadar (qualities that are themselves, in their provisionality, signs of the modern), with an altogether "classical" take—Martin's "more perfection than is possible in the world"—a complex that later, after photography, would be called the optical unconscious.

4 *Paris in Code*

In a short essay of January 1800, "Paris vu dans son ensemble ou le Panorama," chronicler of everyday life in eighteenth-century Paris and psychogeographer *avant la lettre* Louis-Sébastien Mercier shifted his perspective from the streets to the sky, offering a sweeping view of the city from above, a synecdoche for the monumental project of his *Tableau de Paris* and its successor volumes *Le Nouveau Paris*. Tipping his hat to a novel visual spectacle—Robert Barker's panorama, newly introduced to France—Mercier framed his description around an imagined bird's-eye passage down the Seine, taking in Paris's major monuments.[1] Where to begin such a vertiginous voyage? For Mercier, the choice was obvious: "The point of view is taken from the telegraph, that is to say from the upper attic of the Tuileries Palace. It is as if you were there. What an imposing aspect!"[2] If Mercier's sweeping visual glide across the city began at the Tuileries telegraph, it passed other relays along its route. At the midway point: "The telegraphs that seem to walk in the air indicate to you the towers of Saint-Sulpice and the place of its superb portal."[3]

With these aerial strollers—long gone by the time Georges Perec attempted to "exhaust" the place Saint-Sulpice by describing the minutiae of all that passed through it—it is hard to tell "which came first, the semicorporeal allegory of technology or the semitechnological figure of the body" (as Dolf Sternberger once asked).[4] As Sternberger was at pains to point out, "any attempt to exclude the 'telegraph allegory' while retaining the thing itself makes obvious how fast it adheres."[5] To invoke the telegraph was to speak of other things, and sometimes also other times and places. And it was again with the telegraph, as allegory as much as artifact, that Mercier concluded, even though he did not name it as such: "finally, raise your eyes and you will see Montmartre, its abbey built on the foundation of the temple of Mars, its windmills in perpetual activity."[6] Windmills did indeed populate the hills of Montmartre, but the church of Saint-Pierre—that former temple of Mars—was now capped by a war machine of a new order. The Montmartre telegraph became a subject for many artists, from Étienne Bouhot and Georges Michel, whose work I discuss in parts two and three of this chapter, to Louis-Jacques-Mandé Daguerre—who depicted it in a painting made shortly after his success with the diorama, and around the time he began to experiment with the revolutionary photographic technique that came to be known as the daguerreotype (fig. 4.1). In Daguerre's painting, an oil study for a diorama of the same view that opened in Paris in May 1830, the machine can be glimpsed to the right of the largest windmill in the muddy foreground scene, soon to be illuminated by the rising sun that made the operation of both telegraph and diorama possible.[7]

The associations prompted by Mercier remained in place for a long time, and the telegraph found its way into panoramas of all kinds. In 1803, Joseph-Michel Montgolfier reported a device—part entertainment, part demonstration—on show at the new panorama at the Jardin des Capucines. As he described, "its purpose is to transmit thought at a distance, like the machine from which it takes its name, with the difference that its operations can be seen only by the person who directs them and by the one who supervises their effects."[8] This harbinger of electrical telegraphy sent messages between two alphabetized dials, one in the panorama rotunda and one in a cabinet in the gardens. The machine's novelty lay in the lack of visual cues explaining its operation, and it demonstrated the close connection between experimental forms of long-distance communication and the period's culture of spectacles.

In 1852, Edmond Texier's *Tableau de Paris*, riffing on Mercier's eighteenth-century example, opened with a frontispiece that functioned as a kind of squashed panorama, condensing

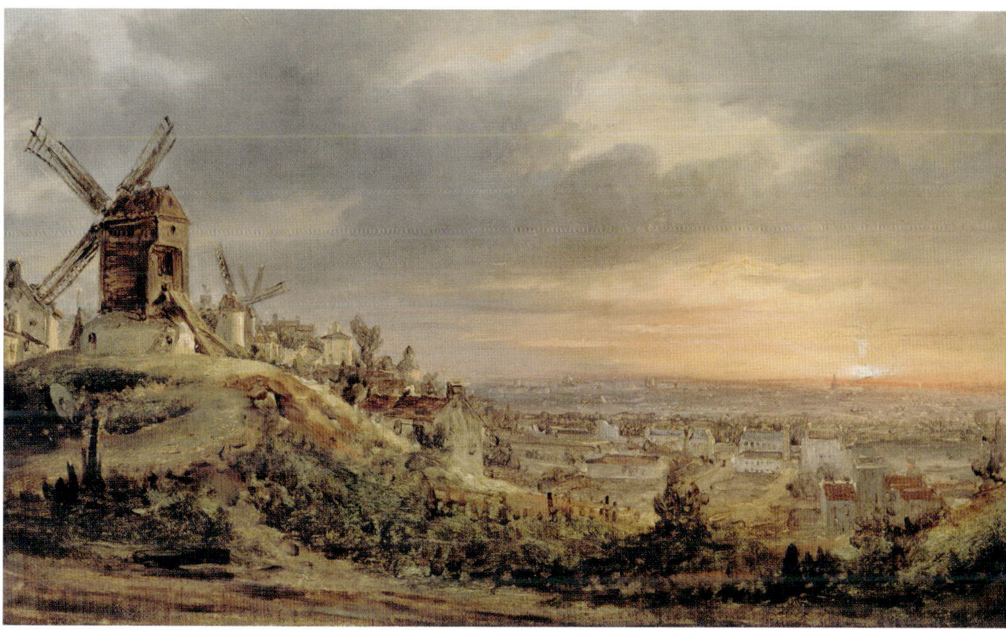

Figure 4.1
Louis-Jacques-Mandé Daguerre, *Paris, vu de la Butte Montmartre*, c. 1830, oil on canvas, 40.8 × 55.6 cm. Musée Carnavalet, Paris.

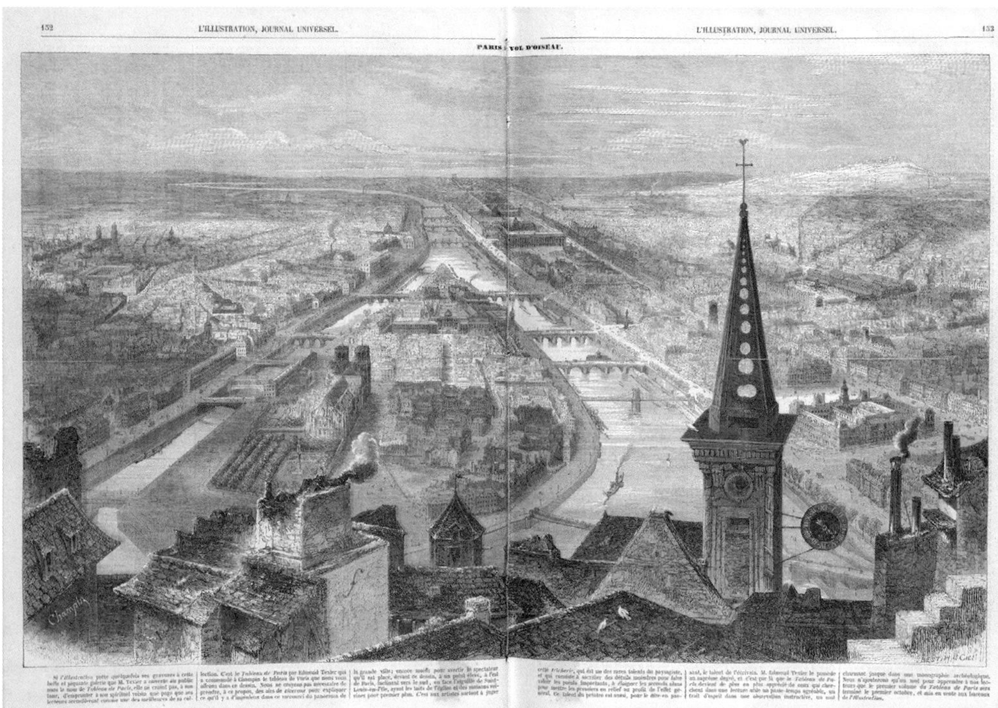

Figure 4.2
Jean-Jacques Champin, "Paris à vol d'oiseau, MDCCCLII," from Edmond Texier, *Tableau de Paris*, vol. 1 (Paris: Paulin et Le Chevalier, 1852–1853). Brown University Library.

the city's monuments into a compact vertical space traversed by solid flows of people and, of course, positioned on the towers of Saint-Sulpice, still the telegraph.[9] A plate from the volume, representing a bird's-eye view of Paris, echoes Mercier's total vision of the city (fig. 4.2). From the crumbling rooftop of an exaggeratedly elevated Saint-Louis en l'Île that provided the artist Champin's vantage point, one might take in a dizzying view west past the Île de la Cité, framed by the Saint-Sulpice telegraphs on the left and the one at Montmartre to the right (the telegraph at Saint-Eustache can also be seen, if less distinctly, to the right of the foreground spire). Later yet, Charles Camille Chazal's panorama of key defensive points around the city, made in 1870 during the Siege of Paris, included, at left, the "tour de l'ancien télégraphe," even though it had been many years since such a device could be seen (fig. 4.3). Writing in 1964 about the view from the Eiffel Tower, Barthes described the dialectic between a "euphoric" sensation of blissful completeness and our drive to decipher, to identify landmarks or objects. Panoramic vision was moreover historical—"it is duration itself that becomes panoramic"—producing a unique consciousness of the strata of the city in time.[10] Distinguishing the presence of the telegraph in the panorama—or even the space where it once stood—satisfied the impulse to punctuate the panorama's "gliding" perspective with recognition of a specific location, while giving free rein to that impulse to optical drift. Telegraphy always designated the total system as well as the individual device; animating landscape, the horizon line, and connections beyond sight, it was of a piece with panoramic literature, panoramic vision, and panoramic time.

It might seem perverse, then, to begin this chapter with a discussion of a single environment, rendered in paint by the genre artist Swebach-Desfontaines, that curtails the wider view. Staging the convergence of infrastructure and information systems within the realm of the infraordinary, Swebach's work drags Mercier's transcendent, frictionless aerial panorama back down into the mud. "My intention," Perec wrote, regarding his meticulous and absurd inventory of the comings and goings on the place Saint-Sulpice, recorded under the sign of the telegraph, or at least in the long shadow cast by its absence, "was to describe the rest instead:

that which is generally not taken note of, that which is not noticed, that which has no importance: what happens when nothing happens other than the weather, people, cars, and clouds."[11] Just as the telegraph's presence had ramifications that were global in their political and geographical reach, it was also a network that bore heavily on more local experiences.

Although not all who described the telegraph discerned an intelligence there—a character in Élie Berthet's 1869 novel *La Tour du télégraphe* called it an "idiotic machine"—the landscape it intellectualized most intensely was found at the center of the network, in and around Paris itself, and it was in the work of artists who focused on the city's changing infrastructures and environments that its effects were most clearly seen.[12] Theodor Adorno wrote that "it is the fortune and limitation of French art never to have entirely eradicated the pride in making little pictures."[13] It was not a compliment. But in the period from the Revolution to the Second Republic, makers of little pictures documented the city's frenetic transformations with considerable facility and sophistication. While history painters subsumed optical telegraphy's signals into a preexisting gestural lexicon, or ignored them entirely, genre painters, landscapists, topographers, and other *petits-maîtres*—all artists brought to greater visibility with the Revolution's democratization of the art world—were more inclined to represent them directly. These artists, three of whom I focus on here, confronted the permutations of the telegraphic imagination in its everyday, locally specific forms, drawing attention to commonplace routines and rituals, the passing of clouds, things forgotten, and "what happens when nothing happens."

Part I

Everything Flows: Swebach-Desfontaines

Strolling along the right bank of the Seine, in the short stretch between the Louvre and the Pont-Neuf, one might be forgiven for missing the place de l'École. Although the space dates to at least 1214, going by this name since 1413, and although it still exists, just about, as a definable urban location, there isn't much to see these days. A curve of pavement, a thickening of the road as one approaches the river from the rue de l'Arbre Sec, and it's gone, barely registering as a distinct area at all. Once a neat three-sided square opening onto the quayside, nestled behind the looming east end of Saint-Germain l'Auxerrois, the place de l'École is now occupied almost entirely by the remodeled frame of the Samaritaine department store, first opened by Ernest Cognacq and Marie-Louise Jaÿ in 1869. It was the expansion of the Samaritaine between 1925 and 1928 that put an end to the place de l'École; its name on the street sign today is a phantom of a more robust space.[14] Taken in 1865, Charles Marville's desolate photograph of the view facing the square seems to prefigure its eventual eradication, as by this point it had been stripped of the central fountain that had dominated it for decades (fig. 4.4). In its combination of geographical centrality and scrappy marginality, the place de l'École seemed to invite reflections of this kind; it is a delicious coincidence that Louis-Sébastien Mercier was born on the square in 1740.

At just over twenty-one by fifteen centimeters, *Vue prise de la place de l'École*, by Jacques-François-Joseph Swebach (known as Swebach-Desfontaines) is, like the square, easy to overlook (fig. 4.5). Still, the painting must have been well known enough to serve as the model for an 1828 lithograph by Jules Arnout of the same scene, which reproduces most of the painting's key features. Best known for his engravings of battle scenes and images of military life and for his designs for porcelain services, Swebach was *premier peintre* at Sèvres from 1802 to 1813, where in 1806 he painted the famous *Service encyclopédique*, its designs based on the printed plates of the *Encyclopédie*, and later the director of Alexander I's imperial porcelain manufactory in Saint Petersburg. He was attentive to overlaps and differences between media. Possessed of an ability to render big themes on a small scale, to see the macro in the micro, he was also attuned to the relationship between the miniature and heroic. Indeed, despite the compressed quality of most of his works, their scaled-down landscapes populated by Lilliputian figures, for his biographer Édouard André "perhaps never before has the art of depicting vast expanses and giving the illusion of distance been pushed to such a degree."[15] The place de l'École provided Swebach with a vantage point from which to consider not only precise transformations in the urban fabric, but connections between local events and history writ large.

Swebach was born into an artistic family in Metz in 1769. His father, François-Louis, was engraver to the comte d'Artois and in the 1790s had produced a series of exquisite mineralogical and

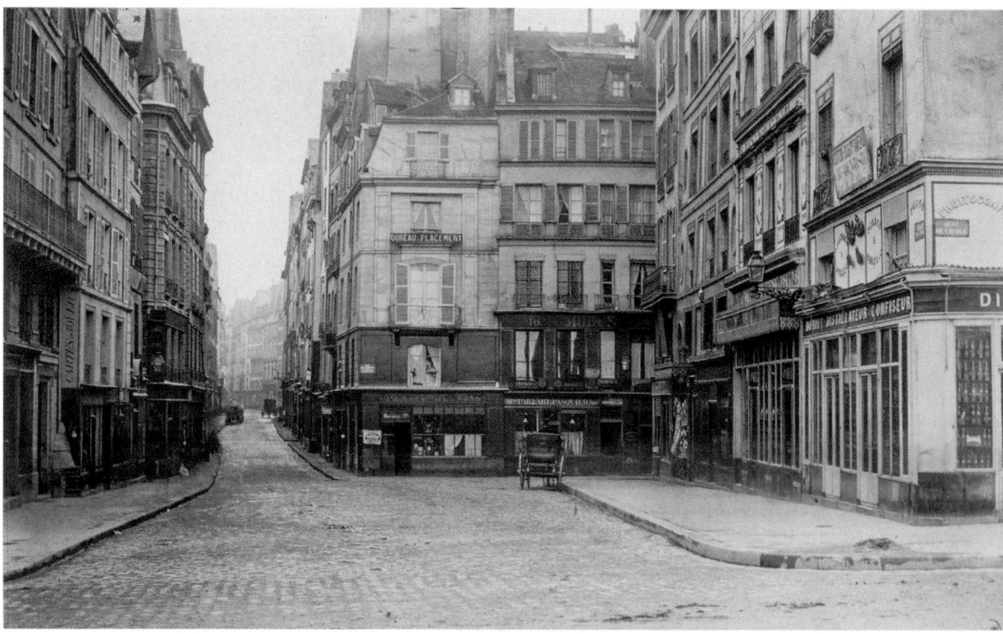

Figure 4.4
Charles Marville, *Place de l'École, vue prise du quai de la Mégisserie, 1er arrondissement, Paris*, 1865–1868, albumen print, 22.5 × 37 cm. Musée Carnavalet, Paris.

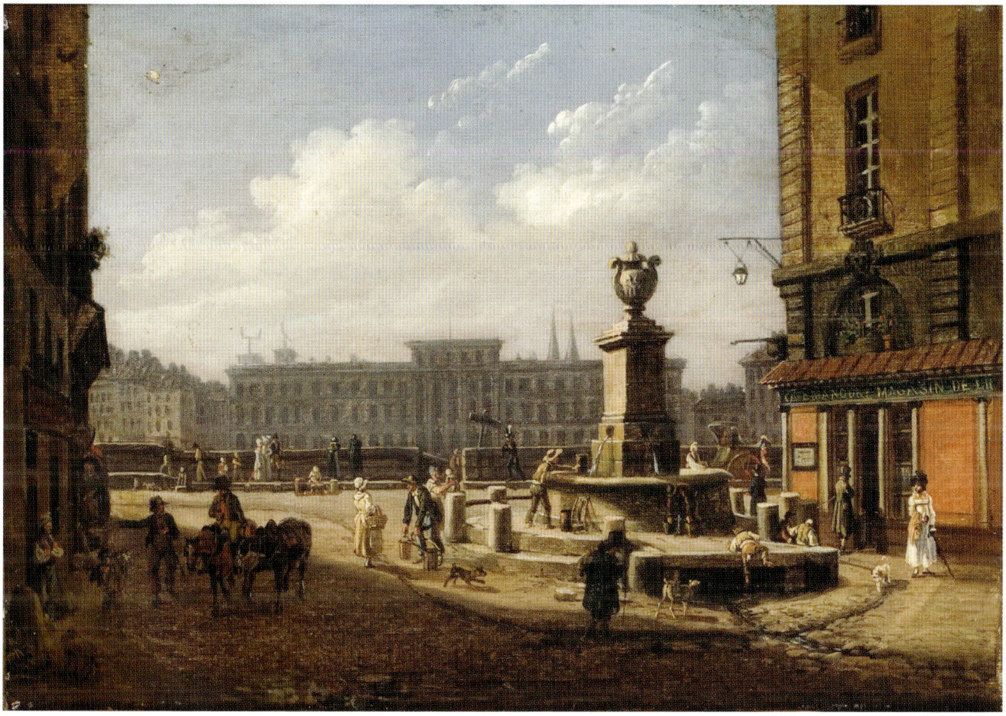

Figure 4.5
Jacques-François-Joseph Swebach-Desfontaines, *Vue prise de la place de l'École*, c. 1815, oil on canvas, 15.5 × 21.5 cm. Musée Fabre, Montpellier.

crystallographic illustrations (such as these for gypsum), the publication of which was thwarted by the Revolution (fig. 4.6). Swebach *fils*, on the other hand, made several successful works in the 1790s, most prominently his contributions to the long-running print series *Tableaux historiques de la Révolution française*, where engravings after his drawings reached a wide audience, and were extensively copied.[16] Outside of this series, Swebach made images of notable revolutionary events, from the festival of Federation to the return of Louis XVI to Paris in October 1789, inscribing his own name, "Desfontaines," on a banner hanging from a window in the latter drawing.[17] He cuts a dashing figure in Louis-Léopold Boilly's *Réunion d'artistes dans l'atelier d'Isabey*, a pantheon of thirty contemporary art world figures painted for industrialist Armand Seguin (the same Seguin whose shoes were admired by Fourcroy, as described in the previous chapter). Shown at the Salon of 1798, the painting offered a cohesive, nonthreatening, and commercially viable template for artistic practice in the new economic climate of Directory France.[18] Boilly, an artist who ultimately gave up a successful painting career to speculate in mortgage loans, was particularly aware of the commercial possibilities presented by the repeated image, returning to the portrait of his fellow traveler Swebach in one of his trompe l'oeil designs.

The Napoleonic Wars gave Swebach ample opportunity to address his favored subject matter, and he became, as Jules Renouvier described him, "the smallest, most efficient military painter of the Empire."[19] Painted around 1815, eight years before Swebach's death in 1823, *Vue prise de la place de l'École* stands out as one of relatively few works in which he addressed the subject of the capital—and, we might add, capital—directly. Economical in every sense, this painting, which prominently features the Paris mint, la Monnaie, proposes something close to Georg Simmel's description of money as "the distilled exchangeability of objects—whose absolute lack of individuality results from the fact that it expresses the *relation* between things."[20] Here, such "things" included the Parisian hub of the optical telegraph network, also visible in the scene. In this work, the differences and equivalences between informational and infrastructural pathways, and what such routes and systems render mobile, are mapped out and measured, their exchangeability assessed. It is also an image that trades in the passage of time and in the ways in which it might be calibrated in, and by, the city.

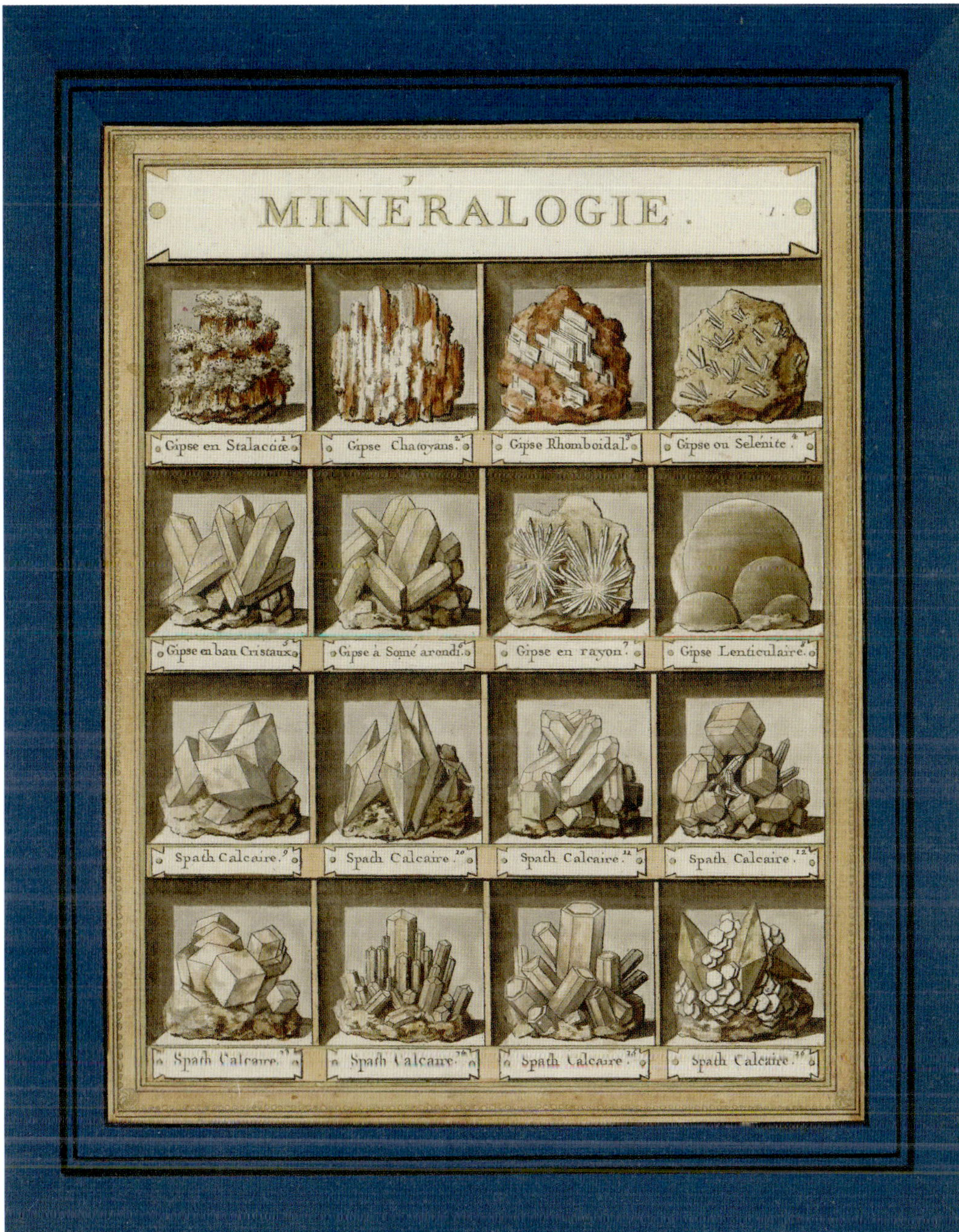

Figure 4.6
François-Louis Swebach-Desfontaines, illustration for *Histoire naturelle: ou, Exposition des morceaux, les mieux choisis pour servir à l'étude de la minéralogie et de la cristallographie*, c. 1789, watercolor on paper, 28 × 22 cm. Natural History Museum, London.

The painting represents a scene of everyday city life: the comings and goings of urban subjects in a public space. In its generic form as well as its ostensible subject matter—the gathering of a community around a fountain, united by a basic human need for water—the painting suggests continuities, in terms of social interactions, exchanges, and ways of life. Swebach's work seems to push this tendency to the point of stasis, transposing a fantasy of the centrifugal village pump to an urban milieu, and recalling the displaced domesticity of his images of military encampments. Yet in *Vue prise de la place de l'École* such continuities confront a more radical reimagining of connectivity and exchange in a modernizing France. Swebach's little genre scene stages, in a concentrated fashion, contemporary debates about transmission and flow in the early nineteenth-century city. Juxtaposing cyclical rural time with the pulsing modernity of Napoleonic Paris, Swebach's densely overdetermined painting shows that modernity to be grounded in a dialectic between grubby materiality and airy abstraction. This was also the dialectic that structured telegraphy's combination of material presence and abstruse codes.

Unlike Marville's photograph, which turns the viewer toward the square, funneling us into Paris, Swebach's painting uses the square as a vantage point from which to confront the river. But it is no less circumscribed in its view. The vertiginously flattened perspective of the buildings on either side of the painting lends this work the character of a peepshow or perspective box—fashionable at the time—wherein a world might be discovered. The sale inventory of Swebach's studio after his death lists several similar devices in his possession.[21] This compressed, ocular effect—telescopic, we might say, or like a projection from one of the magic lantern shows that populated the back streets along this part of the Seine—is enhanced by the fact that at some point the rectangular canvas appears to have been contained within an oval frame, the traces of which are visible on its surface.[22] It seems unlikely that this was its original presentation, as such a frame would have obscured several important details, but the paint here is smudged slightly, as if covered while still wet, and a lingering sense persists of a second order of representation, a change in the atmosphere, a feeling that something is hovering just beyond, or to the side of, visual perception.[23]

The square itself would have been familiar to Swebach prior to his production of this scene, as in May 1788 he had shown some drawings and watercolors at the Exposition de la Jeunesse on the place Dauphine, just off the middle of the Pont-Neuf, at the tip of the Île de la Cité.[24] While these works met with a positive response, Swebach's contribution to the second Exposition de la Jeunesse a year later was less well received. This exhibition took place in the gallery of the painter and dealer Jean-Baptiste Pierre Lebrun, *garde des tableaux* of the comte d'Artois, on the rue de Cléry.[25] This time, with the Revolution just around the corner, the *Mercure de France* of July 4, 1789, remarked on Swebach's flawed use of light and shadow, and the confusing accumulation of figures and objects on the same plane.[26] It is tempting to read some of the *Mercure*'s criticism in *Vue prise de la place de l'École*, although Swebach's use of chiaroscuro effects there seems more strategic than defective. Representing a space just out of shot, from where he would have almost been able to see the place Dauphine, Swebach's painting contains references to his own artistic trajectory, which was punctuated early on by periods of uncertainty and failure.

Swebach's painting frames an opening onto the right bank of the Seine just west of the Pont-Neuf, facing the neoclassical facade of Jacques-Denis Antoine's Paris mint, on the other side of the river. In contrast to Demachy's phantasmatic 1777 view of the newly built Monnaie (fig. 4.7), reproduced to ghostly effect in the improbably tranquil waters of the Seine, in Swebach's painting the production of pristine financial instruments, and the material and symbolic manifestations of economic policy and state power, are juxtaposed with (but also connected to) the muddy, semisubterranean scene in the foreground. Money versus dirt, money as dirt: a familiar story. But also, money in the world, money as power exerted, as a social relation. Hard money—unlike the fragile paper currencies of the Revolution—was, Swebach's painting seems to suggest, in the here and now, and it was real and material. In 1800 Demachy returned to the same scene (fig. 4.8). Time and pollution have left the Monnaie markedly grubbier, its startling whiteness assumed instead by the sheets of the washerwomen that shroud the river bank. Now, on the Louvre, the optical telegraph takes up its deadpan positions.

While the place de l'École no longer exists in the same form, the view across the Seine in the background is still recognizable, and certain features of the square, such as the café on the corner,

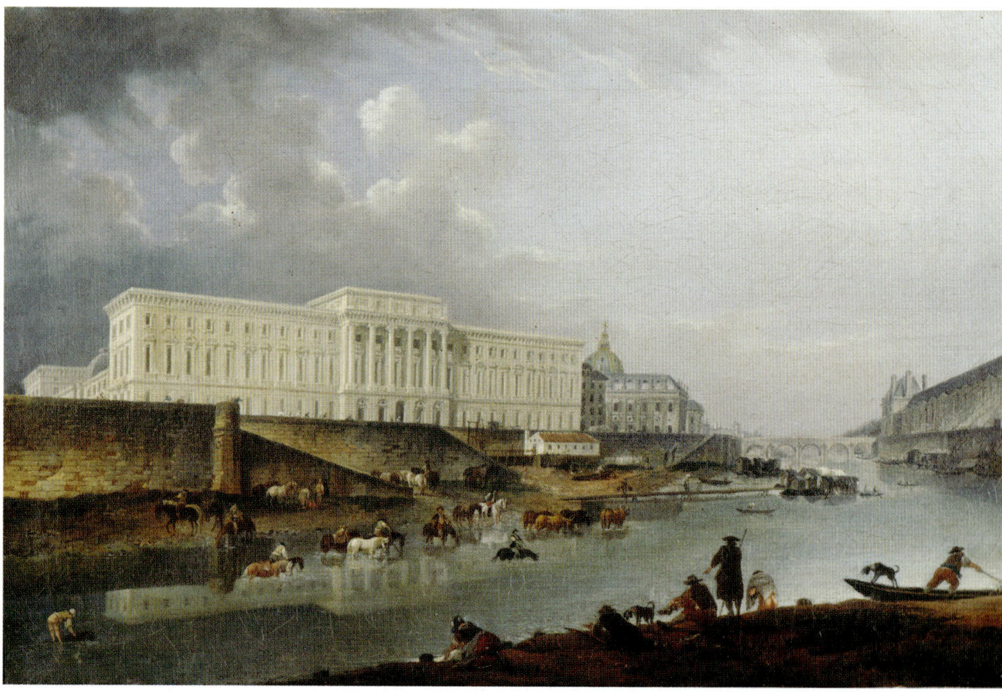

Figure 4.7
Pierre-Antoine Demachy, *L'Hôtel de la Monnaie, le quai de Conti et la Seine, vus de la pointe de la Cité*, 1777, oil on canvas, 41.5 × 64 cm. Musée Carnavalet, Paris.

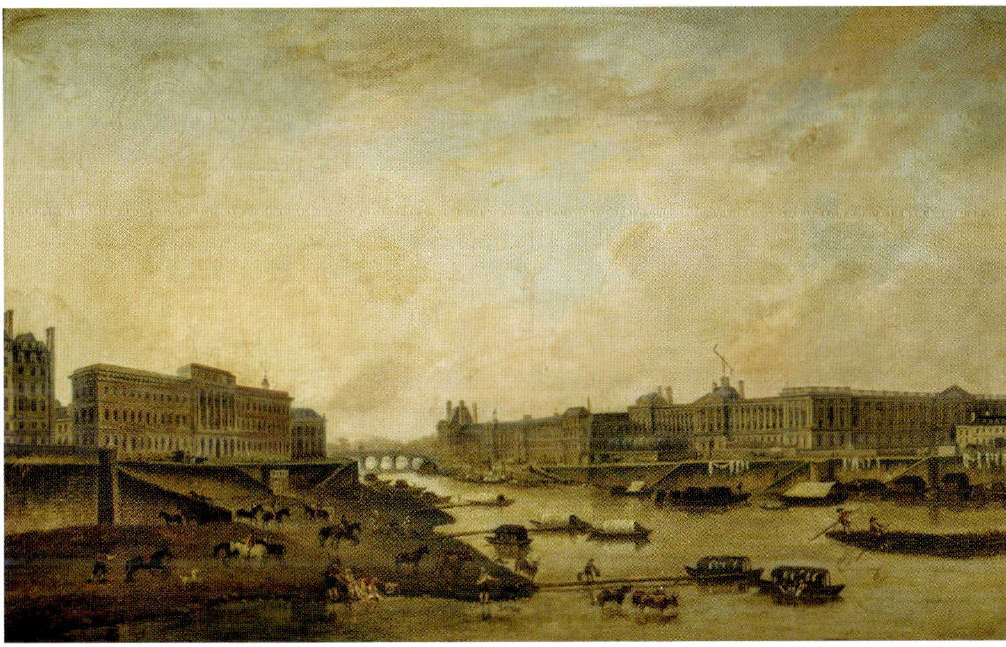

Figure 4.8
Pierre-Antoine Demachy, *L'Hôtel de la Monnaie et le Louvre, vus du Pont-Neuf, vers 1800*, c. 1800, oil on canvas, 60.5 × 95 cm. Musée Carnavalet, Paris.

remain to this day. The Café Manoury, famous for the playing of draughts, was described by both Rétif de la Bretonne and Mercier.[27] Meanwhile, another café on the square, the Café Parnasse, had a noteworthy revolutionary heritage, as prominent Jacobins such as Camille Desmoulins, Jules-François Paré, and Georges Jacques Danton were known to frequent it. It was here that Danton met his first wife, Antoinette Gabrielle Charpentier, whose father owned the café, and whose portrait was painted by David.[28] Her sister-in-law Constance Marie Charpentier made the best-known portrait of Danton, while Gabrielle was the subject of a death mask by Claude-André Deseine, produced after Danton had her body exhumed.[29] It is possible that Swebach had at least some of these associations in mind. His painting adopts a viewpoint from the unseen Café Parnasse's terrace, marking out the place de l'École as a politically inflected space. More forcibly, though, the representation of the café in the image tells us that money made on one side of the river will be spent on the other.

In the center of the square as we see it here, a fountain dispenses an excess of water that pools and tracks in rivulets on the ground. A pumping mechanism can be seen above the balustrade at river's edge, where the famous Samaritaine bathhouses would later stand, although we are spared a view of the river itself. Children and dogs play, washerwomen work and flirt, bourgeois families promenade, and a returning soldier trudges down the quayside. Behind them, la Monnaie sits massive and impassive, while above it on the left-hand side, a pair of optical telegraphs turn silently on the towers of Saint-Sulpice. An airborne object—perhaps a balloon, or some kind of homemade aerostat—floats above. This is a painting of doubles. These take their cue from the Saint-Sulpice telegraphs, positioned to communicate north and south out of Paris. Most obviously, the single spire of Saint-Germain-des-Prés is artificially multiplied to harmonize with the telegraphs, but there are other repetitions: the pails carried by the man in the foreground and the man at the fountain, the wonky form of which echoes the telegraph itself, as well as the two horses on the left, the streams from the fountain, the dogs, and the couples.[30] Everything here needs to be balanced: what goes up must go down, credit must balance debit. Over at la Monnaie, a double-winged screw press called a *balancier* was used to strike coins and medals, its weighted radial arms mirroring these other forms and devices.[31]

The effect is more playful than uncanny, in line with the pun that matches Swebach's name to his work: Swebach "Of the Fountains," rendering the work a subtle self-portrait. Swebach never relinquished his fear that the Germanic origins of his name would prejudice his success, hence the translation into French of his family name "Fountein" as "Fontaine."[32] From this perspective, the painting is an exercise in compensation. Iris Moon, writing about Napoleonic architects Percier and Fontaine, describes how the latter took his name from the family plumbing business, producing a hydraulic sensibility that pervaded the architect's work, a "water magic" that irrupted in Fontaine's loose watercolors and the liquid aesthetic of his writing.[33] Something similar might be diagnosed with his near-namesake's painting, although the flow is reversed: rather than renouncing hydraulics for art, in this painting Swebach's trajectory took him from art to plumbing.

Fluid Knowledge

There are other ways that the painting multiplies its effects. Fountain, mint, telegraph. Water, money, information. This painting thematizes flow, connecting the circulation of money, people, and state secrets, and linking them to the provision and maintenance of public and private structures in urban space. The painting embodies Kittler's description of how, "in a city, networks overlap upon other networks"—bridges cross rivers, gutters traverse electrical circuits, and so on, networks between cities also overlapping, seemingly autonomously, with those of other cities, a media landscape in which, as Kittler puts it, "power thus means occupying at the right moment the channels for technological data processing," even at the most micro level of everyday life.[34] This position echoes that articulated in an 1819 article, "Le pouvoir télégraphique," which identified telegraphic power as distinct from legislative, executive, judicial, municipal, or royal powers.[35] Under the "telegraphic regime," "great arms like mill wings" acted as agents of surveillance, anticipation, repression, and control. Anticipating Kittler's autonomous, militarized mediascape, the author concluded: "this is what telegraphic power is all about. Is the realm where it is exercised and administered inhabited by men?"[36] Such dystopian dynamics are not quite on the agenda in Swebach's image. However, financial, informational, and structural networks are superimposed and shown to be directly exchangeable, in this small and, on the face of it, unremarkable scene. The presence of one of these networks, the telegraph system, points to a world beyond the painting's deliberately restricted purview.

The telegraph at Saint-Sulpice connected, in a route to the north, to an apparatus atop the church of Saint-Eustache. The place de l'École was situated beneath this line. Although inhabitants and passersby would have been unable to decipher the fugitive messages moving above them, they would surely have been aware of them—inconsistently, perhaps, at times even unconsciously, but aware nonetheless. Yet the scene in Swebach's painting also foregrounds another, more comprehensible network. The Samaritaine department store that later appeared in this space was named after the ornate pumphouse introduced by Henri IV (titled for a bronze relief of the Samaritan woman on one side of the pump) that occupied a central position on the Pont-Neuf from 1719 until 1813, shortly before Swebach painted his scene (fig. 4.9).[37] The Samaritaine pump contributed to the soundscape of the area. It housed an astronomical clock with an automated man who hit a bell on the hour, and like the statue of Henri IV on the bridge, it was the focus of political petitions and satire.[38] A pamphlet of 1790 stages a dialogue between Henri IV and the Samaritaine over the current state of France, concluding with "Henri IV" acknowledging the noise of recent events that has kept them both awake.[39] This structure would not have been visible from the perspective selected by Swebach, as the Pont-Neuf is self-consciously just out of the picture. Yet despite its absence here, the Samaritaine is invoked in the attention Swebach pays to the flow of water and to the central fountain, a six-meter-high structure that dominated the space, and it is possible that its recent removal sparked his interest in depicting this space.[40]

Lack of water was a consistent problem in *ancien régime* Paris, and projects to remedy the situation had been in progress for some years.[41] Many people were forced to buy clean water at excessive cost, and it was not until February 1812 that a law establishing the gratuity of water from Parisian fountains was passed, ending such speculation. In 1806, the decree of Saint-Cloud began a campaign to ensure that "water will flow in all the fountains of Paris day and night," ordaining the renovation of Paris's aqueducts, pumping machines, and hydraulic networks, and commissioning the construction of fifteen new fountains. The cost for this work was borne by the treasury, with the city of Paris taking

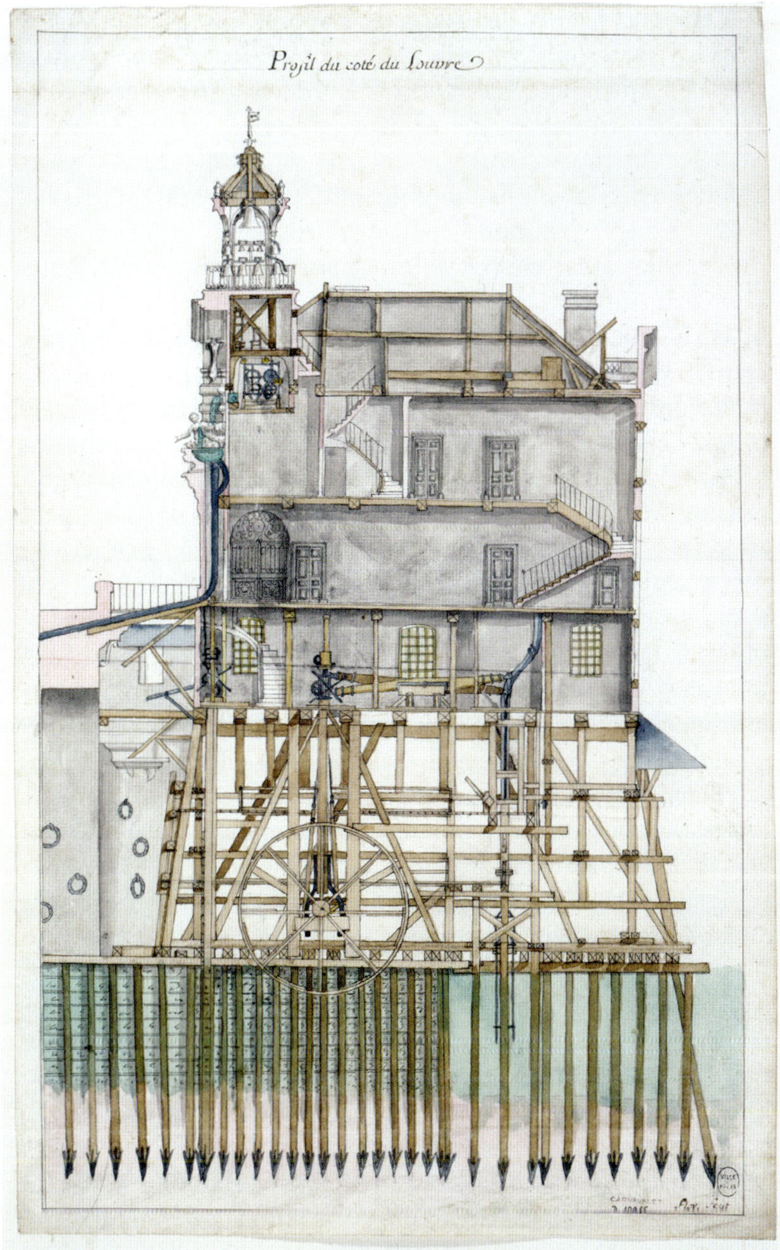

Figure 4.9
Flov., *La Samaritaine, profil du côté du Louvre*, c. 1770–1780, hand-colored drawing, 38.4 × 25.9 cm. Musée Carnavalet, Paris.

on the expense of future maintenance. Between 1806 and 1812, a program of fountain-building took place, headed by the architect François-Jean Bralle. Formally, Swebach's *Vue prise de la place de l'École* shares much with a painting by Étienne Bouhot of the Châtelet square and fountain, which takes a perspective from the next square along the river, looking out over the Pont au Change.[42] Bouhot's painting, exhibited at the 1810 Salon, features another, grander, Bralle fountain, and it seems fairly likely that Swebach had it in mind when he made his work.

In 1812, Charles-Alexandre-Amaury Pineux-Duval, better known as Amaury Duval, published an account of the fountains of Paris, documenting their historical functions and present state.[43] This tome, whose frontispiece dutifully proclaimed its intention to glorify the Napoleonic fountain, was accompanied by sixty engravings, which provided the original impetus for the project.[44] Asked to supply simple descriptions of each engraving, Amaury Duval allowed his aesthetic proclivities to prevail, and he added more complete and partisan observations to his otherwise "dry and tedious" descriptions.[45] Swebach made lively studies of a number of fountains during this period, drawing those on the rue Vaugirard and the place Dauphine, as well as the fontaine des Innocents and the Châtelet fountain; Amaury Duval's text surely informed the production of his painting, which is possessed of what Matthew Hunter has described, after Jeff Wall, as "liquid intelligence."[46] Perhaps he planned his own series to capitalize on interest in the new fountains. The fountain on the place d'École was the first of the fountains to be discussed, at some length, in Amaury Duval's publication, illustrated by Claude Alexandre Moisy's line engraving (fig. 4.10)

Amaury Duval began by recounting the history of the square, which, he claimed, was originally built on land belonging to the nearby church of Saint-Germain-l'Auxerrois, sponsors of the school that gave the square its name (an error, as "école" was in fact an historic misreading of "scala," in reference to the stairs leading to the dock on the Seine, just visible in Swebach's painting). Amaury Duval observed, "there has existed there for a long time a sewer, a receptacle for the refuse of this dirty and muddy neighborhood."[47] This is the tomblike void in the middle of the painting, from which a child clambers.[48] Yet, "The fountain that has now been raised above this sewer is much better placed than one might think. If it had no other use than to continually wash the area around the

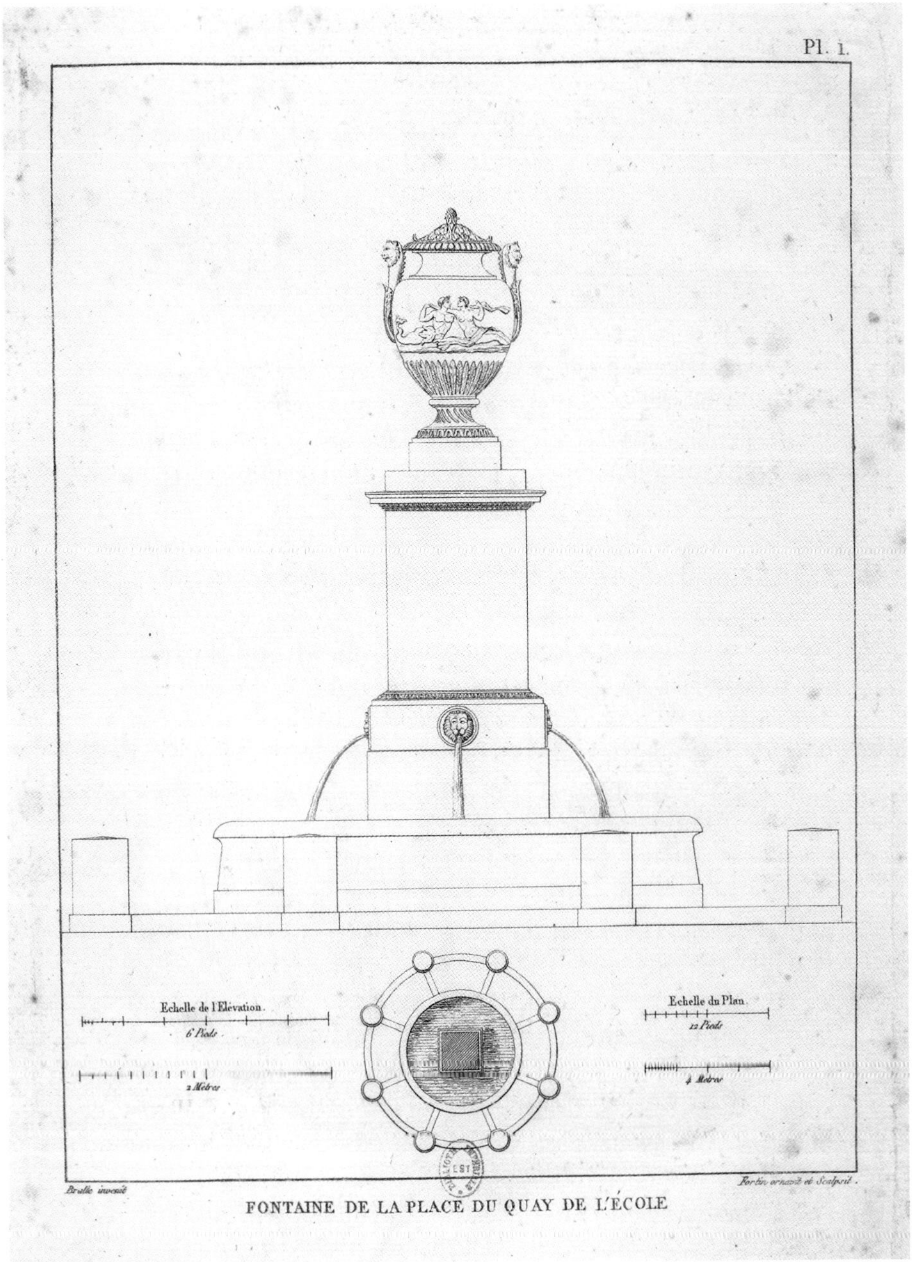

Figure 4.10
Claude Alexandre Moisy, "Fontaine de la place du Quay de l'École," from Amaury Duval, *Les Fontaines de Paris, anciennes et nouvelles* (Paris: Firmin Didot, 1812). Bibliothèque nationale de France, Paris.

sewer, that would already be a very great means to cleanliness and sanitation. Let us add that the water it provides is much healthier than that which one can draw from the Seine, along the quay, where the outfall of several cesspits and the boats of the washerwomen poison this side of the river."[49] Amaury Duval proceeded to describe the new fountain, although his assessment—especially of sculptor Augustin-Félix Fortin's decoration—was not positive, for "This fountain has been condemned, even by those poorly versed in the art of design. The form of the vase, its cover, and above all the size of its handles, the tips of which finish in a chimera's head, seem in very bad taste. From the many antique vases which one finds engravings of everywhere, could we not choose a less disagreeable model?"[50] Situating Bralle's fountain in a contemporary media ecology in which reproductive burin engraving still played an important role, Amaury Duval advocated recourse to commercialized circuits of cross-media image exchange and emulation. Ecology was not just a metaphor, however, for as he continued, "just four years have passed, and already the stone of which this fountain is constructed has become completely black. The reason for this may be due less to the nearby houses and the quantity of chimneys, than to the stay that coal boats make in this basin of the river."[51] Given Amaury Duval's description, one might ask why Swebach chose to make the fountain the centerpiece of his painting. The artist doesn't flinch from representing the fountain's grimy surround or discoloration.

The choice of the place de l'École made sense for Swebach, in part, because the filthy, smoke-stained patina of its fountain and its insalubrious surroundings provided a powerful means to visualize the effects of materials in circulation. It made possible a visual play on the tarnish accumulated by coins once they left the mint, drawing attention to the social life of money, as well as to its abstraction, via the motif of the enigmatic telegraph. Perhaps he also gestures, by way of the floating object in the sky, to the risks of inflation. At the very least, this mysterious dirigible—in common with the telegraphs—belongs to an atmospheric world of clouds and codes, rather than to the earthly circuits below. This may risk overreading a simple scene, but Swebach goes to great pains to compress these things into his painting, and both its inclusions and exclusions are sharply deliberate.

The place de l'École was connected to the place Saint-Sulpice by means other than the telegraph, for Amaury Duval also described the “fountain of peace” in its forecourt (a structure that Perec paid particular attention to over two centuries later). Although he approved of its design, Amaury Duval noted that the fountain was rendered ridiculous by the pace of flow; the fountain looked good with a gushing stream but was less effective with a dribble from its multiple chutes. This was a regular complaint, that fountains serving a primarily symbolic function—for example, the famous elephant on the place de la Bastille—were often next to useless when it came to dispensing water. Flow was not continuous or easily achieved. Like the movement of information via the telegraph, environmental disturbance or faults in hardware curtailed the stream.

Water, here, clearly wasn’t just water. In its mediation between materiality and abstraction, Swebach’s painting is comprehensible in light of contemporary attempts to negotiate between theoretical and practical concerns in the field of hydrodynamics. In eighteenth-century France, the science of flow was increasingly split into two camps, which implied distinct methods and professional identities: hydraulic engineers, meteorologists, and instrument makers, who used empirical methods to deal with practical problems; and hydrodynamic theorists, whose equations were for many years ineffective in practice, and which often contradicted practical observation. This latter group imagined hydrodynamics as a pursuit relevant to a universal system of natural and artificial flows: “the hydraulics of conduits, rivers, and canals; the physiology of blood and sap circulation; aspects of navigation including tides, waves, and ship resistance; acoustic phenomena; the damping of the seconds pendulum in air; atmospheric motion; bird flight; ballooning, and aviation; British theories of the ether as a perfect liquid; and hydrodynamic analogies for electromagnetic phenomena.”[52] We might add flows of information to that list, and the practice and theory of optical telegraphy overlapped with many of these fields. As Olivier Darrigol puts it, for hydrodynamic theorists, “the retardation of water in hydraulic conduits, the friction of water on ship hulls, the subduction of winds near the ground, or the steering of airships, all belonged to the same category of phenomena.”[53] Analogies between fluid movement and telegraphy were made more concrete with the transmission of signals over radio waves later in the century (a revolutionary inheritance here too, in the application of mathematician and first coordinator of the *Description de l’Égypte* Joseph Fourier’s transforms), but the possibility of these things being understood together preexisted such technology. Swebach's painting seems to speak to this broadening of a hydrodynamic perspective to address diverse worlds of flow, while at the same time remaining grounded in the materiality of paint applied to canvas: what Victor Burgin once memorably called "the anachronistic daubing of woven fabrics with coloured mud."[54]

From Steeple to Steeple

Before moving on, we must reconsider the moment at which this painting was made, as some part of the work's meaning rests there. Swebach's painting sides with Amaury Duval's account of the fountain's role in promoting hygiene and celebrating the untrammeled flow of free water (so much so that this once precious commodity can even be spilled freely on the ground). A note of ambivalence remains, however, aligning telegraphy's visual turbulence with forms of atmospheric pollution and environmental wreckage. At one level the painting is a kind of Napoleonic portrait, taking its cue from the portrait factory on the other side of the river, and I have a hunch it was made during Napoleon's return to Paris during the Hundred Days in Spring 1815. From the Treaty of Fontainebleau in April 1814 onward, the coinage of state underwent swift, successive, and no doubt highly confusing changes to its material appearance. Following Napoleon's exile, the stamps and matrices used to make coins bearing his effigy were thrown into a cesspit, leaving just one example of each coin, preserved as comparison pieces to prevent counterfeiting while the coins remained in circulation. With his return to Paris during the Hundred Days, production of new Napoleonic coins was hastily arranged, while royal coins continued to be made alongside them. Following Napoleon's abdication and second exile in the summer of 1815, these coins were, in turn, hurriedly withdrawn, and a report of October 1815 confirmed, again, the successful destruction of all materials needed to make them. For much of this time, then, royal, and Napoleonic coins were in circulation alongside one another. Swebach's painting was made against this backdrop, in which money was politically and iconographically mutable, and in which signs of value could easily find themselves flung into a cesspit, of the kind that according to Amaury Duval populated the bank of the Seine in front of the place d'École. The rapid transformations

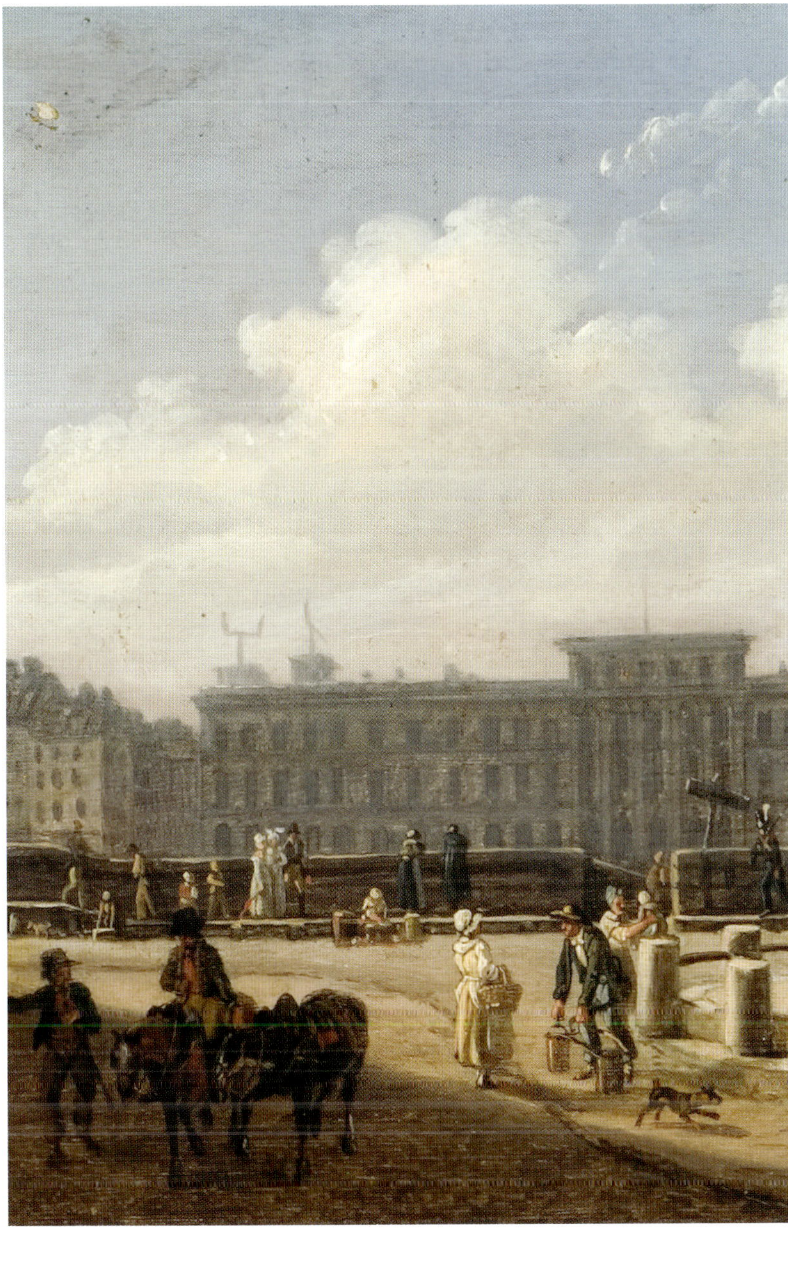

Figure 4.11
Detail of Jacques-François-Joseph Swebach-Desfontaines, *Vue prise de la place de l'École*.

of the "money thing" showed "money" itself to be somewhere else, or perhaps everywhere: in the ground, in the water, in the air, visible and invisible.

The meaning of telegraphy was no less fraught at this time (fig. 4.11). Napoleon's bellicose proclamation to the army on his return from Elba that "the eagle bearing the colors of the nation shall fly from steeple to steeple until it reaches the towers of Notre Dame" can surely be read in light of his embrace of optical telegraphy. Indeed, the restored monarchy of Louis XVIII destroyed telegraph towers in March 1815 to impede the emperor's return.[55] It was via the telegraph that news of Napoleon's definitive abdication was transmitted on June 24, 1815. Telegraphy was, like money, pulled back and forth between political powers, a practical and symbolic necessity at constant risk of obsolescence or transformation. Swebach's vision of overlapping networks in Napoleonic France allowed him to negotiate a relationship between abstraction—value, code, hydrodynamics, or even artistic theory—and the fragile and fungible forms of materiality that accompanied each of these. Combining and conflating the mint, the telegraph, and waterworks allowed Swebach to interrogate the relationship between theory and practice, between the provision of free water and a free market, between bare plumbing and a universal network, between solid currency and abstract money, between ground-level knowledge and encoded information, and, most insistently, between everyday lives and the state.

Part II

Building Codes: Étienne Bouhot

If Swebach's painting offered a relatively positive image of the city's flows, the work of other genre artists focused more trenchantly on telegraphy as a site for recursion, breakdown, and repair. In the "Memorandum" that begins Roland Barthes's *Michelet*, first published in 1954, several images accompany a chronology of the historian's life. Many are identified by the names of the sitters: Michelet's second wife Athénaïs, his daughter Adèle, or her mother-in-law Mme Dumesnil. Others, mainly portraits of Michelet himself, also bear the names of the artists who made them: a photograph by Nadar, a drawing by Thomas Couture, and a medallion by David d'Angers. Among these, a reproduction of a small, unremarkable genre scene goes uncommented upon, and has no title.[56] It is placed adjacent to an image of a letter confirming the 1822 appointment of Michelet as assistant lecturer in history at the new Collège Sainte-Barbe, later Collège Rollin. The work in question, which represents the interior courtyard of the school, with the dome of the Panthéon visible above the rooftops, was painted around the time of Michelet's appointment, by Étienne Bouhot, an artist who had made his name as a specialist in modest, Dutch-influenced scenes of everyday life, and it was shown at the Salon of 1824 (fig. 4.12). The Musée Carnavalet acquired the painting in 1895 from the estate of François-Vincent Raspail, the chemist, physiologist, and republican politician, best known today for the Paris boulevard that bears his name—the longest in the city—built during Baron Haussmann's reconstructions, and later the site of a statue to Claude Chappe.[57] The former rue de Reims, from which Bouhot's work purports to have been painted, was destroyed during the transformation of the *quartier* in the 1880s.

Although it plays the slightest of supporting roles in Barthes's text, there is a logic to the placement of Bouhot's painting here, in this idiosyncratic account of the formative years during the Restoration of a figure who played a key role in the construction of French history, yet whose relation to the revolutionary past was always, Barthes noted, impossibly belated.[58] Despite the ostensibly tranquil quality of this painting—"a seamless world" might be Barthes's description, for this was, he noted, the "great Micheletist theme"—throughout the Restoration Bouhot also made work that spoke to themes of historical dislocation.[59] In their evocation of timeless urban continuities, these paintings were bound up with attempts to portray a peaceful and ordered society, to affirm the stability that, it was hoped, the Restoration would reinstate. Yet, like Swebach before him, Bouhot undercut this message by rendering an image of Paris at a time of political and social flux. While most of Bouhot's work evinced a careful attention to spatial dynamics and a confident handling of architecture, several paintings from this period prefigure a transformation of Paris more usually associated with the latter part of the nineteenth century.[60] More concretely, Bouhot's attention to the fabric of the city and the means used to communicate across and through it were inseparable from, and often in tension with, the Restoration's self-conscious understanding of its historical significance.

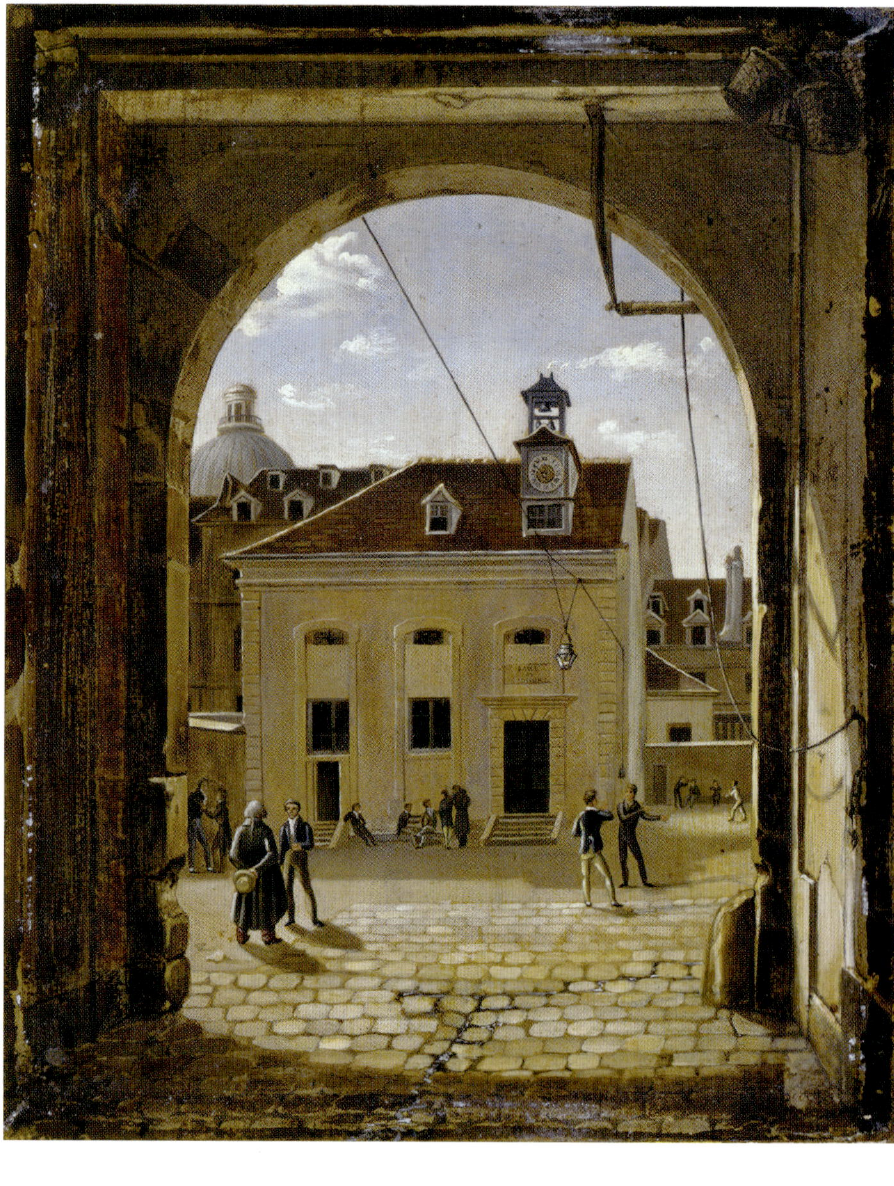

Figure 4.12
Étienne Bouhot, *La Cour du collège Sainte-Barbe, rue du Reims*, c. 1824, oil on canvas, 23 × 19 cm. Musée Carnavalet, Paris.

With the return of the Bourbon monarchy in France after Napoleon's defeat and second exile, a strategy of official forgetting attempted to expunge the recent history of the Revolution and Empire. This policy of *oubli* pervaded the artistic sphere, too.[61] Nonetheless, the Revolution was often more powerfully present because of this repression, returning in unexpected ways. As Stephen G. Nichols puts it, "in many ways, the Revolution is the 'story not told' in the cultural production of history in the Restoration. Not told because the process it began had not yet played out."[62] Interruptions, missed connections, and uneven speeds were intrinsic to the historical process. Moreover, we might add to Nichols's "cultural production of history" the "historical production of culture," for the dialogue between image-makers and historians was two-way. As Nichols observes, during the Restoration, "people still saw themselves as enmeshed in an ongoing political catastrophe, a quest for a workable regime, for which 1789 or 1793 were simply the first manifestations of crisis."[63] Although it cut against the official rhetoric of the Restoration, the present could not in fact be easily separated from preceding events.

A more capacious approach to this issue may be found in a lecture of 1966–1967, in which Hannah Arendt discussed the relationship between the revolutions of the previous two hundred years and her contemporary political situation, developing ideas expressed in her 1963 book *On Revolution*. Arendt noted how the terms "revolution" and "restoration" had initially, in the context of England's Glorious Revolution, been used almost interchangeably, signifying the restoration of monarchical power to its rightful position. Such ambiguities were equally present in the French context, with the French Revolution figuring as, in Tocqueville's account, a "restoration of the *ancien régime* rather than its overthrow."[64] As Arendt noted, "the fact that the word 'revolution' originally meant restoration is more than a mere oddity of semantics. Even the eighteenth-century revolutions cannot be understood without realizing that revolutions first broke out when restoration had been their aim, and that the content of such restoration was freedom."[65]

Restoration, then, functioned in Arendt's reading in a more relational sense, as a moral, rather than strictly historical, course of action, a remedy for a world gone awry (although if revolution itself is analogous with restoration, a call to order for an *ancien régime* that had lost sight of its principles, one might question what value we can ascribe to the Restoration proper). Conversely, Arendt also trades with an understanding of the term that implicates it in a deviation from a rightful course, rather than a setting back on track, observing that "Restoration, the consequence of an interrupted revolution, usually provides not much more than a thin and quite obviously provisional cover under which the processes of disintegration continue unchecked."[66] These intertwined processes of material as well as political dissolution and renewal ("restoration" in a more literal sense) are evident in Bouhot's paintings of this period, which struggle to contain beneath their placid, methodical facades—their "provisional cover"—the contradictions of the era, traces of which were embedded in the buildings and informational networks of the day.

Particularly compelling are paintings made during the 1820s that treat urban infrastructure as a cipher for addressing the Restoration's complex politics of time, and its attempts to manipulate or repress France's recent history. Bouhot's paintings of the Bourse in construction, of fountains and canals, of plaster mines at Montmartre, and, most strikingly, of the operations of the optical telegraph, help fill a lacuna in histories of technology that have focused more heavily on the Empire and July Monarchy. His work elucidates how artists of this period were able to address their political present—and the legacy of the past—in ways that sidestepped direct political reference. These approaches were attached to methods of calibrating and activating time and space that predated the Restoration, yet "continued unchecked," in Arendt's terms, for its duration.

In its careful delineation of precise locations, Bouhot's work conforms to a prominent approach to representation during the Restoration that focused, as Beth S. Wright has written, on "the specific nature of a place, a portrait, a monument" as a means of connecting individual memories to a cohesive "national memory-history."[67] However, unlike many artists active at this time who processed contemporary politics via carefully constructed images of the past, Bouhot's work was often lodged in something like the present. Nowhere outside the Gothic architecture that features prominently in his work might one discern the medievalism, the Anglophilia, or the relic-like effects of the troubadour painting with which he is occasionally associated.[68] Yet Bouhot was an artist who walked a path between current events and a form of subdued archaism. As Jeanne Magnin wrote, in one of few significant essays on Bouhot's work, regarding the Collège Sainte-Barbe painting:

> The little painting is painted with extreme care under the cold light that floods a patch of sky bristling with roofs and chimneys; the touch is meticulous, exact and precise like a *procès-verbal*, all sensibility restrained by a sort of moral modesty which delivers to the public nothing but the decor of the past: the narrow court where the sun fails to linger, the dial of a clock embedded in the fixed shutter of a skylight, a rustic campanile sheltering the bell which sounded in turn for recreation and for study, mute witnesses of the schoolboy life which flowed between these four walls.[69]

Bouhot's attention to effects of light and atmosphere is analogous, for Magnin, to the operations of French bureaucracy. He is "precise like a *procès-verbal*," and the ambitions of his work were attuned to the reconfiguration of time that followed the French Revolution; the way in which innocuous daily events were now replete with the potential to become the substance of law, the matter of a bigger history. At the same time, in Magnin's reading, Bouhot's painting registers temporal calibrations which had remained unchanged for centuries, and which were tied to the lives of the institution and those of the individuals who passed through it.

In due course, Magnin tells us, Bouhot's painting returned as a public artifact of a different kind, for "the view of the Sainte-Barbe institution, which never left Bouhot's studio until his death, reappeared during the Universal Exposition of 1900 in the retrospective section on Teaching."[70] This is an art that even in its posthistory seems suited to the asynchrony, revisionism, and strategic amnesia that characterized cultural policy during the Restoration. Little surprise, then, that the Restoration saw Bouhot's greatest artistic success; he showed his work regularly at Salons in the capital and in the provinces, lithographers regularly reproduced his paintings, and he counted the duc d'Orléans and the duchesse de Berry among his patrons. However, as his image of the Collège Sainte-Barbe demonstrates, meaning is produced not in relation to the figures that populate the scene, who seem something of an afterthought, but is adumbrated instead via a horizon "bristling with roofs and chimneys," and by the ticking hands of a clock. As in many of the works Bouhot made in Paris from 1801 to 1834, there is a sense that broader questions of political and social change—or of stasis, even regression—might be processed, encoded, and transmitted (telegraphed, we could say) by the city itself. Indeed, what better metaphor might there be than the optical system for the snags and backward glances, as well as accelerations and breakthroughs, that characterized this historical period?

Fossils

Born in 1780 in Bard-lès-Époisses, Burgundy, Bouhot was apprenticed to a milliner and subsequently a glazier, although he soon began to fulfill more artistic ambitions. From an early stage he was supported in his art by the celebrated memoirist Mme la Comtesse Victorine de Chastenay, with whom he corresponded for many years. In the late 1790s, Bouhot assisted panorama painter Jean-Charles Langlois in his decoration of the abbey church of Cîteaux. Leaving Burgundy for Paris in 1801, Bouhot soon after joined the studio of Charles Victoire Frédéric Moench, Langlois's master and a student of Girodet. Moench was at that time participating in the Napoleonic restoration of the Tuileries, and consequently Bouhot began a trajectory that involved a significant amount of Bonapartist sponsorship; he later worked for Joseph Bonaparte at Mortefontaine and Lucien Bonaparte at Plessis-Chamant.[71] The announcement of Bouhot's marriage in 1804 listed him as a "peintre en décors," a fitting description for his practice at that time, the vestiges of which may be found in his subsequent attention to the walls and roofs of Paris, the latter by then intermittently speckled with telegraphs.[72]

Sometimes, as in *Toits de Paris*, painted around 1830 (fig. 4.13), we encounter these roofs as a kind of stage set, illuminated for an unspecified display. Such sites form an overlapping series of circuits—of bridges, pipes, and signage—which populate Bouhot's paintings, as they did Swebach's, and whose warp and weft configures the city as a structure for processing data and materials. *Toits de Paris* fits Kittler's description of how "it is common, in the open spaces of a city, to see the skeletal infrastructure on the backside of a building—these are networks."[73] Despite the presence of workers repairing the roof in the foreground, it appears that the city's networks take priority over human agency in Bouhot's painting.

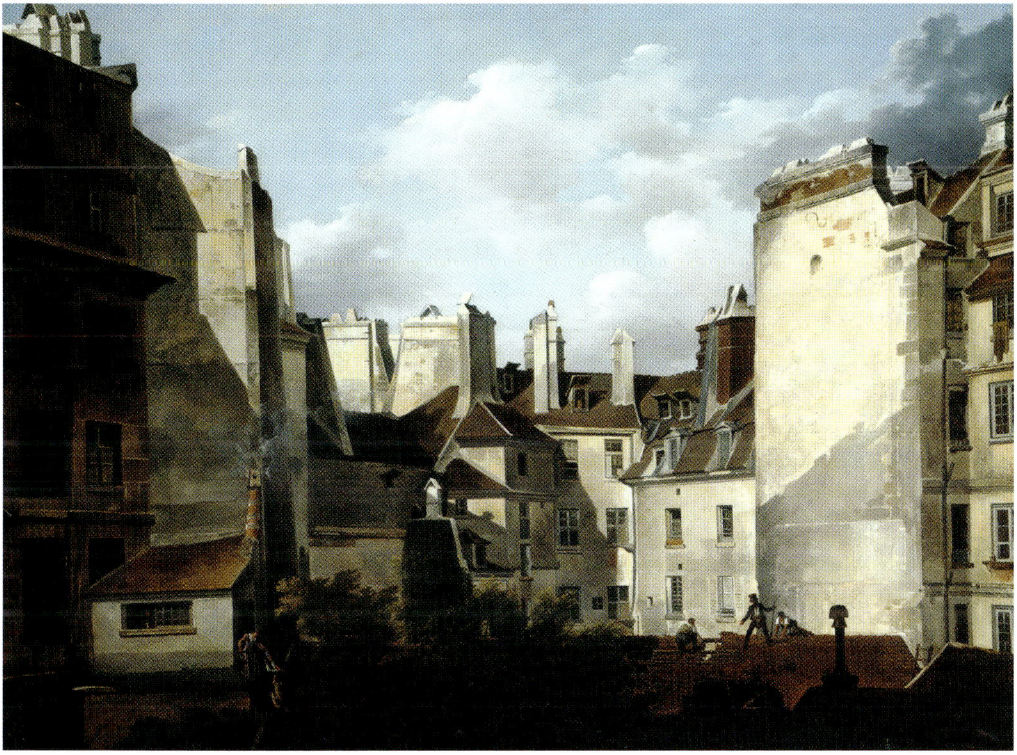

Figure 4.13
Étienne Bouhot, *Toits de Paris*, c. 1830, oil on canvas, 59.8 × 79.3 cm. Musée Carnavalet, Paris.

From 1804, Bouhot worked in the studio of the panorama painter Pierre Prévost, alongside figures such as Charles Marie Bouton, who later developed the diorama with Daguerre. The production of panoramas was to play a key role in Bouhot's career in Paris, until at least 1833, and his career as a panorama painter pervaded his genre scenes. As Sandrine Balan notes, around this time Bouhot repaired the famous panorama of Algiers painted by his former master Langlois, a project whose precisely plotted vistas articulated a vision of French colonial control to a French audience.[74] Expanding his practice from varied forms of interior decoration to the careful, perspectivally tight paintings for which he became known, Bouhot showed his first painting at the Paris Salon, *Vue de la place Vendôme*, in 1808, a work that Ligeret du Cloiseau later described as possessing a "daguerreotype" rectitude in the rendering of its details.[75] Characteristically, the painting juxtaposes bygone fragments—the ruins of the church of the Feuillants—with a scene of harmonious social order, and it was considered a success. Bouhot left Paris in 1834 following the death of his second son, and returned to Burgundy to establish an art academy at Semur-en-Auxois, where he remained until his own death twenty-eight years later. Yet despite the fashionable status of his work during the Empire, the Restoration, and the early years of the July Monarchy, the taste for such paintings was, according to Magnin, swept away by more grandiose forms of Romanticism, for "it seemed outdated, fossilized; it was no longer even discussed, and his works were relegated to some provincial museum, the last refuge of outmoded glories."[76]

Echoing Magnin's summary, Paul Mantz, writing in the *Gazette des Beaux-Arts* in February 1862, described how Bouhot's brand of urban topography and belated *ruinisme* had been the height of fashion around 1816 when Decamps, his more illustrious pupil, was in his studio (in fact, Decamps began working with Bouhot in May 1818), but had been dealt a mortal blow by the advent of photography.[77] This neat account of technological supersession should be taken with a grain of salt, for Bouhot's work was already imbricated within a diverse media landscape that included panoramas and other visual spectacles; however, the reality effects in which his work traded evidently became less convincing as the century progressed. Mantz doubted that Decamps had taken much from the experience of working with Bouhot, apart from learning to accurately represent masonry, and much of his article focused on correcting an earlier piece that had falsely cited the artist as "Bouchot," so far had his star fallen by that point. Yet the quality of outmodedness both Magnin and Mantz identified was, I would suggest, an intrinsic aspect of Bouhot's work from the moment of its production, its presence there freighted with a criticality that exceeded the more mundane forgetting to which art of this kind was subject in later decades. Bouhot's work during the Restoration spoke to issues of time, politics, and media that were absolutely of their moment.

Bouhot made several paintings that addressed elite Restoration politics directly—for example, *Vue de la salle de l'archevêché de Reims le jour du sacre de Charles X*, of 1825, a work honoring the new political regime. Other paintings eschew such firm statements of political and religious orthodoxy. Take, for instance, an earlier work, *Entrée d'une carrière à plâtre à Montmartre* (fig. 4.14), made in 1816 at the advent of Louis XVIII's reign. Stripped of references to contemporary politics, this painting's message is intangible. Bouhot had depicted aspects of urban infrastructure before, such as in two works submitted to the 1810 Paris Salon, *Vue de la place du Châtelet* and *Vue de la pompe à feu de Chaillot*, while his sketches and paintings of Châtillonais forges made around 1823 showed scenes of modern labor. Fountains, pumps, and canals were a common subject for Bouhot during the Empire, his interest in the renovation of Paris's water supply anticipating that of Swebach. The expansion of these hydraulic networks continued during the Restoration, and they were joined by the development of the sewers made famous by Victor Hugo in *Les Misérables*, whose action centered on the period from 1815 to 1832.

Bouhot's image of plaster quarries at Montmartre contrasts with his earlier paintings' positive belief in the regenerated city. *Entrée d'une carrière à plâtre à Montmartre* shows an opening onto a hillside, where a small figure can be seen going about their task, silhouetted against the dark maw of the quarry. The mutilation of the landscape—an abrupt slice through the hillside above—and the striking perspective of the path lend this scene an otherworldly flavor, exacerbated by an uncharacteristically loose handling of paint. This sense of spatial uncertainty, at odds with the forensic specificity of the majority of Bouhot's urban scenes, is heightened

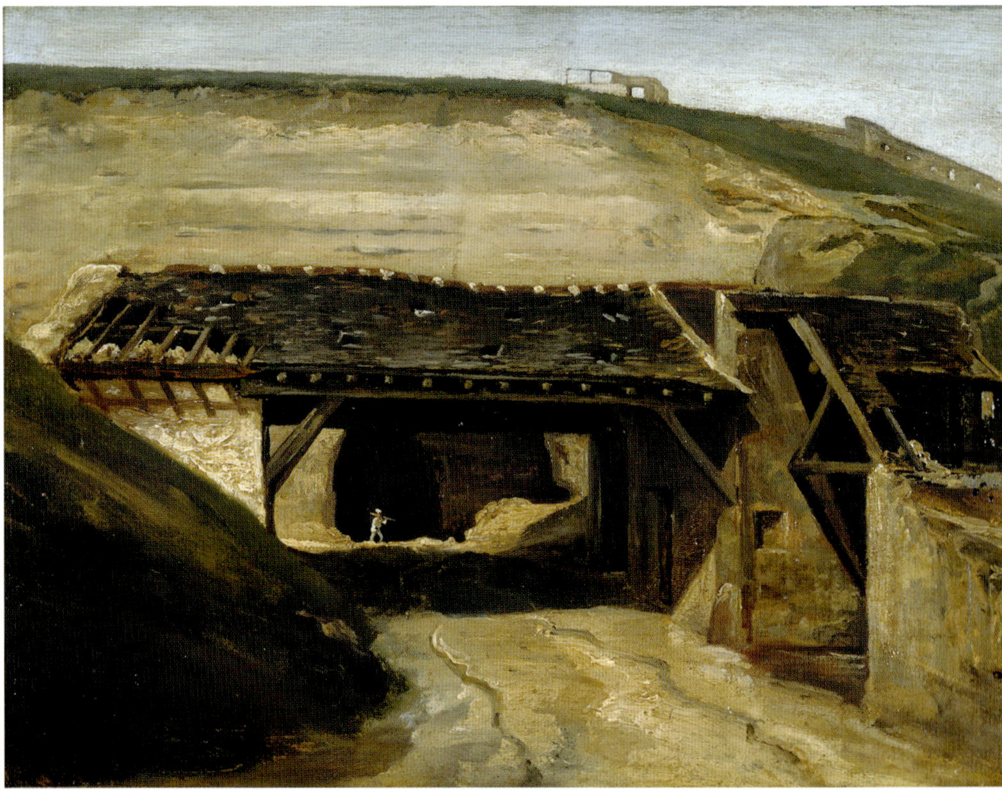

Figure 4.14
Étienne Bouhot, *Entrée d'une carrière à plâtre à Montmartre*, c. 1816, oil on canvas, 28 × 37.9 cm. Musée Carnavalet, Paris.

by the structures in the foreground and the ridge of indistinct shapes surmounting the bluff on the right-hand side. Missing some tiles and seemingly uninhabited, this technology is hardly cutting-edge but rather a site of mostly invisible labor suffused with a semirural sensibility that belies Montmartre's location just outside Paris. The mining of gypsum from the hills of Montmartre and Ménilmontant had taken place for many centuries and was the foremost source for plaster of Paris, while during the Directory, the Montmartre quarry was the site of the discovery of fossil specimens that facilitated Georges Cuvier's early paleontology. Montmartre's main quarry operated until 1862, when it joined areas that already comprised the cemetery based there. Bouhot pursued this archaeological mode after his return to Semur-en-Auxois, where, in addition to painting Semur's picturesque walls and towers, he played a key role in establishing the town's museum and made trompe l'oeil images of ex-votos and other objects excavated from the source of the Seine and from Massingy.[78]

Bouhot returned to the quarries in a work shown at the Salon of Arras in 1822—*L'Entrée d'une carrière au plâtre près Pantin* was bought by the Société des Amis des Arts de Bordeaux—the same year that Théodore Géricault also completed a memorable image of plaster production.[79] While Géricault represented plaster as a dusty cloud, puffing into the air from the factory building, rather than as a solid substance, Bouhot elided the material entirely, focusing instead on entrances to subterranean structures, and largely effacing the work of its extraction. In 1816, an image of a quarry called to mind the provisional materials that dominated attempts at a monumental visual culture during the Revolution and Empire, when temporary materials such as plaster were frequent substitutes for bronze and marble. The fragility of such objects, and the risk that their political message might be equally delicate, remained a source of anxiety into the Restoration. The material imaginations of previous regimes were difficult to obliterate, even if their objects frequently were not, and the Restoration's recourse to signifiers of distant royal or religious history might be read, in part, as an attempt to evade the articulation of contemporary politics by flimsy symbolic means. In the context of these works, one wonders if such anxieties extended to the transitory visual acts of the telegraph, a "future fossil" whose fugitive mannerisms eschewed tradition, the labor of its operation similarly concealed.

Bouhot's Montmartre is only partly comprehensible in terms of the city's underground networks. As significant is what he chose to leave out. An anonymous watercolor and pen drawing from 1826 fills in some gaps (fig. 4.15). Above the quarry was the church of Saint-Pierre de Montmartre, the site of the optical telegraph that communicated in and out of Paris to the north. The hub of the first line to be opened, the telegraph at Montmartre was closely associated with the area and was the last to close when electromagnetic transmission replaced the optical system starting in the mid-1840s.[80] During the time the telegraph was installed at the Louvre, it was decided that Ménilmontant was no longer usable, and that visibility to Écouen was restricted, so a new station was constructed at Montmartre, on the Paris-Lille line. Saint-Pierre was not strong enough to hold the machine, so it was bolstered with a huge beam and masses of stone—much of it broken tombstones—that were placed in the middle of the choir.

The administration of the telegraph entered a period of crisis toward the end of Charles X's reign, and it was withdrawn from the supervision of the Chappe family in September 1830, with the retirement of the last of the brothers.[81] This is the context into which Ignace Chappe's history of "l'art télégraphique" appeared in 1824.[82] Chappe clarified what was acknowledged by omission in Bouhot's painting: that communication networks were inherently political, that their visible acceleration of information was often at odds with the Restoration's backward look to the *ancien régime*, and that to invoke such a device during the Restoration was, potentially, to call to mind the structural transformations wrought by the Revolution and the Empire. This point is emphasized in an anonymous satirical image, *Les Royalistes visitant les travaux de Montmartre* (fig. 4.16), from 1815. As quarry workers toil in the background and the mysterious shapes of the telegraph appear at the summit of the hill, the returning royal family are appraised, as if through a telegrapher's telescope, and found wanting.

The absence of the telegraph in Bouhot's painting of the Montmartre quarry is, however, corrected in his later work *La Bourse en construction* (fig. 4.17), which represents a building site in front of the Parisian stock exchange, shortly before it was completed, in 1826, under the supervision of Éloi Labarre (the lateral wings of the

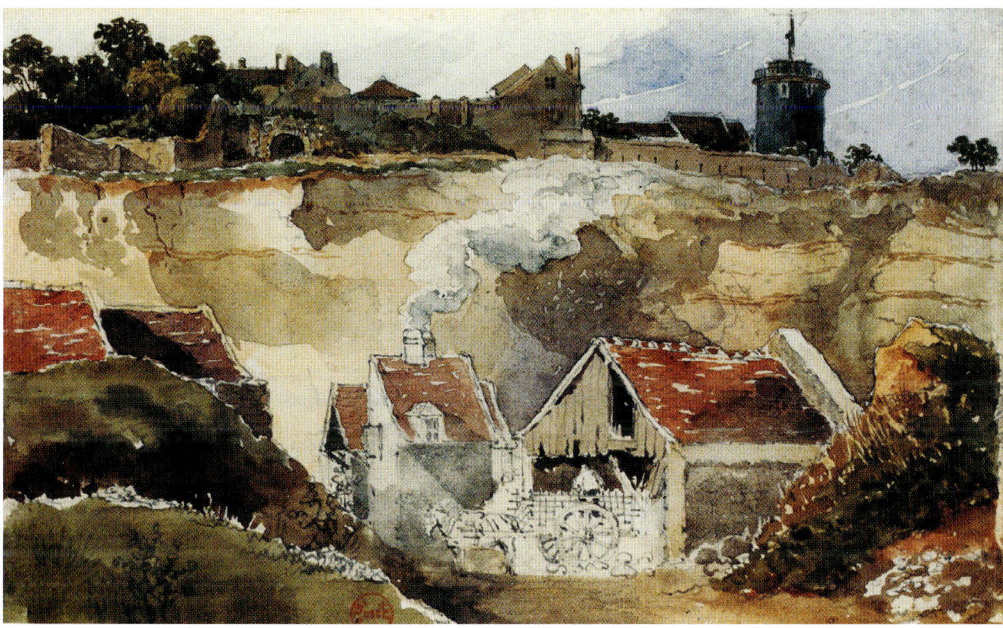

Figure 4.15
Vue de Montmartre, carrière et tour du télégraphe, 1826, watercolor with pencil, 12.5 × 20 cm. Bibliothèque nationale de France, Paris.

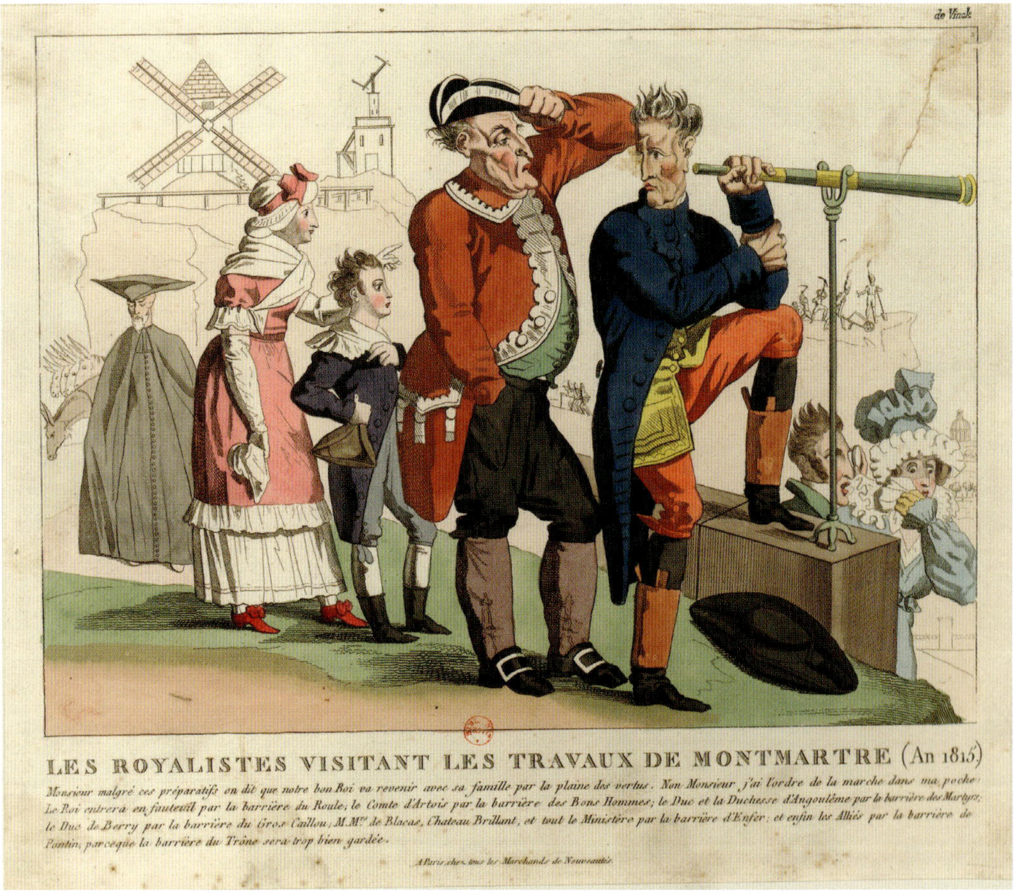

Figure 4.16
Les Royalistes visitant les travaux de Montmartre, 1815, hand-colored etching, 19.9 × 27.3 cm. Bibliothèque nationale de France, Paris.

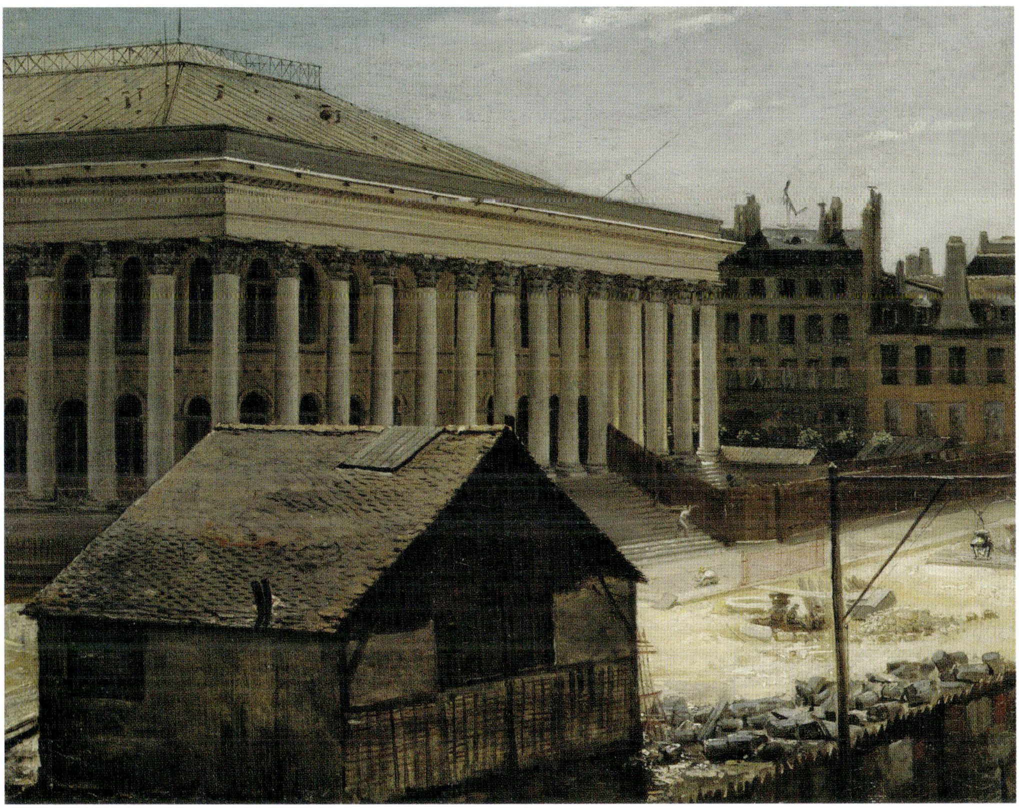

Figure 4.17
Étienne Bouhot, *La Bourse en construction*, c. 1820, oil on canvas, 47.5 × 55.1 cm. Musée Carnavalet, Paris.

building's present cruciform plan are a twentieth-century addition).[83] In this work, a minor detail animates the whole: barely visible above the houses to the right-hand side of the image, a Chappe telegraph mechanism rotates on the church of the Petits-Pères, Notre-Dame-des-Victoires, a building which had only relatively recently been restored as a place of worship after serving as the base for the national lottery and as a temporary stock exchange during the Directory.[84] These days, the prominent radio masts of Agence France-Presse occupy the same horizon.

Begun in 1808 following Alexandre Brongniart's design, the Bourse was, by the time of its completion, already a relic of Napoleonic reconstruction. Emerging in finished form during the Restoration, it was a belated, rather stiff, reminder of the previous regime. In Bouhot's painting, faceless workers in the distance carry away scaffolding and apply finishing touches. In the center, partly obscured by a streetlamp, laborers hunched in a semisubterranean space work at an unidentifiable task. Pieces of stone and wire reinforcements lie about the place. The foreground is occupied by a dusty, fenced-off area scattered with the bric-a-brac of construction, featuring a remarkable depiction of a looming, decrepit workers' hut, whose patched-up, shoddy character contrasts sharply with the triumphant monument to speculation glistening behind it. It makes for an interesting contrast with Swebach's image of an implacable Monnaie, although that too acquired its gloss from juxtaposition with the lived reality of the modern city. This "taste for the provisional," as Pierre Pinon terms it, links a number of Bouhot's paintings from this period.[85] The tattered posters visible on the hoardings marking off the place de la Bourse find echoes elsewhere in Bouhot's work, notably in his *Vue intérieure du porche de l'église Saint-Germain-l'Auxerrois* of 1831—a location yards from Swebach-Desfontaines's place de l'École—where scraps of text on the scarred wall advertise a "maison à louer" (Magnin transcribes this incorrectly as "maison à vendre," a more secure concept). Bouhot's *Bourse* may be a site of construction rather than extraction, as in his quarry images, but it is hard to tell the difference, and one might as easily read it as an emblem for Arendt's "processes of disintegration." Here, in the Paris of the arcades, center of a bustling, confident modernity, Bouhot flips the picture to expose that modernity as always already ruined.

An astonishing painting from the mid-1820s works as an effective pendant to this image (fig. 4.18). It shows the repurposed

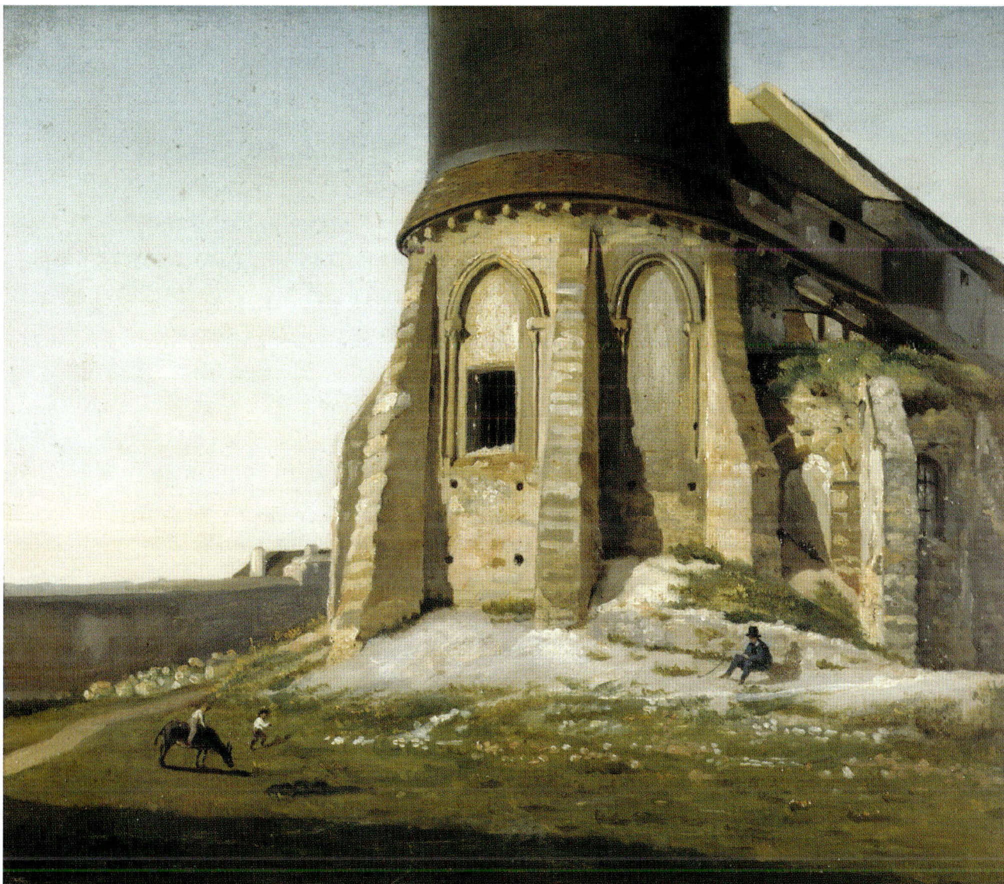

Figure 4.18
Étienne Bouhot, *L'Église de Montmartre, avec la tour du télégraphe Chappe*, c. 1825, oil on canvas, 29.7 × 35 cm. Musée Carnavalet, Paris.

church tower at Montmartre, amputated at the top of the frame at the point where the Chappe apparatus was positioned. Even when cropped brutally, the shadow of the telegraph was powerfully present. Bouhot's paintings highlight the uncanny effects of modern technologies of communication, rendering them alternately visible and invisible, and they do so in a way that largely occludes people and the work involved in their operation. Yet if, at Montmartre, windmills and sheep partly defused the shock of the telegraph's placement there, in Bouhot's painting of the Bourse new communication systems are shown to overlap with other networks—financial, architectural, and social; ground-level, airborne, and subterranean—although where the old ends and the new begins is anyone's guess. In that sense, at least, it is a consummate Restoration painting. If Bouhot sometimes provides a vision of Napoleonic reconstruction as belated, vainglorious, and run-down, he also shows us that the networks that mark, and define, a city function as sites of memory on their own terms. These paintings are telegraphic not only in their subject matter, but in the way they reach across time. The purposeful anachronism of the Restoration city was harnessed to telegraphy's historical layering, and even to its obsolescence. Bouhot's paintings are often equivocal, but what remains striking is how, in their quiet way, even after the Restoration of monarchy in France, these works refuse to forget the past. One might add, they remember the present too.

Part III

A Storm Is Coming: Georges Michel

"Let's not forget the dead!"[86] With these words, in April 1873, dealer and critic Alfred Sensier summoned the specter of landscape painter Georges Michel, an artist who had indeed been all but forgotten until that point. Michel, who died in 1843, was little known in his lifetime but achieved some small renown later in the nineteenth century as a forerunner of the Barbizon School, primarily thanks to Sensier's necromancy. Sensier had vested interests in promoting Michel, as he hoped to sell the artist's work to Paul Durand-Ruel and other dealers, an ambition that was successful.[87] However, his rehabilitation of this marginal revolutionary artist in the name of reactionary political values strikes an odd chord.

Sensier's retrieval of Michel from historical neglect was nostalgic, even deathly, in tone: his preface described Théophile Thoré's early attention to the artist in the 1840s as "clairvoyant," and he cheerfully termed his own study an "exhumation."[88] Although it was published only thirty years after the artist's death, Sensier's *Étude sur Georges Michel* represented him as an inhabitant of an entirely different world.

Michel was, in Sensier's telling, the archetype of a uniquely French kind of *petit-maître* active at the end of the eighteenth century and in the years after the French Revolution. Sensier talks of Michel's works appearing among sellers of bric-a-brac, where they dwelled unsigned, undated, and mysterious; only long-term inhabitants of the streets and a few fellow artists knew the secret of his name. Although Michel struggled throughout his long career to find buyers for his art, he supplemented his income by working as a *brocanteur*, opening a shop on the rue de Cléry in 1813; by giving drawing lessons; and, allegedly, by restoring paintings.[89] Yet Sensier did not read Michel's outsider status as typical of conditions experienced by many artists after the French Revolution. He interpreted it instead as the marker of a distinctive virtue, one which by the time of his text's publication had become part of the ideological apparatus of advanced art. In contrast to the measured professionalism of Swebach-Desfontaines and Bouhot, Michel's "wild and irritated" paintings, "full of temerity and insolence," were made by someone "who had regard neither for the traditions of the school, nor for the proprieties of the salons, nor for the measures to take for mercantile industry."[90] They were, in short, the product of a modern artist, albeit one recast in a conservative light.

Michel's landscapes—focused, as they were, on his local milieu—did not emulate the sights and atmospheres of the Roman campagna familiar to artists who had pursued an established academic path. Nor, unlike several other landscape painters who came to prominence in the 1780s, such as Flemish artist Simon Denis, was he aligned with a tradition linking him to prestigious forebears such as Pierre-Henri de Valenciennes or Joseph Vernet.[91] He did, however, receive training from the history painter Leduc, and joined the studio of Nicolas-Antoine Taunay, as well as, possibly, that of Carle Vernet. Michel did not show his work in the prerevolutionary Salons, nor at the Académie de Saint-Luc. The first major public exhibition of his work was at the democratic Salon of 1791, where he showed three paintings. From the early 1790s he was associated with the genre and landscape painters Jean-Louis Demarne and Lazare Bruandet, as well as with Swebach-Desfontaines, collaborating with the latter on at least one painting.[92] He exhibited regularly at Salons from 1793 until 1814, yet although he was prolific, success was limited. Despite Michel's purported marginality he became something of an artist's artist. Notably, Vincent Van Gogh wrote admiringly of Michel's work in letters to his brother Theo, bemoaning his own inability to live up to such a "maître," an accolade delivered wholly without irony.[93] Even here, the fact of Michel's being forgotten, his simultaneous centrality and eccentricity, became, in the absence of much other information, a defining feature. Van Gogh is known to have owned a copy of Sensier's biography, and he had no doubt metabolized the critic's account of Michel's obscured place in the history of art, alongside the formal lessons of his paintings.

Michel was, for Sensier, a primitive whose unfettered artistic impulses connected him to the greats of the past, and to more ancient, primal forces: "He is rough and fierce to the point of excess; he shows human forms as if a creator of the first worlds; he has in him the gigantic, wild movement, the archaic or barbaric animal gesture. As humane as Michelangelo, his 'confidential studies' are so many monuments that demonstrate how his first thoughts were unalloyed and unmitigated by the all too delicate taste of the world."[94] Although they were made by a man whose extra-artistic career involved varied humble means of making do and getting by, Michel's landscapes and genre scenes were imparted a terrible, prophetic power. Unsettling the hierarchy of genres, within which landscape was accorded a lower rank, Sensier's description of Michel's "savage" work and "archaic" painterly gestures appropriated the spartan gravitas accorded to neoclassical painting as it had evolved in the 1780s and 1790s, while introducing him as an artist working apart from, even in conflict with, such models.[95]

What then, given these elemental claims, are we to make of Michel's careful attention to a modern technology such as the telegraph? Not only did Michel depict the machine repeatedly, but he also made it disappear, severing it from the church below in a manner that recalls Bouhot's version of the same scene (fig. 4.19). He was attentive to the ways in which it might come into view above a hill or escarpment, or from behind a cloud, and to how it retreated from sight. Indeed, Michel's works—read through Sensier, as well as on their own terms—are profoundly telegraphic images, for they speak to connections across time and to relationships between environment, locality, and history.

Although its details are frequently difficult to corroborate, Sensier's *Étude sur Georges Michel* has influenced Michel's reception to the extent that it has been hard to separate fact from fiction. Due to the emphasis on recuperation and stylistic genealogy, attention to the way his works themselves slip in and out of visibility, the way they conjure an atmosphere that proceeds from, but does not begin and end with his paintings, and the way they allude to broader social and political issues, is often deferred. Instead, commentators on Michel have focused on cementing his place within a tradition of nineteenth-century French landscape painting, and connected him to sources in Dutch art—Rembrandt, de Koninck, and Ruisdael—as well as to British artists who drew similar inspiration, such as Crome, Turner, or Constable.[96] There was, wrote Sensier, a "contagious air" that linked these artists otherwise unknown to one another, a "community of sensations" that connected disparate practices regularly excluded from the academic mainstream.[97]

It is worth pausing over Sensier's wording here, to read another history in his conjuration of Michel's spectral presence. Rather than seek connections that discern stylistic or iconographical precedents for Michel's work, or which investigate him as a harbinger of later artistic movements, Michel's practice reveals a historical dynamic of its own, in which nebulous and often dangerous evocations of atmosphere played a crucial role. This dynamic comes into play through Sensier's encounter with Michel, although such issues are also immanent in the works themselves. Michel's work and its interpretation are eloquent, when read together, about the freighted processes of recall and forgetting that attend the study of "minor" artists in the postrevolutionary period. Additionally, in Michel's ostensibly quite homogeneous corpus, which brings together discourses of environment, urban transformation, technology, and social upheaval, the newly mediatized world of early nineteenth-century Paris can be glimpsed. Rather than standing alone, a "savage" proto-modern extracted from a primordial

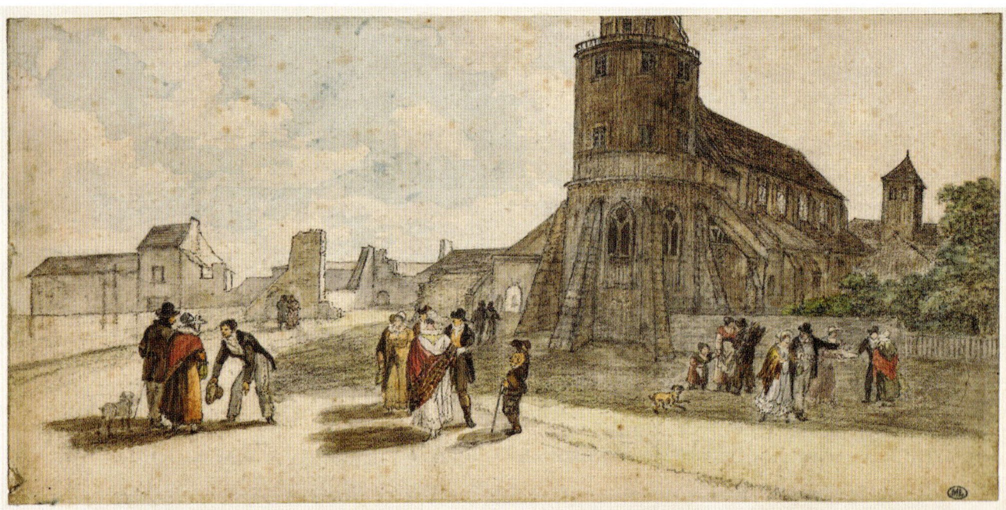

Figure 4.19
Georges Michel, *Vue de l'église Saint-Pierre de Montmartre*, early nineteenth century, watercolor, pen, brown ink, and black chalk, 14.2 × 29.2 cm. Musée du Louvre, Paris.

past, Michel might be better understood as an artist who laid bare diverse forms of mediation current in France just after the Revolution and, moreover, as an artist whose work and its interpretation offered insight into the political history of media in the period from the Revolution to his posthumous reclamation in the early years of the Third Republic.

Proposing an "infrastructuralist" epistemology of media, John Durham Peters has observed how "the history of media is the history of the productive impossibility of capturing what exists. . . . Clouds that vanish yield some of our most beautiful paintings, and clouds that obscure give us some of our most precious meteorological knowledge."[98] "Above all," Peters notes, "media capture and fail to capture time, whose fleetingness is the most beautiful and difficult of all natural facts."[99] Marville's studies of clouds over Paris, made at midcentury in the immediate aftermath of the optical telegraph, illustrate well this exquisite ambivalence, although photography had no absolute privilege in this regard, and painters such as Michel anticipated the ambition of these studies (fig. 4.20). Thinking about the material properties of media or artworks can only take us so far, if we remain blind to those phenomena—often mundane or trivial—that fail to register, that resist apprehension, whose temporality transcends that of a given work, or whose history slips in and out of view. Such phenomena included clouds, and sometimes telegraphs too.

Despairing and Gesticulating in the Wind

If Michel was known for anything, it was for his clouds. Michel specialized in the representation of bleak, windswept heaths, often beneath an incoming storm. Such subject matter was hardly unusual in the nineteenth century, but his intensely gloomy and turbulent skies were a topic of concentrated artistic investment. Michel was a painter of atmospheres, of meteorological drama considered from a panoramic distance as well as up close. In some works, such as *The Gathering Storm (Landscape with Quarry and Figures)*, a gauzy mist drifts in.[100] In paintings such as *Landscape in a Thunderstorm*, fulminating clouds loom menacingly (fig. 4.21), while elsewhere, inkblot skies meet country roads, ragged semirural suburban sprawl, or industrial production—a lime kiln, for instance—with murderous intensity (fig. 4.22). The best of Michel's skies have a centripetal force that seems to suck all the air out of the picture,

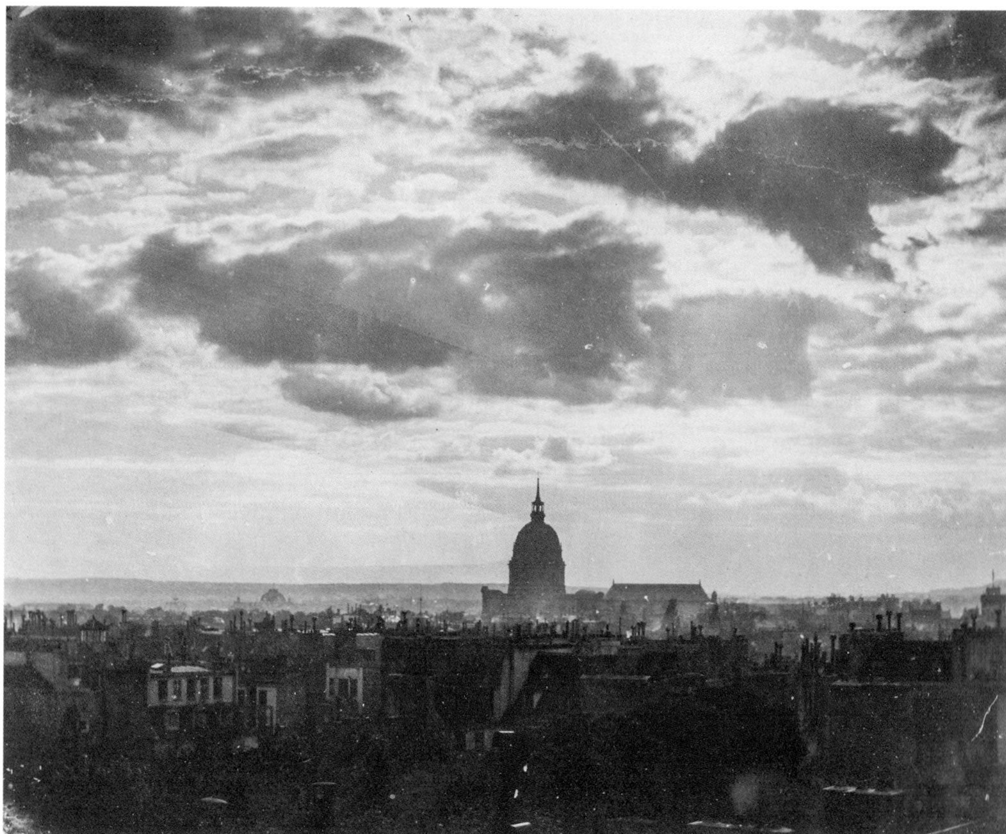

Figure 4.20
Charles Marville, *Cloud Study over Paris*, 1856–1857, albumen silver print, 16.7 × 20.6 cm. Metropolitan Museum of Art, New York.

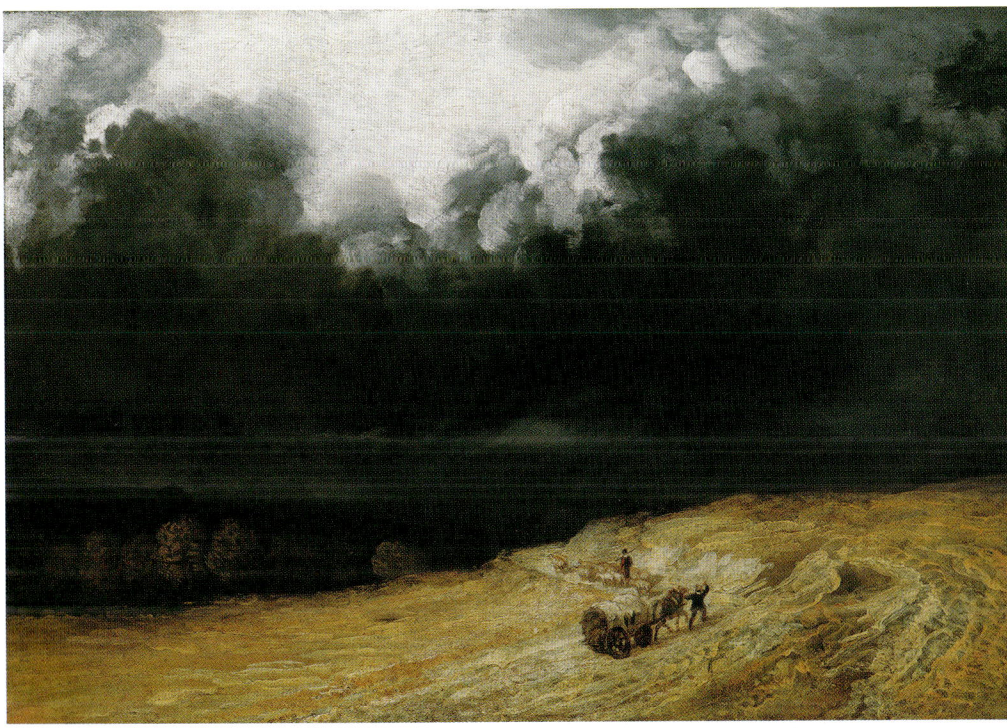

Figure 4.21
Georges Michel, *Landscape in a Thunderstorm*, 1830–1840, oil on paper mounted on wood, 51.5 × 73 cm. Belvedere, Vienna.

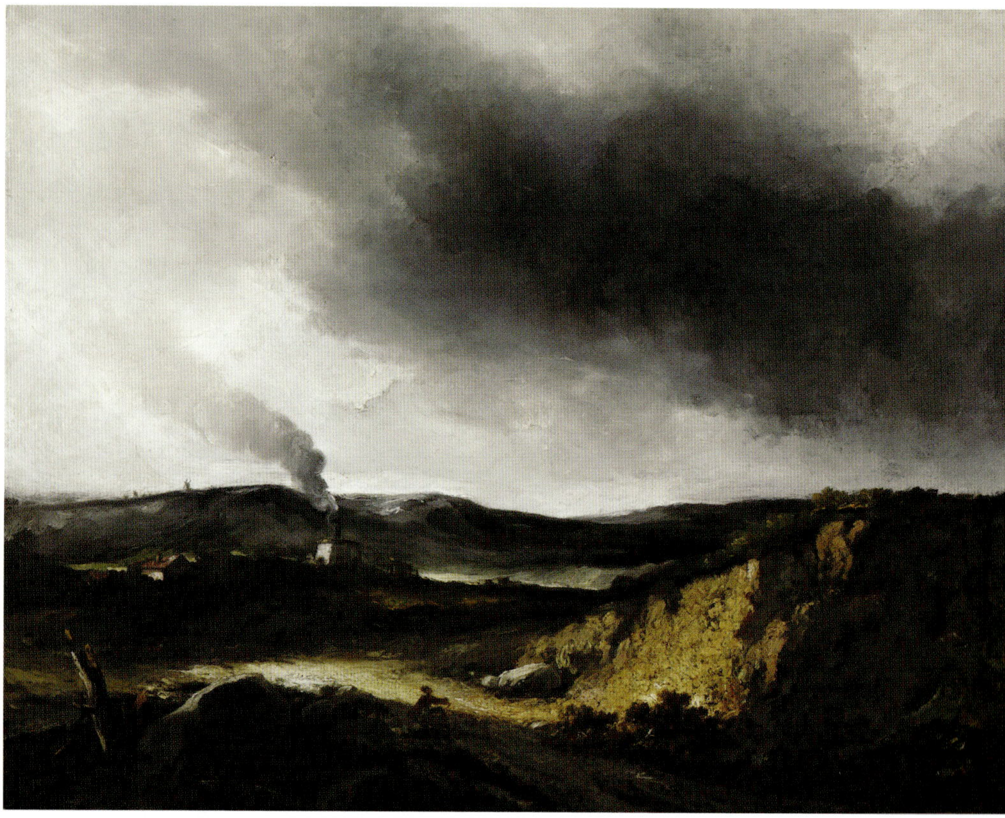

Figure 4.22
Georges Michel, *The Lime Kiln*, c. 1790–1810, oil on canvas, 55 × 70 cm. National Galleries of Scotland.

registering as an acute change in pressure such as that accompanying an impending tempest. Indeed, this is an art that foregrounds not only gathering disruption, but a state of suspended, imminent transformation: the storm is usually right on top of us. Michel's humble protagonists and we as viewers are held in a state of grace before the skies collapse. Perhaps this is partly what Sensier had in mind when making a claim for Michel's antediluvian qualities, as his art seems to bear witness to an incoming deluge and its accompanying destruction. And so, it goes without saying that time is from the outset a central feature of Michel's paintings. Calibrated in seconds, Michel's storm clouds provided an intelligible if polysemic language with which to comprehend the passing of epochs: the imperceptible progress of geological change beneath the landscape, as well as more abrupt human interventions, and on its surface, the transient mechanisms of the political order, as one by one regimes ended and others took their place.

Michel's works betray an acute attention to the specific tonalities and temporalities of different atmospheric formations. In a converse maneuver, they also exploit a Romantic preoccupation with clouds as an index of subjective experience, and mobilize key aspects of the sublime. While in this respect Michel's work is representative of broader aesthetic concerns, to think of it solely in terms of its sympathy with such categories obscures aspects of his practice which had distinctive resonance in his Parisian postrevolutionary milieu. Although the word "atmosphere" had been current for many years, its meaning as ambience, mood, or mental environment in fact appears for the first time during the French Revolution: in 1797, in John Foster's journal, and earlier in French, in 1793, in Antoine Barnave's indictment of the "false atmosphere" present in the 1791 National Assembly.[101] Etymology, of course, can only take us so far. All the same, it is notable that Michel's work gained currency at the exact point when "atmosphere" became a matter for meticulous if occasionally expressionistic scientific method, in the work of Lamarck and others, as the infinite variety of clouds were processed via observation to fit existing categories of genera and species, while serving at the same time new metaphorical uses as signifiers of broader, morally inflected structures of feeling. As the example of Barnave demonstrates, "atmosphere" swiftly garnered a political coloring, and then as now, it was not always positive in tone. These tensions were, as we have seen, legible in the stilled machine-engraved skies made by Conté in Egypt, which worked to generate a sense of atmosphere without clouds for assistance, and in contemporary struggles to ensure the optical telegraph's visibility in extremes of weather or pollution.

In this light, the windswept quality of Michel's landscapes deserves scrutiny. For Michel's skies were set in motion, the buffeted clouds offering tangible evidence of forces that would otherwise remain unseen. The atmospheric disturbance that characterizes many of Michel's paintings and drawings undoubtedly served the practical painterly purpose of imbuing his scenes with a sense of dynamic energy, providing the genre of landscape—which otherwise struggled to emulate the grand themes of history painting—with drama and purpose. However, the windswept landscape acquired a distinct political valence in the years after the Terror, articulating the emotional and physical destruction wrought by the Revolution. Iris Moon describes how the panoramic seascapes and barren windswept landscapes on a pair of Dihl and Guérhard porcelain vases were linked to attempts to process the revolutionary past.[102] Connecting the blustery motif of these exquisite objects to diverse and often more demotic media that emerged in the Revolution's wake—from printed silhouette imagery to Robertson's phantasmagoria shows and Carmontelle's transparencies—Moon details the interpenetration of different images and objects that became possible in the freer cultural climate of the late 1790s. The grisaille landscapes on these vases may have been painted by Demarne, who with Swebach-Desfontaines and a number of other genre painters, including Martin Drölling, worked for Dihl and Guérhard.[103] There are, consequently, robust reasons to consider such objects alongside Michel's practice, even though it was directed predominantly toward oil painting, for it evolved in the same climate during the Directory and after; we have firm evidence that Michel knew, and worked with, several of these artists. The "narrative indeterminacy" Moon diagnoses in the Dihl and Guérhard vases, which she associates with the contingency and uncertainty of the years following the Terror, is discernible in Michel's landscapes, which frequently lack clear focus or intent.[104] Yet shipwrecks and storm-tossed landscapes were not merely generic motifs; in the context of postrevolutionary France, they were also specific markers of a lived turmoil, the experience of being swept up in currents beyond one's control, and of historically distinctive intersections between media.

The context for such objects as the Dihl and Guérhard vases—which presupposed an elite audience, if not even an actively antirevolutionary one—was also different from that for Michel's paintings. Not all storms are the same, as Lamarck's research was making clear, and neither were they equal from a political or social perspective. During the Revolution, for instance, a cloudy sky pierced by light would signify the obscurantism, ignorance, or deceit of the former regime, rather than the traumatic effects of the new one.[105] As the Empire gave way to the Restoration, and in turn ceded to the July Monarchy, Michel's localized yet extreme and uneasy atmospherics became harder to read as signaling a unique political standpoint. Rather than dispensing with one or another of these meanings, they seemed to represent a confluence of them all. The sense of traumatized desolation that characterized the stormy skies of Directory-era reflections on the Terror was parlayed in Michel's paintings into a swirling free-for-all of signifiers. This was exacerbated by the fact that few of Michel's works are dated, making it hard to track a clear chronology.

Michel's fretful skies and abandoned, marginal plains do not appear to trade in the politics of aftermath or recrimination, which makes sense given his likely revolutionary politics, but they figure the movement of the wind as a cipher for broader, if vaguer, historical changes. This lack of specificity allowed observers to discern clearly in Michel's paintings their inheritances from Ruisdael and other Dutch masters. And it was by way of a technology prominent in those same Northern paintings—the windmill—that Michel was better able to register the intensity of his winds (fig. 4.23). However, to a Parisian audience viewing these works, the recurrent presence of these structures in his paintings would have returned to the scenes a particularity of place and time that was only partly clouded by their art historical associations. This time was set too by the telegraph, that "mute and ungainly interlocutor who despairs and gesticulates, in the wind."[106]

Underground Overground

A signal characteristic of Michel's work was his predilection for a particular site, just outside Paris: the hills of Montmartre, then an austere escarpment dotted with windmills and gypsum mines (used, as we have seen, for the manufacture of plaster of Paris), bordered by the abattoir of Rochechouart, and, from the mid-1790s,

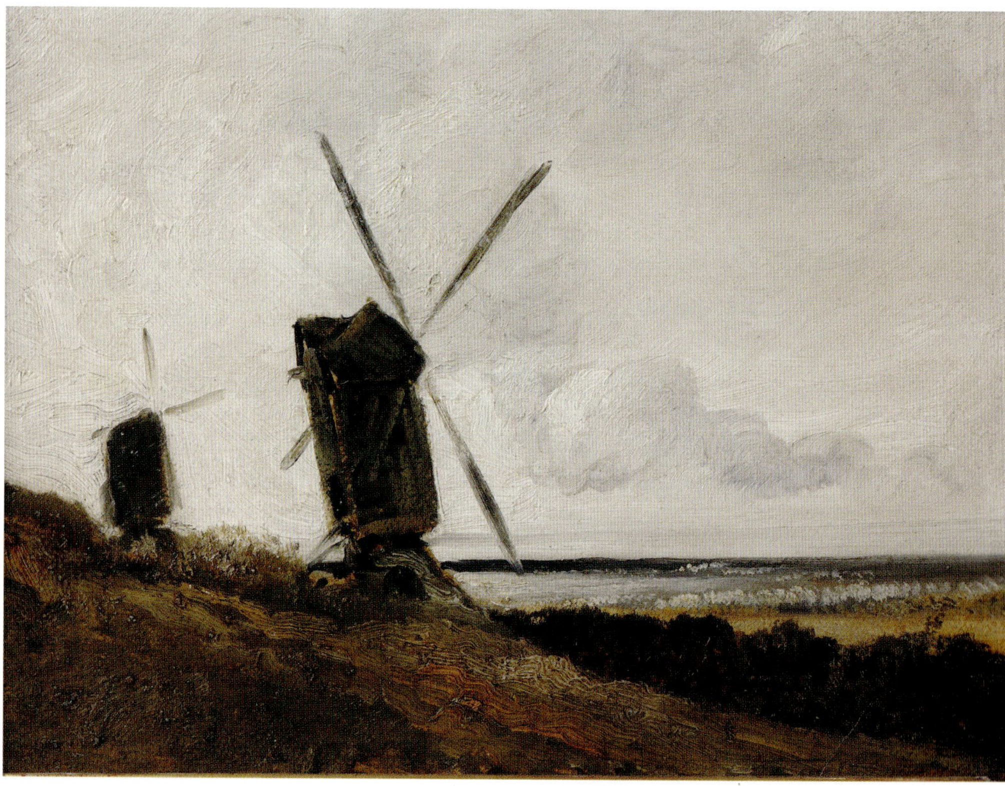

Figure 4.23
Georges Michel, *Paysage avec moulins à vent*, n.d., oil on card, 25.5 × 33.5 cm. Musée des Beaux-Arts de Strasbourg.

topped by the telegraph. This was a landscape in the process of transformation, cut up and butchered, whose protean qualities were both acknowledged and stabilized in Michel's work. Here, wrote Charles Sellier in 1893, channeling Sensier, this "poor and bizarre artist, whose success was posthumous," finally "found his Roman campagna."[107] Michel returned, again and again, to this place, which was the subject not only of finished works, but of expressive drawings, made on the spot, often on blue tobacco wrapping or papers for binding candles, where smoke, clouds, and graphite became one (fig. 4.24). Sometimes this sooty, particulate sensibility found other, equally self-referential, outlets: the 1842 sale of Michel's work and studio materials, conducted the year before his death, included "a snuffbox belonging to M. Michel adorned with seven reverse paintings on glass, of genre subjects painted by him."[108]

Montmartre was brought officially into the territory of Paris in 1790, as part of the National Constituent Assembly's reorganization of France's administrative districts. At the Paris Salon of 1791, only one painting was exhibited that represented this former village, a view by Lefebvre of the caverns there, which were rather oddly described as being inhabited by "Arab thieves."[109] However, in ensuing decades the area became increasingly popular among artists, and views proliferated. Well before it became associated with avant-garde circles in the later nineteenth century, from the 1820s artists—such as Bouhot—were attracted by Montmartre's panoramic views, by its picturesque buildings, and, strikingly, by its folding together of landscape and industrial production. This included the mines that perforated the district, rendering the whole area unstable and ensuring that boundaries between subterranean space, the ground upon which one walked, and the air above were never absolute.[110]

As noted, it was at Montmartre in the late 1790s that Cuvier conducted early paleontological investigations, discovering fossil remains in the quarries that led to important reflections on catastrophic extinction.[111] Surely, Sensier had such a local history in mind when he invoked Michel's archaism and his primordial qualities, and when he deployed metaphors of "exhumation" to describe his own scholarly task? Indeed, the area had been connected to the revelation of such secret knowledge for several decades by the time Sensier wrote. In 1812, Johan Meerman's *Montmartre*, a lengthy poem in Dutch, published with a French

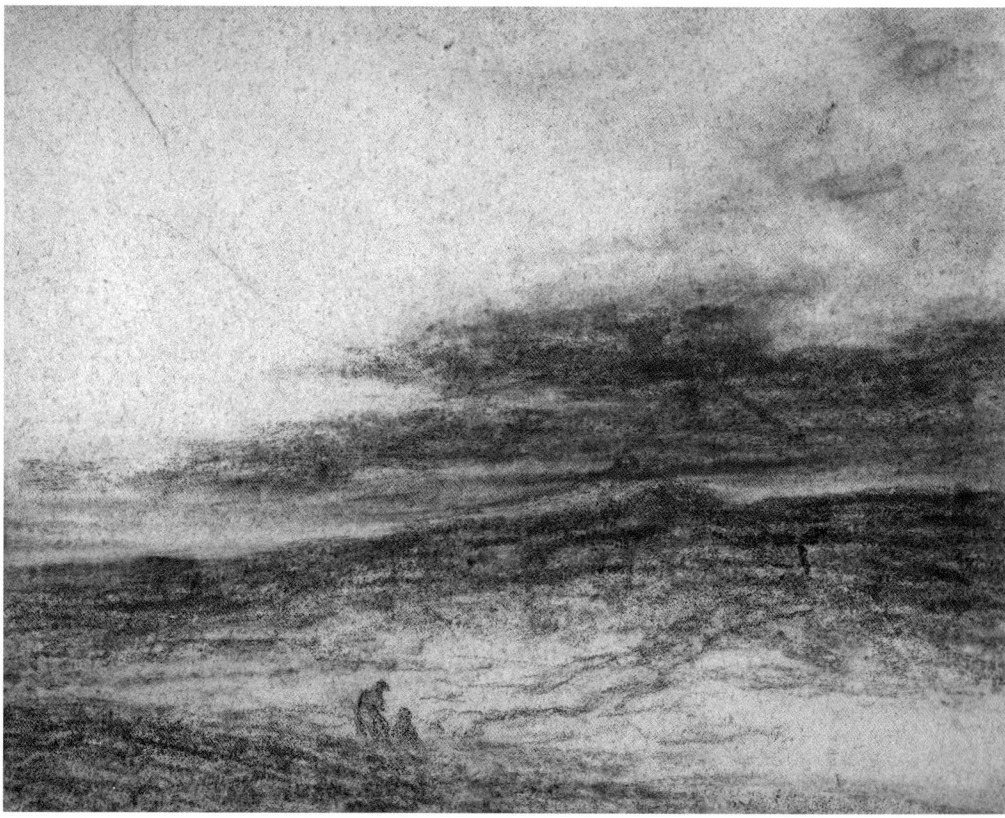

Figure 4.24
Georges Michel, *Landscape*, c. 1830, charcoal on laid, faded blue-gray, Ingres d'Arches paper, laid down on Japan paper, 41.2 × 52.2 cm. Yale University Art Gallery.

translation, read the site as an emblematic locale for the "thundering storms" that had transformed French politics, religion, and society.[112] However, Meerman began his poem by eulogizing Montmartre as a "vast museum, filled with fragments of the mineral and animal kingdoms." Drawing together the atmospheric and the subterranean, Meerman's elemental history exploited a dramatic contrast between unimaginably distant life forms and the marvels of modern science. At Montmartre, "Fire, water, or the steam that sighs after expansion, make the earth tremble . . . and the remains of a thousand species of animals, accumulated here without number, are the crushed debris of a million skeletons that Nature under construction deprived, as if by a lightning strike, of breath and movement."[113] As Meerman concluded, "few animals existing are made of such bones; but a primitive world presents these ones again to the genius and contemplation of the CUVIERS of our century."[114]

Michel stalked this area. Sensier's vivid account drew heavily on the testimony of Michel's second wife, Anne-Marie-Charlotte Claudier-Vallier, whom he married in 1828, one year after the death of his first wife, Marguerite Gros, and eight years after the death of the last of his eight children (Michel, like Sensier, had overwhelming reasons not to forget the dead). Claudier-Vallier was herself presented as a kind of apparition, having been discovered by chance in 1849 living in obscurity with her daughter on the rue Breda. Her account, relayed in the second chapter of Sensier's biography, began as follows: "Never tired, he made journeys that the famous runners before the Revolution could never have done, and he said of his legs that they were like good steel: the older they got, the more they were tempered by the fresh air."[115]

Claudier-Vallier described a typical day in the Michel household later in his career: Michel would rise at seven each day and work until three, at which point he and his wife would set out from their lodging at 2, rue des Fontaines, near the Temple, and walk until evening, without fail and in all weather. Michel rarely painted from nature, but he would stop to make drawings as the pair progressed. They would go to Belleville, to the bois de Romainville, to the plaine Saint-Denis, to Saint-Ouen, to the buttes Chaumont, to Pantin, to Bondy, and even to Vincennes. Return journeys, however, were always made via Montmartre.[116] Such was his fascination with his local environment that Michel would travel, always on foot (he had no time for carriages), to sketch inauspicious sites such as the rubbish dumps at Pantin or the emptying of a reservoir, but it was from Montmartre's combination of the sublime and the mundane that he drew his greatest inspiration.

Many figures in Michel's landscapes seem to function as avatars of his perambulations. Often, they face away from the viewer, hunkered down against the storm or dodging the lightning strike invoked by Meerman. Always, however, they are on the way somewhere, following paths both timeworn and new through the landscape. These routes could change in their meanings: in Year III, the boulevard Montmartre was renamed Mont-Marat, in honor of a more recent martyr than those from which the area had originally taken its name (Marat had, supposedly, taken refuge there in December 1789).[117] Only infrequently are Michel's figures personified fully. More usually, they appear as stock characters, although in a category of more finished drawings, they reach a more complete realization (see fig. 4.19). Such drawings largely focus on the boulevards and streets of Paris and are more formal in execution than Michel's rough landscape sketches. Michel was preoccupied here with perspective, which as Ger Luijten observes, he never quite mastered.[118] But individuation of figures was not really the point. Far more significant was that these characters were, like Michel himself, travelers through a shared landscape. Michel's paintings and drawings show us spaces that are intricately networked, crisscrossed above and below ground by circuits and meandering tracks. In this sense, they are informational spaces, densely coded with knowledge both deeply personal and local, as well as national or universal. The clouds and winds that fulminate above are bearers of information too, although they also threaten to impede any clear message. With all this in mind, Michel's attention to the most spectacular means by which information might be passed from and through Montmartre is striking.

Windmills and telegraphs seem to stand at opposite ends of a history of technology, one pointing to the past and the other to the future, one grounded in place and dealing with transformations of matter and the other suggesting dematerialized and invisible flows. Yet although the windmills at Montmartre in fact hindered the clear operation of the telegraph, the visual connection between the two technologies, both of which involved rotating arms atop a tower, was pronounced (acknowledged, as we have seen, by Daumier in *Le Moulin du télégraphe*). Some of the first witnesses to the telegraph had confused it with a windmill, while fifty years later, an anonymous correspondent described how "the telegraph has, in its gait, all the sadness and bad grace of a windmill. It is all elbows and angles; it wants to work well, because every good citizen must serve his country, but it seizes the slightest pretext to withdraw and return to its rest." For this author, the optical telegraph's principal failure—to be remedied by electricity—was its sensitivity to atmospheric conditions: "Has the sun gone down? Good evening. The telegraph has the principles of the honest man who goes to bed early to get up late. He seems to fear colds at the slightest fog, the slightest rain, the slightest snow, he puts his head under his wing and falls asleep."[119] There was a longer history here, as an early attempt to develop a telegraph system in France, by Guillaume Amontons in 1690, proposed using windmills with characters attached to the wings, and indeed, Chappe did use some windmills to support his apparatus, albeit only as base structures, with the wings removed.[120] Devices powered by the wind, and which registered its force and movement, were connected irrevocably to those which passed information through the air.

While Michel's steel legs pounded the trails of Montmartre above the fossilized bones of countless animals, the metal limbs of the telegraph moved quietly above, the sinister messenger of unknown dangers, of victory, invasion, or intrigue. At Montmartre, through which the Paris Meridian line also passed, multiple calibrations of time were at play simultaneously.[121] The deep time of the geological past being exposed by Cuvier contrasted with the accelerating technological modernity above; one expanded time almost beyond comprehension while the other compressed it. Everyday life at Montmartre, which moved to the rhythm of biological time, offered a counter to both. The Montmartre telegraph appears in several drawings by Michel, seen from different distances and viewpoints, its wings folded and inactive, and occasionally caught in use, a scratchy "Z" against the sky (fig. 4.25).[122] A larger *Vue à Montmartre* shows it silhouetted against the horizon, as workers, families, and animals make their passage down the hill in the foreground (fig. 4.26). Some of these may be workers who, during the Revolution, defended themselves against complaints from residents of Paris about delays in completing the roads constructed across Montmartre.[123] The telegraph here joins the other

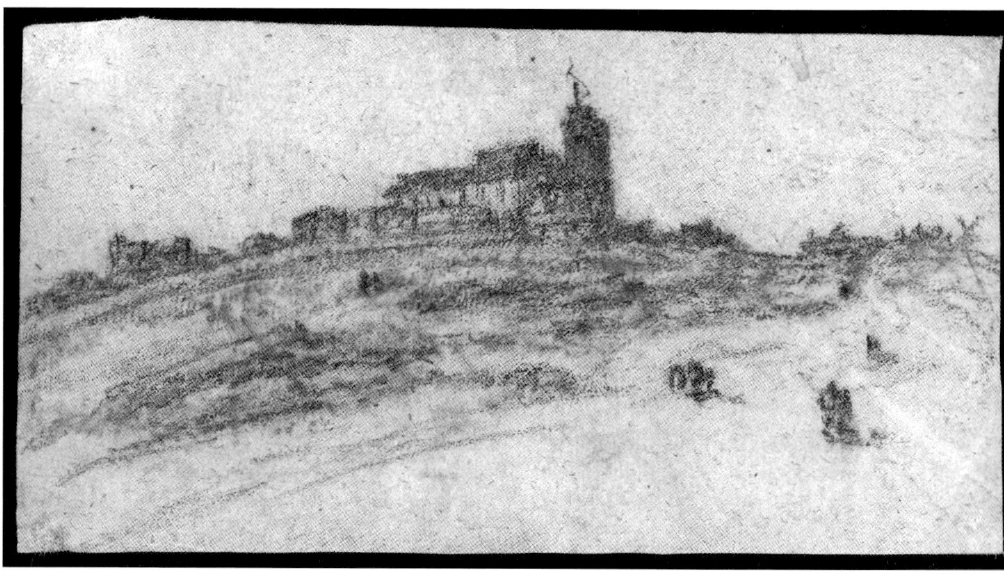

Figure 4.25
Georges Michel, *View of the Butte Montmartre with the Church of Saint-Pierre*, n.d., black chalk and stumping on beige paper, 90 × 169 cm. Fondation Custodia, Collection Frits Lugt, Paris

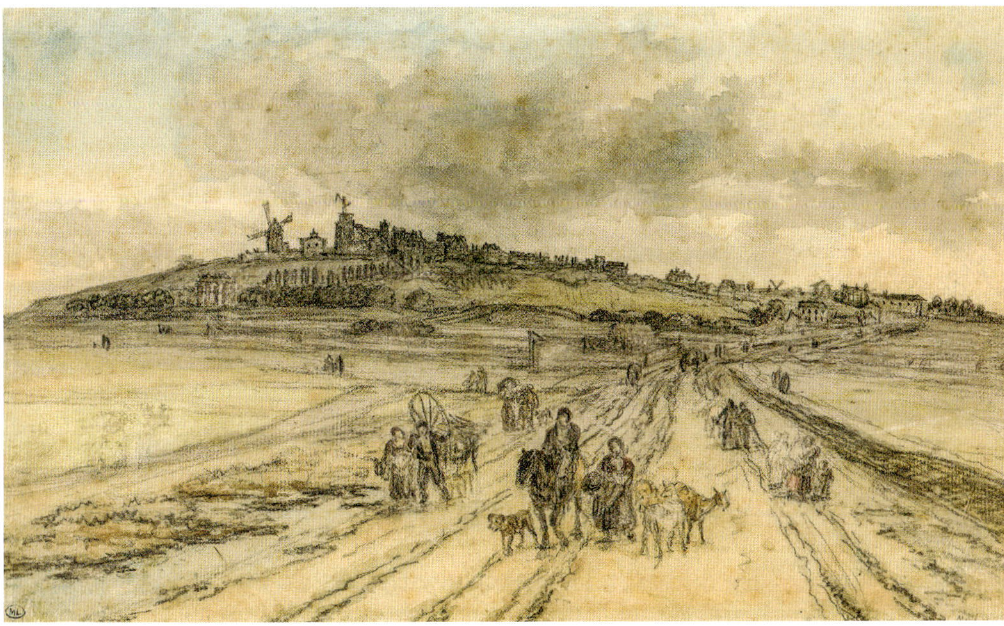

Figure 4.26
Georges Michel, *Vue de Montmartre*, n.d., black chalk and watercolor, 18.1 × 29.8 cm. Musée du Louvre, Paris.

networks that lattice Michel's paintings and drawings, as part of a layered archaeology of media, simultaneously bound together and torn apart by his stormy skies. In time, it was returned to by one of Michel's lauded Barbizon successors, Jean-François Millet, who made a black chalk drawing of a derelict telegraph tower on the Plain of Chailly, although by then the machine itself had vanished and the tower was little more than a ruin (fig. 4.27).[124]

Servants and Parrots

Sensier's text, geared as it was toward the sale of Michel's work, is revelatory of the ways in which postrevolutionary French art reentered the market in the late nineteenth century. Michel was a malleable product, as his work did not seem explicitly political in character, and his long career confused attempts to identify him closely with a particular regime. By 1888, Lucien Rhéal, writing in the satirical newspaper *Le Tintamarre*, could report on the exploitation of amateurs by an unscrupulous dealer, who rigged the sale of a painting by Michel, "whose canvases frequently sold for fifty francs, ten years ago, and cost five thousand francs today."[125] Two years later, Étienne Arago, recounting the acquisition by the Musée du Luxembourg of two paintings by Michel, followed Sensier to the letter, in describing this belated acknowledgment as "reparation posthumously made, without blind admiration, to this poor child of Paris, to this special painter of Montmartre and his surroundings, an unsung talent, almost unknown in his lifetime, and yet a virile strange brush, from his grounds to his skies, and considered today the first milestone planted solidly on the road by the English and French landscapists of the naturalist school."[126] Despite such rhetoric lauding Michel's singularity and verve, there were many artists working in a naturalist vein in postrevolutionary Paris who, as Steven Adams has pointed out, were not "subjected to Sensier and Durand-Ruel's publicity machine" and who consequently were never "exhumed."[127]

Sensier's artistic biographies were certainly projections of the author's sensibilities and ideological outlook.[128] His writing, as Parsons and McWilliam recount, articulated a position in relation to contemporary French society, whereby "rural France came to symbolize an unchanging society preserving the values of an organic community which had been sacrificed to the atomized environment and decadence of urban life."[129] Sensier's accounts

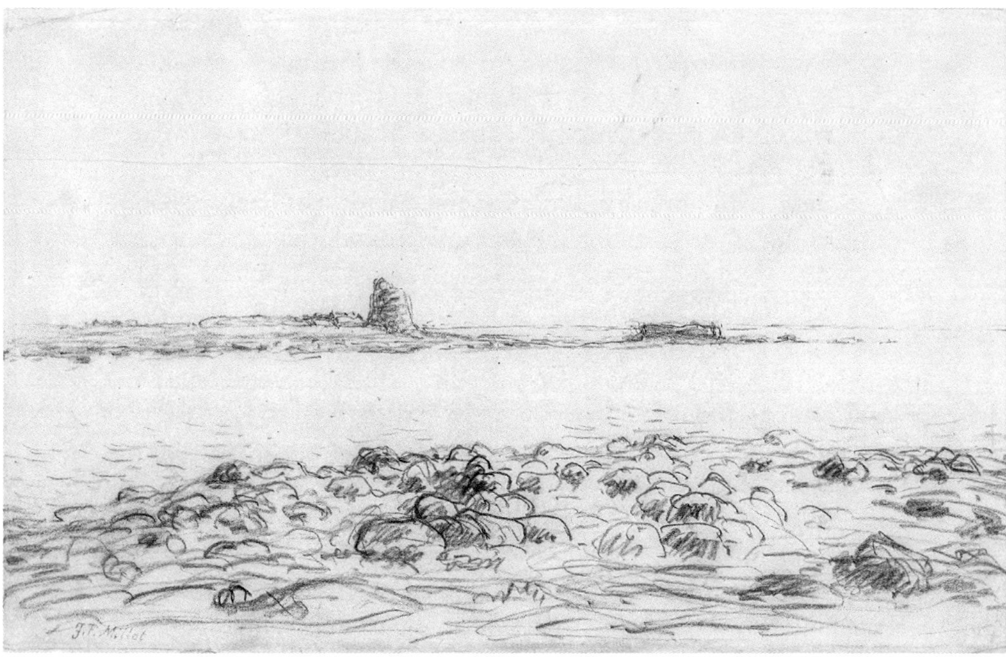

Figure 4.27
Jean-François Millet, *The Plain of Chailly with a Ruined Telegraph Tower*, n.d., black chalk on paper, 19.05 × 36.2 cm. The Morgan Library and Museum, New York.

of Rousseau and Jules Breton are replete with anxieties about modernization, yet his essentially conservative politics did not preclude discussion of artists such as Millet, who offered a less flattering view of the French peasantry.[130] The peasantry occupied a central role in Sensier's critical thought, affirming an unchanging vision of France unblemished by values of rationalism or social and political revolution, and grounded in a fantasy of a sacrificial moral order that stood in stark contrast to the capricious and violent urban crowd.

In this mythic present, technology and industrialization were destructive forces. An animosity to the city linked Michel not only to more recent figures such as Millet but also to appropriated forebears such as Claude and Poussin, for Sensier thought Michel excelled as an artist only when he removed himself from the "corrupting atmosphere" of Paris. In his Michel book, Sensier gave full vent not only to his distaste for industrialization but to "the depredations of the Commune and the Franco-Prussian war, the evils of absinthe and moral laxity, the uniformity of suburban expansion and the blight of factory building."[131] All of these, Parsons and McWilliam continue, "are associated with images of corruption, decay and death. Paris itself becomes a leprous growth, transforming the remnants of unspoilt nature which surround it into a wasteland of mud, brick and human degradation."[132] Here, we might understand Michel as a kind of time traveler in Sensier's worldview, a figure retrieved from the past to affirm values that he sought to identify in the present. Michel was, in Claudier-Vallier's testimony, an artist who represented the modern world in front of him, telegraph and all, but by 1873 Sensier had parlayed this version of the artist into a sense of loss, a yearning for a utopian past to which one might aspire to return. In this sense, Sensier, quite despite himself, offers us a version of Michel as a romantic anticapitalist.[133]

This is also how Michel—the "peintre excentrique"—makes sense as part of a tradition of outsiders, by virtue of his alleged self-taught status (an incorrect assumption) and his separation from official academic culture. Michel was ranked alongside artists such as Simon-Mathurin Lantara and Lazare Bruandet as part of a French tradition of amateur practice that blurred the boundaries between them and their peasant subjects.[134] These "tutelar deities" were united by their perceived ability to avoid the dictates of the market—Sensier approvingly quotes Thoré's description of Michel as an irremediably unfashionable painter in the *Constitutionnel* of November 25, 1846—and by an individualism praised as a source of unique authenticity.[135] It is striking that these claims for the power of Michel's unfashionability, the recoding of his untimeliness as timelessness, were made of an artist whose career evolved at a time when fashion itself became a particularly complex barometer of political meaning. To place Michel out of time in this way not only neglected the milieu within which his art was made and displayed, but also obscured—in plain sight—the extent to which the work is itself all about time: from the fossilized remnants embedded in the hills of Montmartre to the constantly changing permutations of the clouds and the flickering now-time of political affairs heralded by the telegraph signals passing overhead.

This brings us, finally, to a central paradox in Sensier's admiration for this artist, given the writer's antipathy toward revolutionary politics and the urban working class in general. For, according to Claudier-Vallier, Michel had been highly sympathetic to the French Revolution. She described how, learning of the convocation of the Estates General, Michel returned to Paris from a spell teaching drawing in Switzerland, arriving a few days before the storming of the Bastille, and participated in its liberation. "From that point on," Claudier-Vallier recounted, "he became acquainted with the fashionable patriots, frequented the Jacobin club, and became a great admirer of Mirabeau, Pétion, Robespierre, Camille Desmoulins, and other famous speakers of the Revolution."[136] During the Revolution, Michel lived on rue Neuve-de-l'Égalité (now part of rue d'Aboukir), located in the Bonne-Nouvelle section, a sansculotte stronghold. After Thermidor, his militant Jacobinism led to his incarceration at Le Plessis prison for four and a half months in 1795. Michel supported Gracchus Babeuf, and there is evidence that he subscribed to radical publications and maintained his revolutionary sympathies for decades.[137] Later, with the arrival of the 1830 Revolution, Michel "thought he was dying with joy." As recounted by Claudier-Vallier and transcribed by Sensier, Michel believed that "it was not enough to hunt kings," but one must also "get rid of their servants and parrots." However, the events of July 1830 disillusioned him, and he discovered that all republicans were not, like him, "men of courage, probity, and good intentions."[138] Throwing himself back into his painting, he said no more about politics in the remaining thirteen years of his life. Were the telegraphs in Michel's work—which shared his inscrutability—also drawn and painted with their revolutionary origins in mind, or with the feeling that their political purpose had been subsequently corrupted?

Sensier presumably did not exaggerate Michel's radical politics, given that they seemed so antithetical to his own beliefs. Michel's alleged renunciation of republicanism no doubt helped in this regard, and admitting the artist's earlier revolutionary sympathies may have been merely an *effet de réel* that lent authenticity to Sensier's text, operating as further evidence for Michel's untamed nature. Yet, even if this were the case, Michel the revolutionary still seems a strange subject for Sensier's pen.

There is, however, another testimony that might be brought into play. The publication of Sensier's book on Michel coincided with the writer's retirement in 1873, following a bureaucratic career as Chef de Bureau des Musées and as Chef du Secrétariat for the Ministry of the Interior from 1850, positions he held while amassing his own large collection, dealing, writing, and pursuing his art world contacts. Much of Sensier's initial collection had, however, been built from that of his father, Jean-Baptiste-Théodore Sensier, a notary who rose to Commissaire Général during the July Monarchy.[139] Born in 1779, Sensier *père* had, like Michel, witnessed the storming of the Bastille, which made a lasting and profound impact on him. Later, he recalled the terrible spectacle of the condemned being led from the Conciergerie to the scaffold: from his window on the quai d'Horloge, Jean-Baptiste-Théodore witnessed the liquidation of the Girondins, as well as the executions of Mme Roland, Mme du Barry, Charlotte Corday, Marie-Antoinette, and finally Robespierre and other members of the National Convention in Thermidor Year II.[140] He also attended, and participated in, patriotic festivals, and began his career as a clerk in the mid-1790s corresponding with Thermidorian luminaries such as Santerre, Fréron, and Tallien. Later, in 1830, he was one of the *commissaires* entrusted by Lafayette with the task of collecting and publishing the official account of the July Revolution and totaling the number of dead.[141] Like his son, Jean-Baptiste-Théodore could not forget them.

By then, as the baron de Girardot observed in an August 1863 issue of Gabriel Charavay's *L'Amateur d'autographes*, the elder Sensier was acknowledged as one of the foremost bibliophiles of the Empire and Restoration, having amassed a formidable collection of books, medals, and autographs.[142] Many of these betrayed his fascination with the revolutionary political process and its material culture, and he accumulated many objects from the period. Paul Mantz, in his foreword to Sensier's posthumously published book on Millet, remarks on the son's fascination with revolutionary history in the wake of 1848 and his desire to trace the history of this period back to the National Constituent Assembly and National Convention in the 1790s.[143] Consequently, Alfred Sensier's collecting and critical practices—both of which in due course included his "production" of Michel—were bound, in highly personal ways, to his inheritance of a postrevolutionary reckoning with the radical past that was not, in fact, however hard Sensier might have tried to repress it, wholly negative in character. Like the optical telegraph that populated Michel's works—defunct when Sensier wrote—this history signaled to the present from the past, in a kind of generational relay. It is difficult not to think that Sensier's sympathetic reading of a figure such as Michel was made possible by his family history. Yet it was also bound to its moment. What, in fact, was Sensier writing about when he wrote about Michel *when* he did?

Dead Men's Souls

In the first of two lectures delivered at the London Institution in February 1884 under the title "The Storm-Cloud of the Nineteenth Century," John Ruskin recalled a dismal light encountered one morning in Matlock, Derbyshire, in midsummer 1871. This was not normal weather, but something more dreadful, more pestilent, "For the sky is covered with gray cloud;—not rain-cloud, but a dry black veil, which no ray of sunshine can pierce."[144] This, Ruskin, recounted, was a "plague-wind," characterized by darkness, intermittence, tremulousness, and malignancy. There was, he admitted, a scientific explanation, for the evil cloud comprised poisonous smoke emanating from the two hundred furnace chimneys compressed in the surrounding two square miles. But such rationalization could not account for the wild movement of the plague-wind. Rather, as Ruskin noted:

> It looks more to me as if it were made of dead men's souls—such of them as are not gone yet where they have to go, and may be flitting hither and thither, doubting, themselves, of the fittest place for them.

> You know, if there *are* such things as souls, and if ever any of them haunt places where they have been hurt, there must be many about us, just now, displeased enough![145]

Immediately after transcribing this unusual comment, Ruskin proceeded to reflect, from the contemporary moment, on his 1871 observations: "The last sentence refers of course to the battles of the Franco-German campaign, which was especially horrible to me, in its digging, as the Germans should have known, a moat flooded with waters of death between the two nations for a century to come."[146]

While Ruskin, looking back fourteen years later, in dialogue with his earlier thoughts, is clear about the consternation the Franco-Prussian War caused him, it is notable that the events that followed it go unmentioned. For if one were to look for ghosts, the Paris Commune of March to May 1871 would be a fine place to start. Its presence here is implicit, neutralized in Ruskin's recollection as a matter of European discord rather than civil war. Although he elsewhere decried the Commune as a "Kakotopia," Ruskin had some sympathy for its communitarian ideals, which he pursued via his Guild of St George.[147] His intuition in 1871 that the plague-wind of atmospheric pollution might harbor phantasms of political sacrifice demonstrated that the most scientifically minded analyses of cloud formations were sure to be accompanied by more ineffable effects, and that the movement of winds forced reflection on historical process. One of Camille Corot's 1871 drawings on lithographic transfer paper, made while he was in exile from Paris during the Commune, and published the following year, registers something of this ambiguity, and seems to channel the spirit of Michel in its depiction of a windblown traveler (fig. 4.28). That year, the coming storm depicted in Michel's paintings raged and raged.

The outbreak of the Commune was signaled by the people's capture of cannon at Montmartre, reenacting almost play by play a scene on July 15, 1789, when the cannon of Paris had been brought to Montmartre for safekeeping to prevent them being taken by the king's troops. And here, the specter that haunts Sensier's text rattles its chains. For it is truly striking that Michel's compulsive returns to the hills of Montmartre were reimagined by Sensier in an immediate postrevolutionary climate, in 1873, only two years after the traumatic repression of the Commune, with ruined buildings still standing and the dead populating the social and political

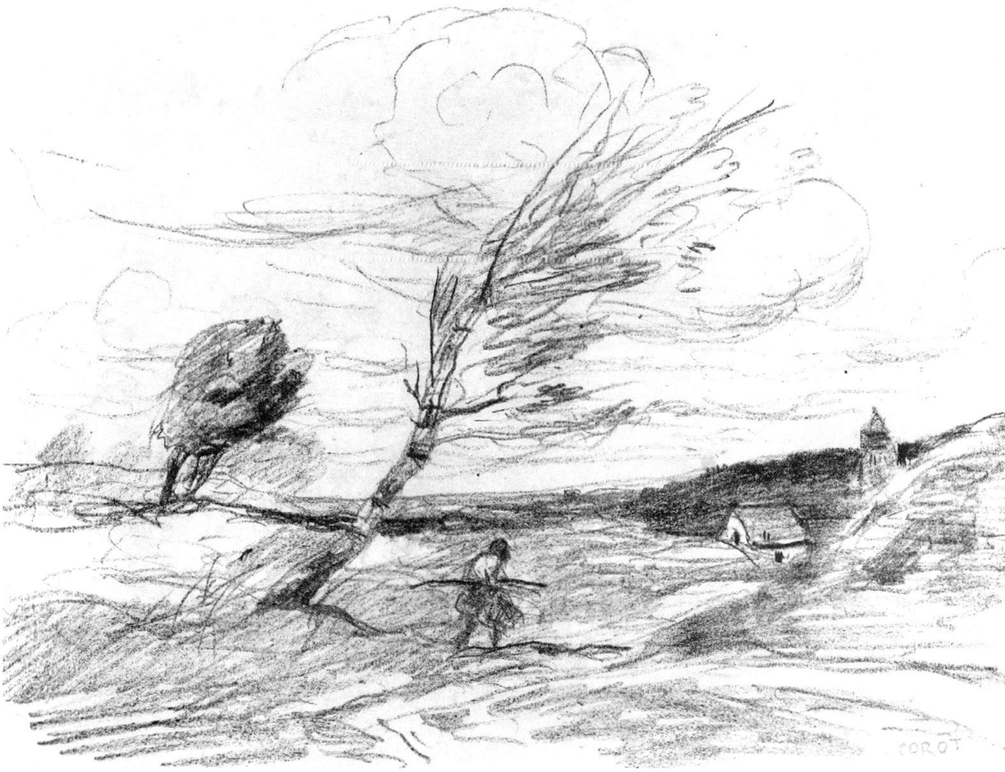

Figure 4.28
Camille Corot, *The Gust of Wind*, from *Douze croquis et dessins originaux sur papier autographique par Corot*, 1871, transfer lithograph, 21.8 × 28 cm. The Metropolitan Museum of Art, New York.

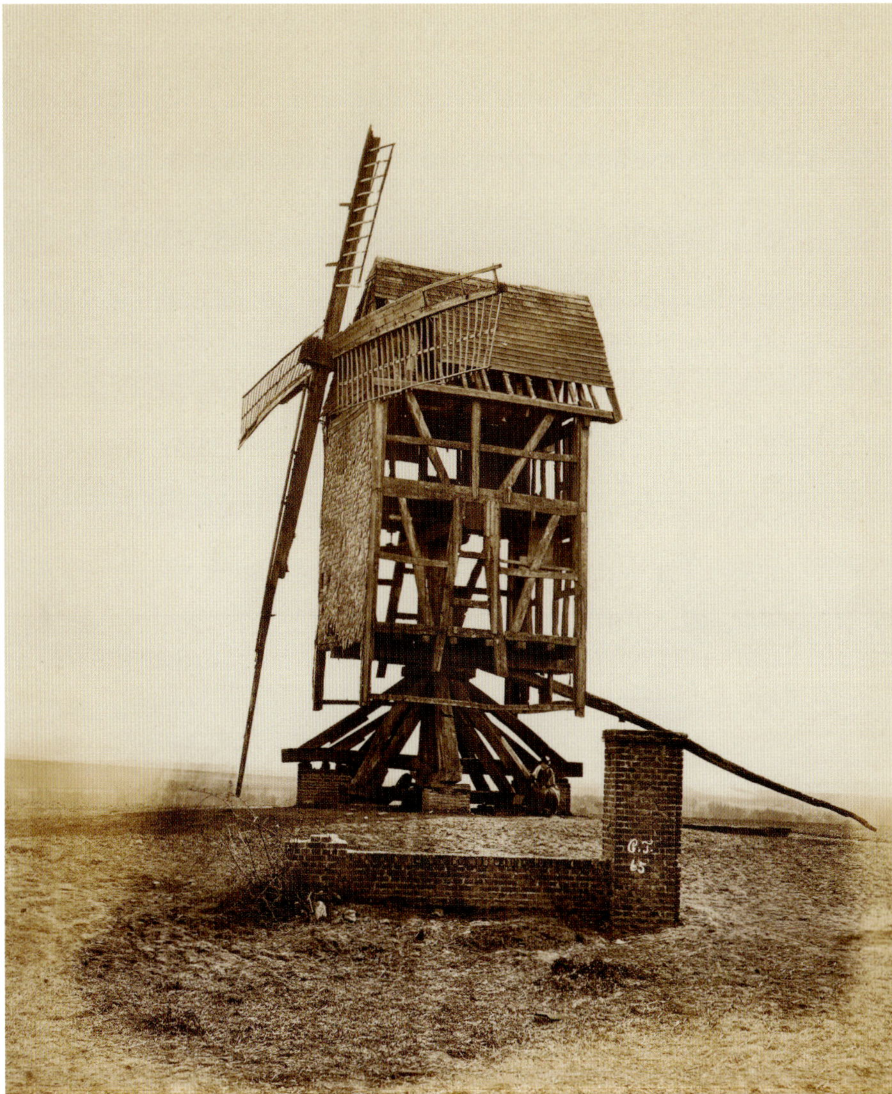

Figure 4.29
C. Jung, *View of a Windmill and a Soldier after the Paris Commune of 1871, St. Quentin [?]*, 1871, albumen silver print, 21.4 × 17.2 cm. Canadian Centre for Architecture, Montreal.

imaginary (fig. 4.29). Michel's radical politics were not a matter of passing relevance here, at a time when the forgetting of the Commune was a matter of official policy, and neither was his attention to informational routes into and out of the capital. Rather, in Sensier's account, Michel's artistic persona and his turbulent skies operate as the connective tissue that links the Revolution of 1789 to that of 1871, allowing Sensier to process his own conflicted relationship to the revolutionary past and its present iterations, as well as to diverse forms of postrevolutionary settlement.

The dead that Sensier urged his readers not to forget—the revenant *petit-maître* artists of the Revolution, Empire, and Restoration—were, then, closer than might be imagined to the dead men's souls conjured by Ruskin, flitting hither and thither in the atmosphere, and to the bodies of slaughtered Communards hastily buried in mass graves around Paris. Sensier recalled Michel for the present, yet by accentuating his archaic qualities, he held the artist's political convictions and his work's close connection to a uniquely fraught period in French history firmly at a safe distance. Attempts to bring back an artist such as Georges Michel, to collapse historical time, or to reanimate his sensibility in the living, could never—*must* never—be fully realized, and he remained, for Sensier, as for many subsequent art historians, the object of an irrevocable, unnamable loss, obscured by a dark cloud of unknowing.

During the Commune, feminist Paule Minck operated a free school at Montmartre in the choir that had earlier served as base of operations for the telegraph above. By century's end, however, Saint-Pierre de Montmartre—one of Paris's oldest churches—was dwarfed by the rising basilica of Sacré Coeur, a conservative Christian monument to the Franco-Prussian War and a repudiation of the Commune.[148] Its lofty parvis now the site of a popular "natural" panorama of Paris for the thousands of tourists who visit each year, we can read another telegraphic history of nineteenth-century France through such politically layered spaces. In the half-century or more that followed the French Revolution, Swebach-Desfontaines, Bouhot, and Michel all saw something interesting in the telegraph, and chose to depict it. They, like it, were later understood as outmoded or anachronistic. Their

Parisian careers all took form beneath it, and only Bouhot outlived the optical system, dying in 1862, although by then he had long departed the city. These are just three: there were other artists who responded similarly, although each one would tell a different story of French art in the age of telegraphy. Rather than seeing these artists solely in relation to the telegraph, a longer examination of their careers as they developed around and beneath the machine shows how it moved in and out of view, to different effect, and at different times, in their work, exposing new capacities of both genre painting and landscape in the process. Much of the time, of course, they ignored it: no one *just* painted telegraphs. But by taking surreptitious glances at the telegraph, staring it down, or turning away, these artists—and their interpreters too—found ways to think about their changing panoramic city: to orient themselves in relation to its fabric, structures, inhabitants, and ecologies, and above all its place in French history. The gimlet eye of the telegraph looked back—who knows what it was thinking.

5 *When I Was a Telegrapher*

Discussing his adopted city of Paris, where he lived from 1830 until his death in 1837, German writer Ludwig Börne described it as "the telegraph of the past, the microscope of the present, and the telescope of the future."[1] Börne's technological analogy is both familiar, consonant with the infrastructural comprehension of the capital's long history we encountered in the previous chapter, and at the same time oddly discordant. City-as-microscope makes a certain amount of sense as a vivid descriptor for a world becoming accustomed to new forms of spectacular realism, even if microscopy itself was hardly a concept or technique specific to the nineteenth century. Telescope-town also works, as the predictive powers of an urban milieu that drove not only industry but also novelties in art and fashion align with our expectations about the modern city's future-oriented *telos*. But "the telegraph of the past"? Why would telegraphy look backward, and what kind of past might it communicate?

While microscopes and telescopes clearly operate by visual means, telegraphy resonates today as a technology grounded in a turn away from representation, a document of the modern world's gradual drift toward immaterial, virtual presence. Viewed in retrospect, it seems to mark a victory of writing over image-making, visual culture abandoned by the speed with which information could now be transmitted and by the inability of the telegraph to convert a visual field into code.[2] However, Börne's allusion was not to electrical telegraphy, which had not yet been implemented successfully, but to the Chappe optical system. Via association with the telegraph, the city figures in Börne's neat one-liner as a site where the ghost of the 1789 Revolution might reappear in the guise of 1830. The longevity of the Chappe telegraph—still current through the July Monarchy—meant that it was, arguably, able to offer a more convincing metaphor for historical thought than the technological and social caesura later suggested for electrical telegraphy. But perhaps Börne had an eye to the fact that the optical telegraph would within the decade start to be replaced, because well before the retirement of the system, many rival systems attempted to supersede it. As the decades passed and the optical telegraph ceded to electrical machines, Börne's telegraphic comparison retained its relevance, and the metropolis remained a scene for the reanimation of specters of all kinds.

Honoré de Balzac, as recounted by Nadar, believed that all physical bodies were composed of ghostly layers.[3] Balzac never shed his fear of the daguerreotype, for he thought that each photograph would strip away one of these layers from the photographed subject, which would pass through the ether and fix to the surface of the image instead. The telegraph, on the other hand, prompted fewer qualms, although its workings were likewise otherworldly and obscure. In *Les Marana*, first published in 1834, Balzac was attentive to how the sensorium of optical telegraphy incorporated both frantic chatter and uncanny silences; divorced from the machine itself, his telegraph metaphor permeated social life. Balzac found his telegraph in the perplexing operations of Parisian gossip ("all those ears hear, all those tongues say, all those minds know"), from which nothing could be hidden: "Do not ask what mysterious telegraph it was which conveyed to him in the twinkling of an eye, at any hour, in any place, that story, that bit of news, that scandal; do not ask what prompts him. That telegraph is a social mystery; no observer can report its effects."[4] Balzac gave the example of the assassination of the duc de Berry at the Opéra in February 1820, news of which had reached the Île de la Cité just ten minutes after the crime. In Balzac's version, the "social mystery" of this telegraph—as everyday as gossip, but bewildering in its secret machinations and discomfiting in its shocks—exposed everything to the light and at the same time exceeded any visual record.

The "sun telegraph," as Richard Cobb called it—a technically incorrect but intuitively photographic description—had a prolonged and complex relationship with its visual history, and with other media, old and new.[5] The demise of the Chappe system was as fuzzily open-ended as its beginning, and for a few years electrical telegraphs replicated in miniature the semaphoric code of the Chappe telegraph, seemingly unable to commit to the obsolescence of a familiar technology and unwilling to embrace fully the potential of electric telegraphy as a form of communication with devices and visual vocabularies of its own. The wonderfully sci-fi Pouget-Maisonneuve receiver, all whirring parts, glass domes, and cranking arms, was one of several to domesticate the telegraph in this way, combining all the components needed to operate a telegraph station in a single device (fig. 5.1). In the interests of visibility, a much earlier generation of telegraphers had experimented with placing the arms of the Chappe machine flat against a white wooden screen, rather than rotating them atop a pole (so-called *télégraphes à cadran*). However, such versions of the device had retained the telegraph's original scale, as well as its mechanical animation. By encasing the device in ornately turned wood and placing the little telegraph arms behind a glass screen, later electrical models such as this one, or the 1848 Cacheleux system, which used an electrical pointer to indicate static Chappe signs on a circular dial, required only the focused attention of their operator, and their effect was reduced to their primary use as transmitters of code.

Given its innovation in the field with the Chappe machine, France was a relative latecomer to electrical telegraphy, despite long-standing experimentation with electrical systems (the work of François Arago, André-Marie Ampère, and Antoine César Becquerel, among others).[6] The achievement of Charles Wheatstone, Carl August von Steinheil, and Samuel Morse had not been matched in France, in large part because the optical network was so successful and so well established, with 534 stations already operating across the country. Under pressure from the rail companies, an experimental electrical line was installed in 1842 from Saint-Lazare station to Versailles, and the inaugural Rouen line followed three years later. Yet the long and embedded history of telegraphy was one that needed to be contended with in France as nowhere else. Pouget-Maisonneuve's rather literal remediation

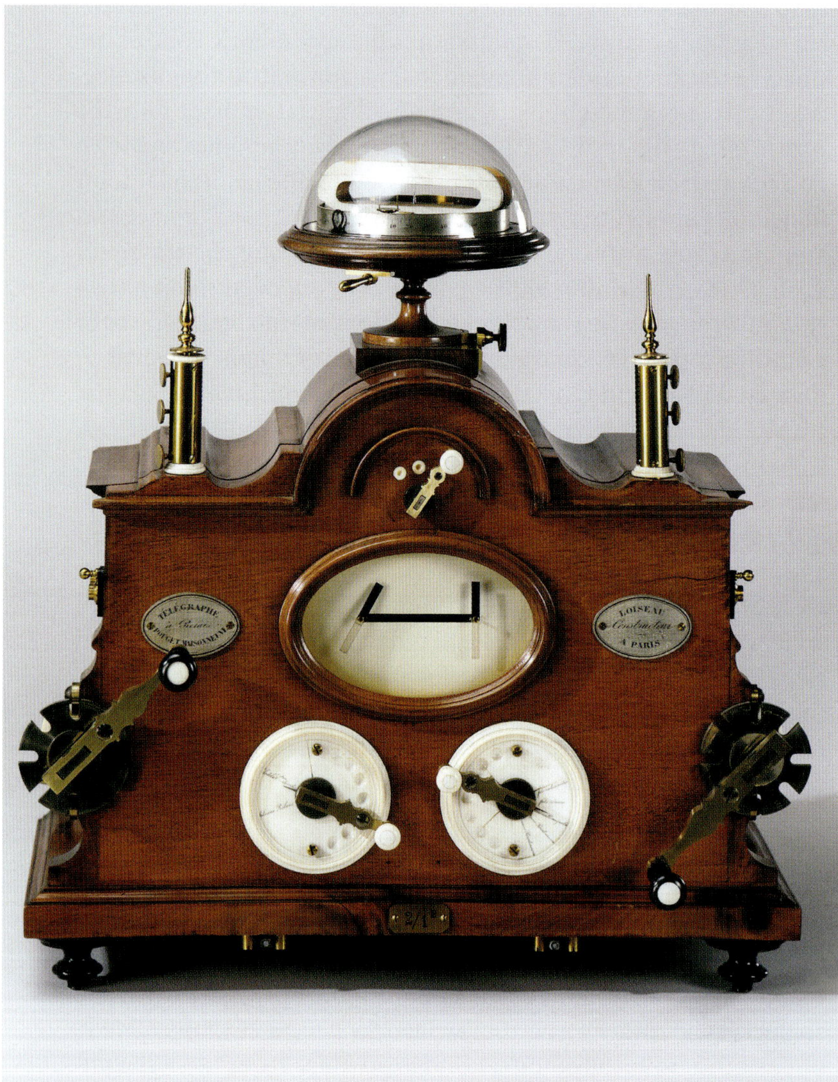

Figure 5.1
Louis François Achille Loiseau and Louis Alexandre Pouget-Maisonneuve, *Poste télégraphique type français complet à indicateurs de signaux Chappe système Pouget-Maisonneuve*, 1853, steel, wood, copper, ivory, glass, and iron, 45 × 42 × 27 cm. Musée des Arts et Métiers, Paris.

of the Chappe machine was but one direction in which the technology traveled.

It is to the long afterlife of optical telegraphy that this chapter turns, less to trace a linear technical history characterized by patterns of evolution and decay, rupture and regress, than to suggest that visuality continued to inflect the subject of telegraphy in France after the 1850s and to draw out some of the ways in which telegraphy provided a means of conceptualizing the historical meaning of diverse media, particularly in their ability to materialize images at a distance. The time of telegraphy did not only look forward, but demanded, as Börne intimated, and as the work of artists discussed in the previous chapter demonstrated, a reckoning with history. As the century progressed, the emergence of other media with an ability to conjure the past—most notably photography, in its various forms—did not initiate a decline in telegraphic metaphors: rather, it gave them new life.

Sic Transit!

In *Paris, ses organes, ses fonctions et sa vie dans la seconde moitié du XIXe siècle* (1875), writer and photographer Maxime du Camp provided for his readers a meticulous anatomy of a singular institution: the *bureau central* of communication on the rue de Grenelle-Saint-Germain in Paris.[7] In this "palace of electricity," the noise of Morse and Hughes telegraph machines created a constant, repetitive din, which, combined with the austere barracks-like interior, made for an overwhelming and punishing environment.[8] Yet the fierce modernity of the space and the contemporaneity of its operations existed alongside a trail of obsolete technological artifacts and quietly moldering archives of earlier correspondence. In addition, the building itself, constructed with an older, pre-electric form of communication in mind—Chappe's telegraph—accentuated the troubling presence of the outmoded. At the *bureau central*, a tower at one end of the building had accommodated the Chappe apparatus for ten years, from the time the rue de Grenelle had replaced the Hôtel de Villeroy as the center of the network in 1831 until 1841 (fig. 5.2). By the time du Camp wrote, optical telegraphy had been replaced wholly by electrical signals. Yet, despite disappearing from view, the Chappe telegraph continued, in du Camp's reading, to haunt the future path that telegraphic communication might

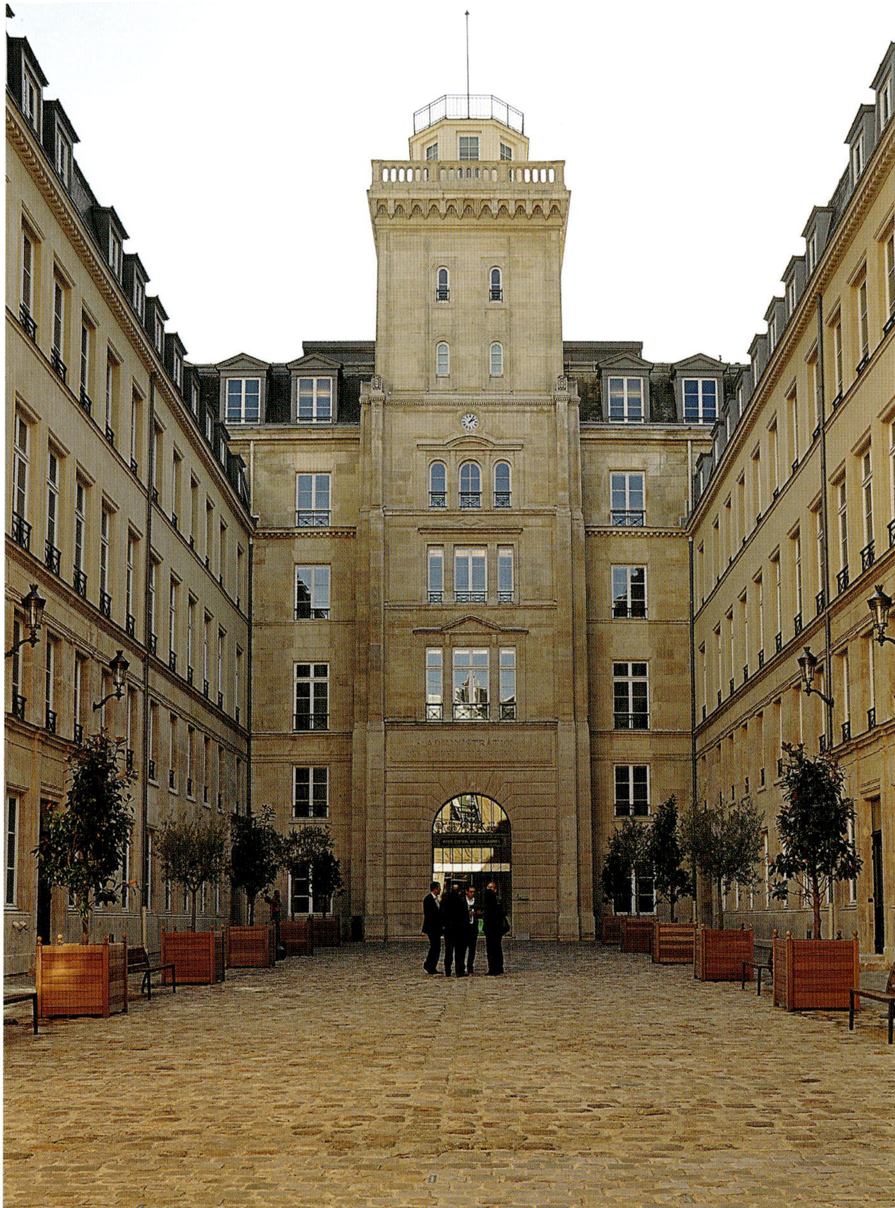

Figure 5.2
103, rue de Grenelle, Paris, 75007. The Chappe telegraph tower is visible at the top of the building.

take, and the traces of its hardware and administration could still be found at the *bureau central*. Although in its lifetime the Chappe telegraph had increasingly appeared a symbol of state repression, the utopian possibility of telegraphy's earliest incarnation, as a means of universal communication between citizens, seemed to remain in the 1870s as a reproof to its current formulation.[9] Napoleon III's opening of the telegraph system to the public, pursuant to a law of 1850, had offered a route to realizing this promise, although telegraphy's commercial operation produced new shortcomings of its own.

"Eccentrically" located far from the Bourse, the avenues, the Interior Ministry, and the Tuileries, the *bureau central* in the 1870s was, du Camp claimed, both "shameful" and "absurd," a fact reflected in the haphazard interior arrangement of the building, with the current telegraph rooms situated impractically on the second floor.[10] As du Camp put it:

> This old fortress of telegraphy is stripped of her splendor; she makes one think involuntarily of those medieval castles on which we put wings and which became mills. We removed the Chappe machines that gestured toward the four cardinal points; we took away the telescopes that searched the horizon; the employees climb two hundred stairs, grumbling increasingly, and in the post office, where all the news of France and the world ended up, we stacked cartons, old registers, piles of paper; the mice walk about in peace, spiders spin their webs unconstrained: *Sic transit!* The central cubicle has become an attic.[11]

So much for communication's inexorable forward march. Instead, technological progress has reconfigured the present as a retrofitted Middle Ages, burdened now by a jumble of cobwebbed papers. Against this image of decay and diminution du Camp set out a detailed description of the physical, technological, and social organization of the *bureau*, stressing the frantic labor that took place in its departments (fig. 5.3). Echoing Hugo's complaint, in *Le Télégraphe*, that the secretive movements of the Chappe telegraph on the twin towers of Saint Sulpice were distracting him from the task of writing, du Camp compared the sparse, atomized workspaces of the telegraphic clerks to the garret lodgings inhabited by hard-up poets. Something, however, had changed in the fifty-six

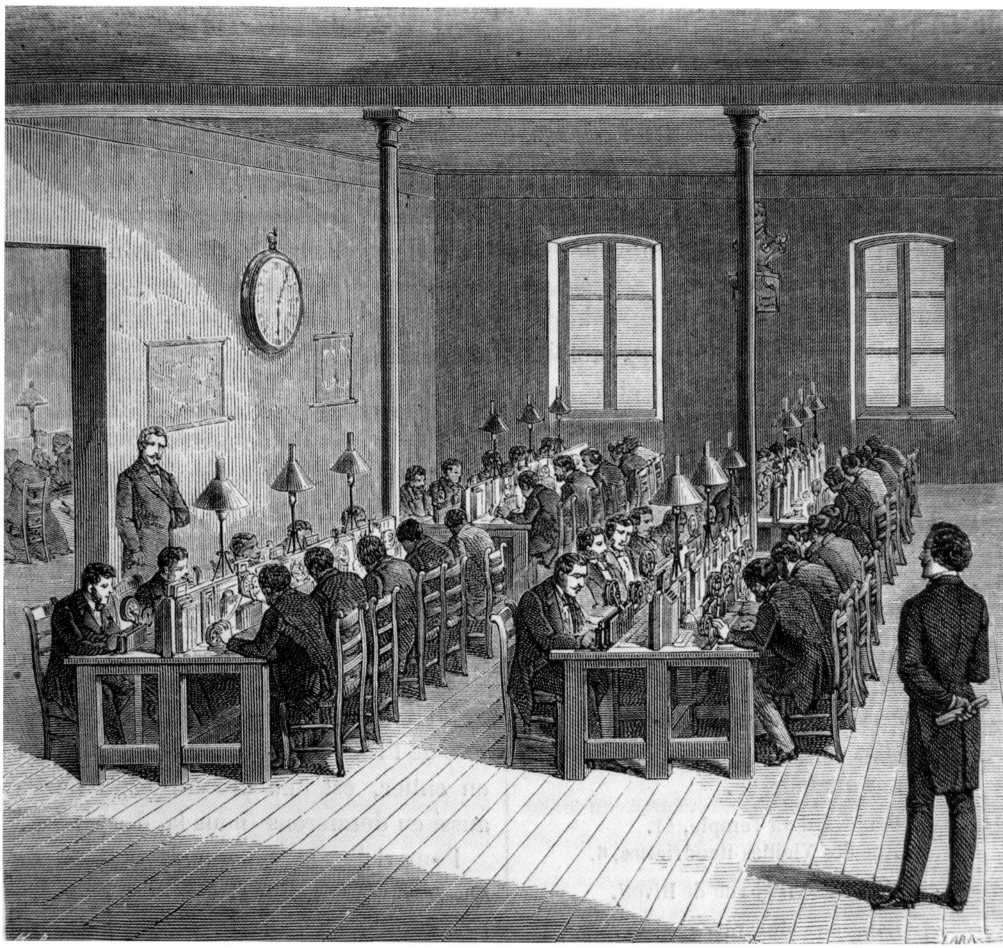

Figure 5.3
Crété et fils [printer], "Une des salles des instruments à la station centrale des télégraphes de Paris," from Louis Figuier, *Les Merveilles de la science, ou description populaire des inventions modernes*, vol. 2 (Paris: Furne, Jouvet, et Cie, 1868).

years between the publication of Hugo's and du Camp's texts. Although he suggested analogies between the young clerks' ceaseless writing and the production of literature, du Camp presented the unending labor of the clerks as an obstacle to a creative interior life: "Some of these young people, whose pale features announce their tiredness, have a book with them, hoping to read if their apparatus remains motionless for a few minutes. None of them, I am sure, could finish the paragraph started; a dispatch arrives, then another, then another, and so on and forever, and with it a job that changes with each new telegram."[12]

Du Camp's diagnosis of the exploitation and alienation at the heart of telegraphic communication was compounded by his discovery of the "illusory" pay structure employed at the *bureau*, where many of the workers, he failed to note, were women.[13] A description of the *bureau* in *L'Illustration* of September 20, 1888, is similar in tone to du Camp's, but relates that 542 men and 414 women worked there. While many images of the era show rooms wholly occupied by men, an arresting cover by Louis Sabattier for the March 20, 1909, issue of *L'Illustration* redresses this gender blindness in the context of the first general strike for all services covered by the administrative umbrella of the Postes et télégraphes (from 1921 known officially as the PTT, or Postes, Télégraphes, et Téléphones) (fig. 5.4). For the first time in the domain of communications, women played a prominent role in this strike, resisting attempts by those loyal to Clemenceau to cast them as dangerous revolutionaries. *L'Illustration*'s cover shows Julien Simyan, the under-secretary of state for the Postes et télégraphes, and Louis Lépine, a politician, lawyer, and innovator in forensic science who was at this time prefect of police, engaged in animated conversation with a representative of the striking women. Her apparatus, and a fainting comrade, are visible behind her. This unnamed woman leans in to make her point, and from Lépine's alarmed response—surprising in a noted disciplinarian known as the "little man with the big stick"—we are led to judge her appeal unfavorably, as aggression.

In its gestural polemic, the image reproduces, across the decades and in a stiffer format, aspects of the "telegraphic" humor of Gavarni, Cham, Daumier, and other lithographers of the July Monarchy, in which women's semaphore frequently signified adulterous correspondence (fig. 5.5). Gavarni's cuckolded husband falls asleep with, tellingly, his newspaper, while his wife signals to a

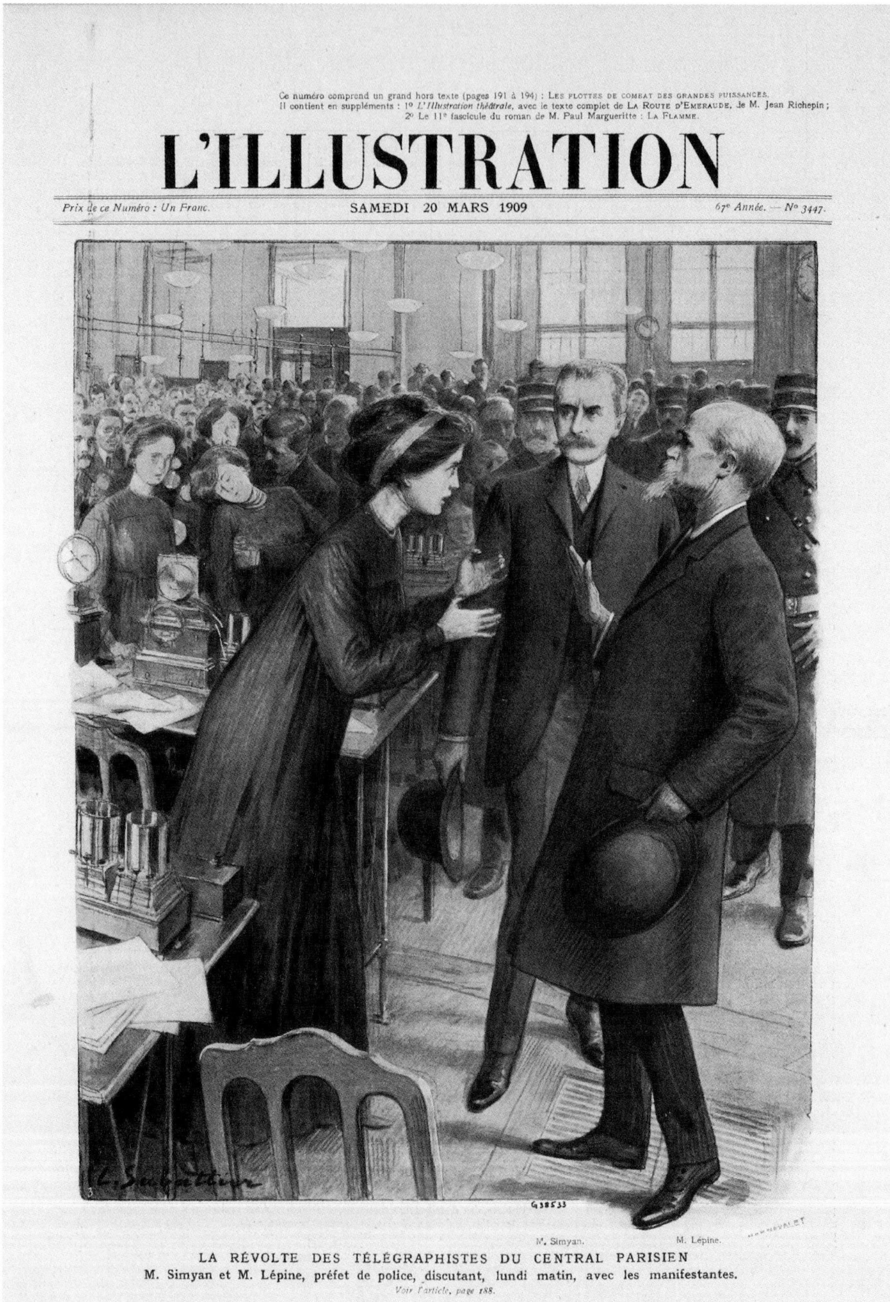

Ce numéro comprend un grand hors texte (pages 191 à 194) : Les flottes de combat des grandes puissances.
Il contient en suppléments : 1° *L'Illustration théâtrale*, avec le texte complet de La Route d'Emeraude, de M. Jean Richepin ;
2° Le 11° fascicule du roman de M. Paul Margueritte : La Flamme.

L'ILLUSTRATION

Prix de ce Numéro : Un Franc. SAMEDI 20 MARS 1909 67° Année. — N° 3447.

M. Simyan. M. Lépine.

LA RÉVOLTE DES TÉLÉGRAPHISTES DU CENTRAL PARISIEN

M. Simyan et M. Lépine, préfet de police, discutant, lundi matin, avec les manifestantes.

Voir l'article, page 188.

Figure 5.4

Louis Sabattier, *La Révolte des télégraphistes du Central Parisien*, published in *L'Illustration*, March 20, 1909, 38.7 × 27 cm. Musée Carnavalet, Paris.

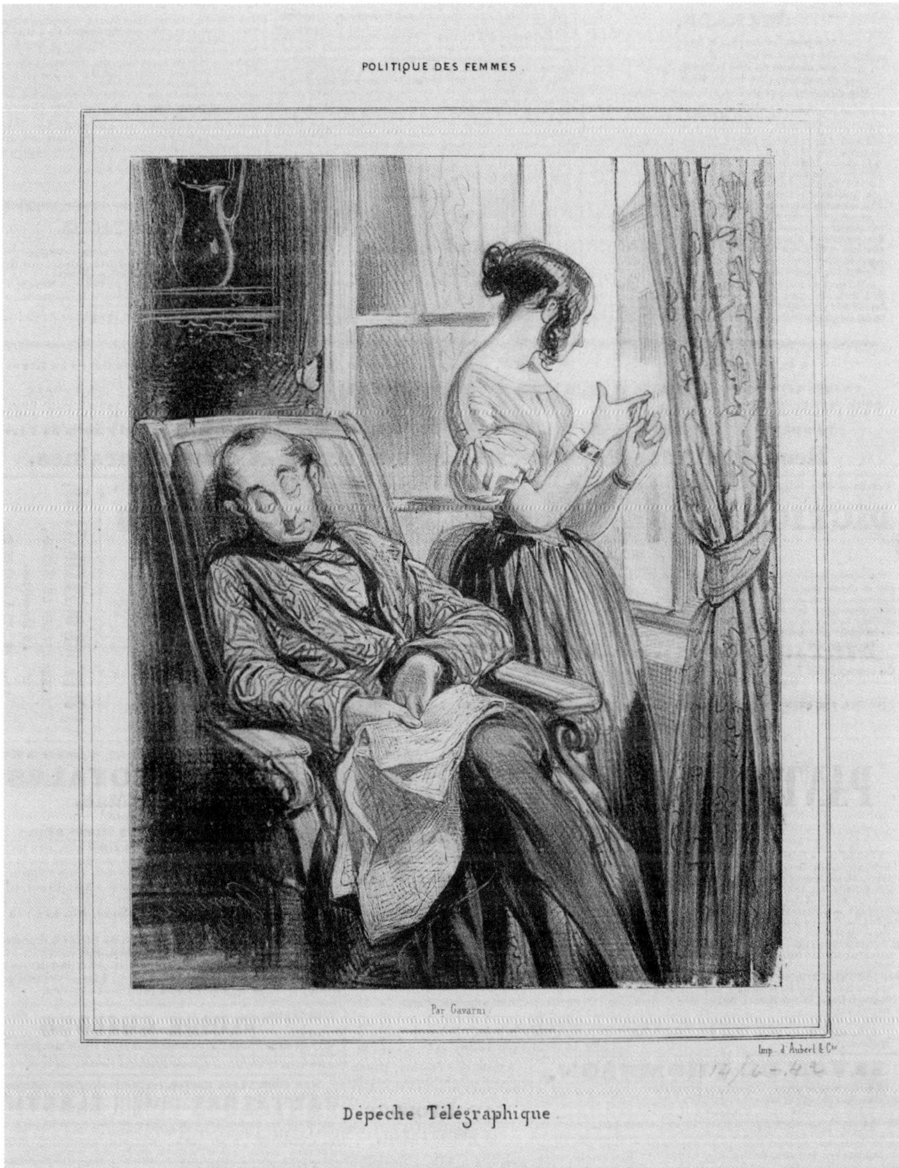

Figure 5.5
Paul Gavarni, *Dépêche télégraphique*, from *Politique des femmes*, 1839, lithograph, 20 × 15.7 cm. Harvard Art Museums.

lover beyond the window. He may read the news secondhand, but she makes it, and his world of print falls a step behind telegraphic sign language—two-timing was a matter of consequence for both dozy husbands and their media avatars. As with these earlier prints, Sabattier's cover image takes its cue from optical telegraphy's long-vanished sign systems, which still served the purpose of rendering intelligible electrical telegraphy's unseen transmissions. *L'Illustration*'s stance on the strike was negative, yet the newspaper's subtle invocation of the Chappe system's semaphore also resurrected (seemingly inadvertently, like the return of the repressed), the technology's long-held associations with exploitation and labor-induced madness—the central figure's fingers are, perhaps, deformed by long service at her machine, although they recall too the hysteric's clawing hands. Women here are presented as dissimulative—like the figure in Gavarni's print, not to be trusted—and at the same time subject to irrepressible passions they cannot conceal.

The disguised reference to the older technology made sense in the context of a strike. With electrical signals stopped, the simple efficiency of gestural semaphore was all that remained as a means of communication, but any sense that this might truly substitute for messages sent by wires was undermined by the archaic associations of the Chappe system. *L'Illustration*'s front page therefore conveyed a sense of journalistic immediacy while aligning the *télégraphistes*' withdrawal of labor with a defunct technology. The clockfaces that frame the space indicate time suspended, rather than the smooth homogeneous time of global communications.

Back in the 1870s, du Camp had observed that at first sight employment as a telegraph operator at the *bureau* appeared easy, for "to be seated on a chair in the presence of an intelligent machine which seems to function for itself, following by sight the features it draws, and to unroll a strip of paper, is all the apparent work." However, "to be well done, it requires a quick hand, a fixity of gaze, a tense mind, and often even a deployment of considerable strength."[14] Far from creating a leisurely atmosphere for the worker, technologies of transmission produced detrimental physiological and psychological effects in those who operated them, people who had, du Camp reported, largely given up on speaking to one another, but who now conversed solely in code, like the "intelligent machines" that surrounded them. At the *bureau central*:

> There is not a second of rest, all nerves are overexcited; the sheer diversity of news which follows relentlessly leads to more weariness: family matters, bank intrigues, commercial operations, political news, coded letters, English, French, Italian, Spanish, Dutch, German, arrive one after the other, like the ticking of a clock, regularly and tirelessly in the space of the same quarter hour. To this we must add the continuous noise of devices, nervous noise, staccato, almost as sour as it is dry and which, by dint of being reproduced without discontinuity, ends by undermining the most vigorous of natures. If one ever comes to write the history of diseases special to each trade, I am convinced that the electric telegraph will furnish a remarkable and quite particular share.[15]

The frustration apparent in Victor Hugo's experience of the aerial telegraph in 1819—its stubborn industry mocking his "sterile brain"—vanished in the closed interiors of the palace of electricity, where cyborg operators communicated in code, their bodies twitching to constant stimuli. The madness Hugo discerned as a by-product of telegraphy had invaded the system itself. Communication had become pathological. We can see here, perhaps, vestiges of an early harbinger of telegraphy playing themselves out: the optical telegraph, so clearly anthropomorphic, had often appeared a scaled-up, dumbed-down extension of eighteenth-century automata, with their feedback loops, programmable operations, and simulations of consciousness. Scholars of automation have noted how the prestige and significance of automata waned in the nineteenth century, as they moved from the realm of legitimate scientific inquiry to fairgrounds, shows, and other spectacles. Different iterations of telegraphy, however, seem to provide an alternative trajectory for some of the ideas contained in the dream of artificial life.[16]

The telegraph's early evocation of a machine human body, and its animation by invisible powers, also suggested that it might contain the potential to outthink or subjugate its human masters. In an intriguing image, dated 1800, a multifaceted diagram of the French telegraph's movements is overlaid with a likeness of the god Mercury (fig. 5.6). An elegant rendering of the machine's semaphoric possibilities is combined in this print with a cross-section of the cabling system as well as a detailed breakdown of

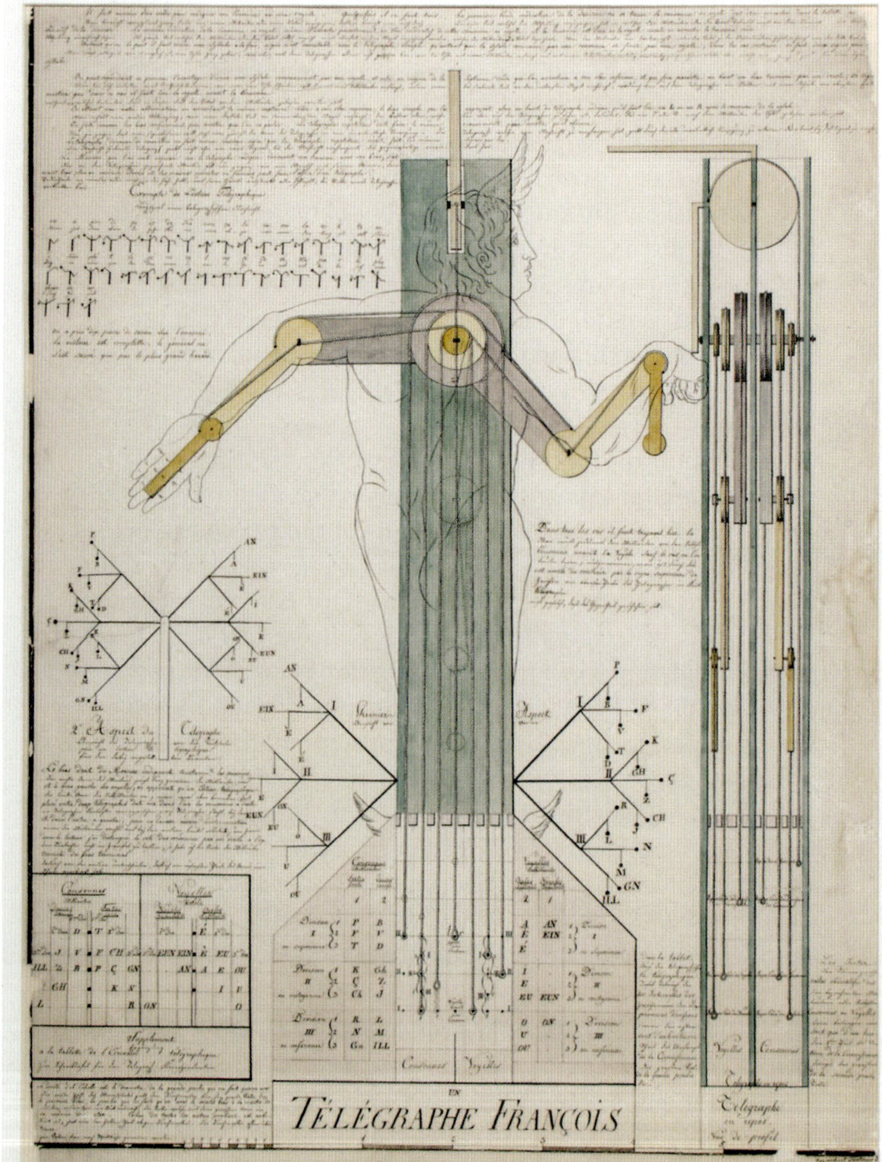

Figure 5.6
Fartempas, *Télégraphe François*, c. 1800, pen and ink on paper, 57.5 × 43.5 cm. Museumsstiftung Post und Telekommunikation.

the logic of the code. Despite this rational diagramming of the system's workings, the force of the image comes from the telegraph's doubling with Mercury's body. Here, mechanistic philosopher Julien Offray de La Mettrie's conception of the human, published clandestinely as *Man a Machine* in Amsterdam in 1747, reached full realization. "The human body is a machine which winds its own springs," wrote La Mettrie.[17] "The body," he continued, "is but a clock."[18] La Mettrie anticipated a future in which a reverse engineering process might produce marvels superseding those of Jacques de Vaucanson—a mechanized talking man was no longer to be thought impossible.[19] Further, as Sabine Arnaud describes, for La Mettrie "machine" was more than a metaphor for the body; rather, "he made the machine the point of departure for thinking about both the body *and* the mind, and understood their relationship in plastic terms as interdependent."[20] Something of this total intelligence can be glimpsed in the Mercury-Machine's sinewy and synaptic form. By the 1870s, the radicalism of eighteenth-century mechanism had turned into something closer to a technological nightmare, enacted directly on the bodies of the *bureau central*'s overwhelmed workers, rather than on classical gods. In many texts and visual images of the early nineteenth century, the waving limbs of the telegraph made an impression on casual observers, even if they did not understand its nefarious messages. However, that machine operated in plain sight and could be avoided. In contrast, the collateral effects of the 1875 telegraph were absorbed in full by those poor souls who operated the Hughes and Morse machines, whose dissatisfaction would, several decades later, come to a head with the 1909 Postes et télégraphes strike.

Perhaps to get away from the infernal, driving cacophony of the transmission room, du Camp retreated to the lower depths of the *bureau*, to the realm of mice and spiders. Here he passed a "mysterious" door, barred to the public, where, as if in one of his earlier photographs of Egyptian monuments semiconcealed by drifting sand (fig. 5.7), the remnants of a previous age were gradually revealed, obscured partially by the sediment of time's passage. If the transmission room was dominated by an accelerating, disciplining clock time, here time went backward but was no less configured by established power dynamics.[21] This is where news came that was not meant for "little people like you and me"—the deaths of emperors and kings, revolutions, abdications, peace treaties, declarations of war, assassinations, royal

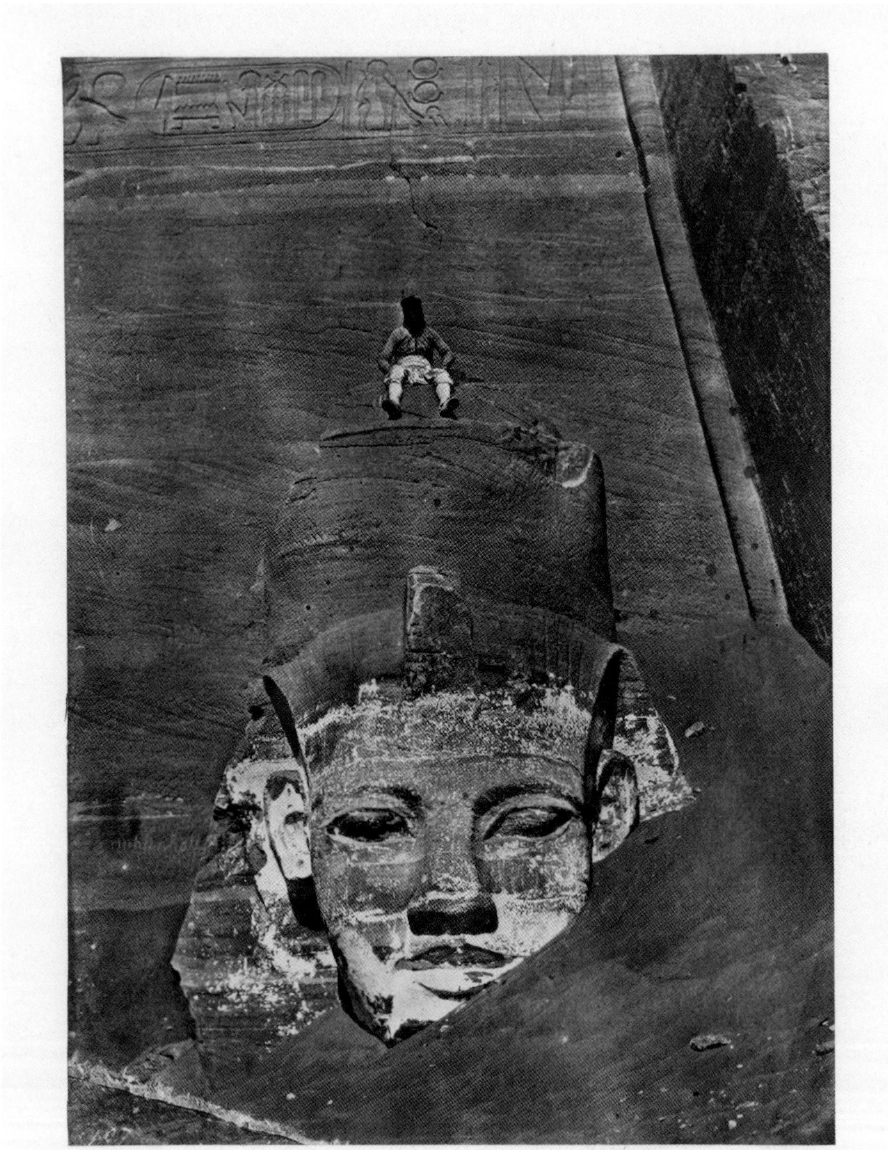

Figure 5.7
Maxime du Camp [photographer] and Louis Désiré Blanquart-Evrard [printer], *Westernmost Colossus, the Great Temple, Abu Simbel*, 1850, salted paper print, 22.7 × 16.2 cm. The J. Paul Getty Museum, Los Angeles.

marriages and princely births—information about world historical events that required extensive mediation before it could be let loose on an unsuspecting public, and which, du Camp observed, speculators at the Bourse would have killed to get their hands on.[22] This room harbored the secret information so intrinsic to the telegraphic enterprise since its systematization in the 1790s.[23] This is where fantasies of capitalist accumulation came to be realized, although only retrospectively, impotently, in the form of privileged information transmitted by a continually changing technology, situated always on the precipice of the outmoded. All that is solid melts into air. And here, in a little room on the ground floor, far away from the open, cloudless sky so necessary to the aerial forerunners for which the building was constructed, lurked four machines which for five years in the 1860s had promised to send images through time and space, but which, by the time of du Camp's investigation, had already begun their slide into disrepair and neglect.

Facsimiles and Other Apparitions

These machines were pantelegraphs. The invention of an Italian priest, Giovanni Caselli, they were registered in 1861 and brought into official service in 1863.[24] The term itself was not exactly new—it appears in 1829 in an article in *La Mode*, where it is used to describe a universal form of telegraphy—but the technology was.[25] "Everyone knows," du Camp writes, "that this device, which is electrochemical, reproduces in *facsimile* everything that one can draw on paper: a portrait drawn in pen, submitted to the influence of the machine in Lyon, will be photographed, so to speak, by the apparatus in Paris."[26] Located between the subheadings "Mystère" and "Sorcellerie" in du Camp's text, this machine was itself a hybrid. Formed from an etymological conjunction of "telegraph" and "pantograph," its name imparted a confident universality to this kind of transmission that was not supported by the range of different technologies that ran alongside it at this time, competing for a share in the market.[27] Unlike Chappe's aerial semaphore, the pantelegraph's operation did not depend on an operator discerning a visual sign and conveying it to the next outpost. Rather, it offered, for the first time, the miraculous ability to send pictures down a telegraph line. Contained in the dark rooms of the telegraphic *bureau*, the pantelegraph did not figure *as* image, but instead gave rise to images of its own making.

The pantelegraph took the form of an elegant A-shaped frame, approximately two meters in height, bisected by a heavy pendulum that hung its length (fig. 5.8). The device transmitted messages to an identical machine at the destination, to which it was connected by an electrical cable. On one side of each device a pair of curved copper plates provided a support for the transmission of dispatches, one to send and one to receive incoming messages (fig. 5.9). The user drew or wrote their message on a sheet of tin or metallized paper in nonconducting ink. Clips attached this sheet to one of the curved plates. The swinging pendulum animated a stylus that scanned the message by moving across it in a series of parallel lines, while on the other plate incoming messages were inscribed. One movement of the pendulum corresponded to the movement of one line. Extremely accurate clocks, functioning independently of the electrical current of the telegraphic cable to minimize atmospheric variation, ensured that syncopation between the two machines was perfect. Each time the stylus passed the nonconducting ink, it broke the signal, enabling an exact replica of the message to be produced at the other end, as long as the timing was correct.

At the destination apparatus, a sheet of paper saturated with potassium ferrocyanide was attached to the receiving plate. Those parts of the paper that were subject to an electrical current passing through the stylus were marked in Prussian blue, by virtue of a chemical reaction with the paper—a specimen sheet demonstrates its graphic possibilities in contrast to more limited coexistent and subsequent systems (fig. 5.10). Earlier attempts at electrochemical telegraphy—the Davy machine of 1839 or Alexander Bain's device of twelve years later—had been limited to the transmission of figures and preset signs.[28] The pantelegraph was the first such device to transmit faithfully other kinds of inscription: portraits, signatures, plans, or in fact any image that could be drawn on the surface of the tin. It was followed soon afterward by competitor devices or adaptations, such as Étienne Lenoir's *électrographe*, granted a patent in 1866. Yet, over a decade before the implementation of telephone lines, the pantelegraph realized the possibility that images might transcend their rootedness in a single place and time to appear, almost simultaneously, at another location.[29]

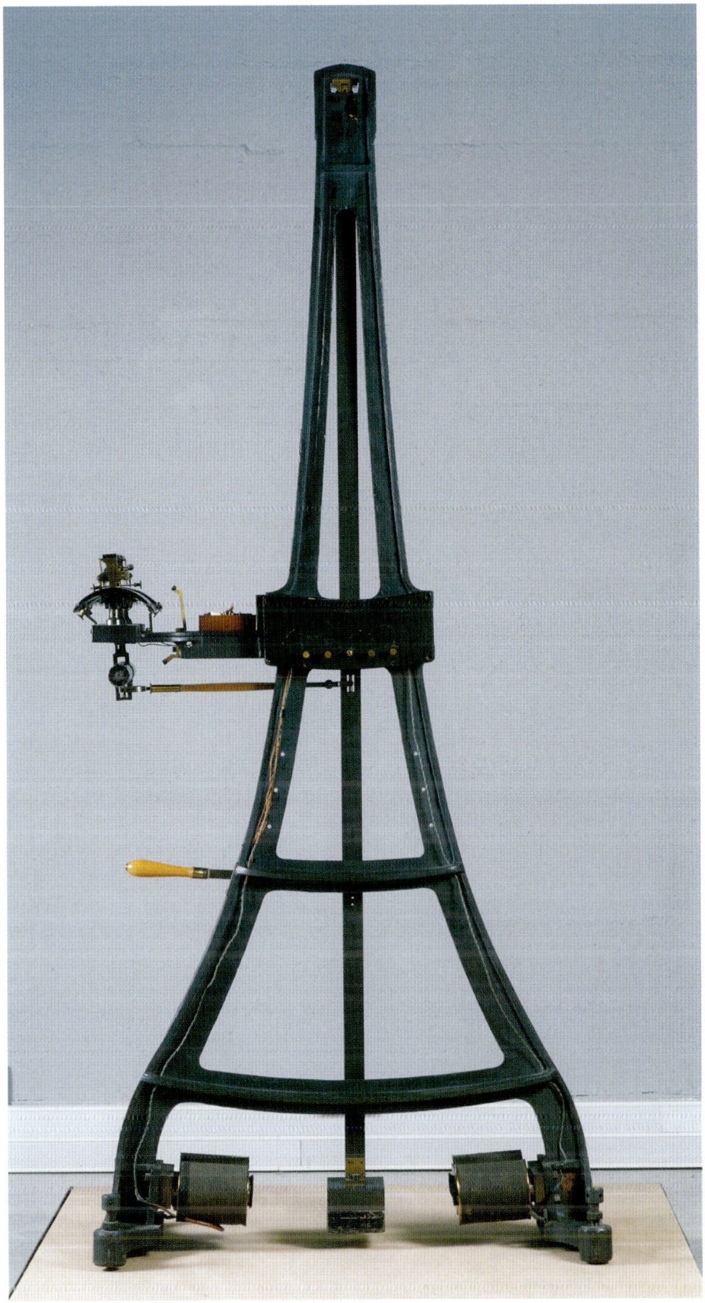

Figure 5.8
Giovanni Caselli and Paul Gustave Froment, *Télégraphe autographique système Caselli dit Pantélégraphe*, 1861, wood, brass, iron, copper, lead, 216 × 52 × 94 cm. Musée des Arts et Métiers, Paris.

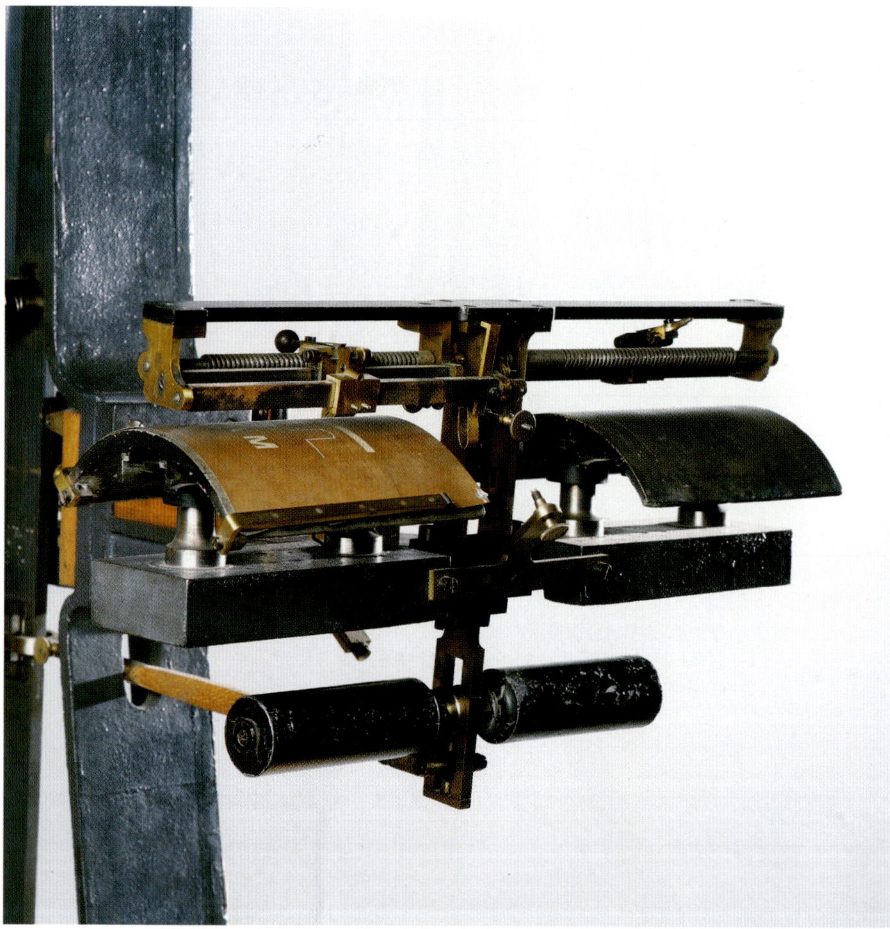

Figure 5.9
Detail of Giovanni Caselli and Paul Gustave Froment, *Télégraphe autographique système Caselli dit Pantélégraphe*.

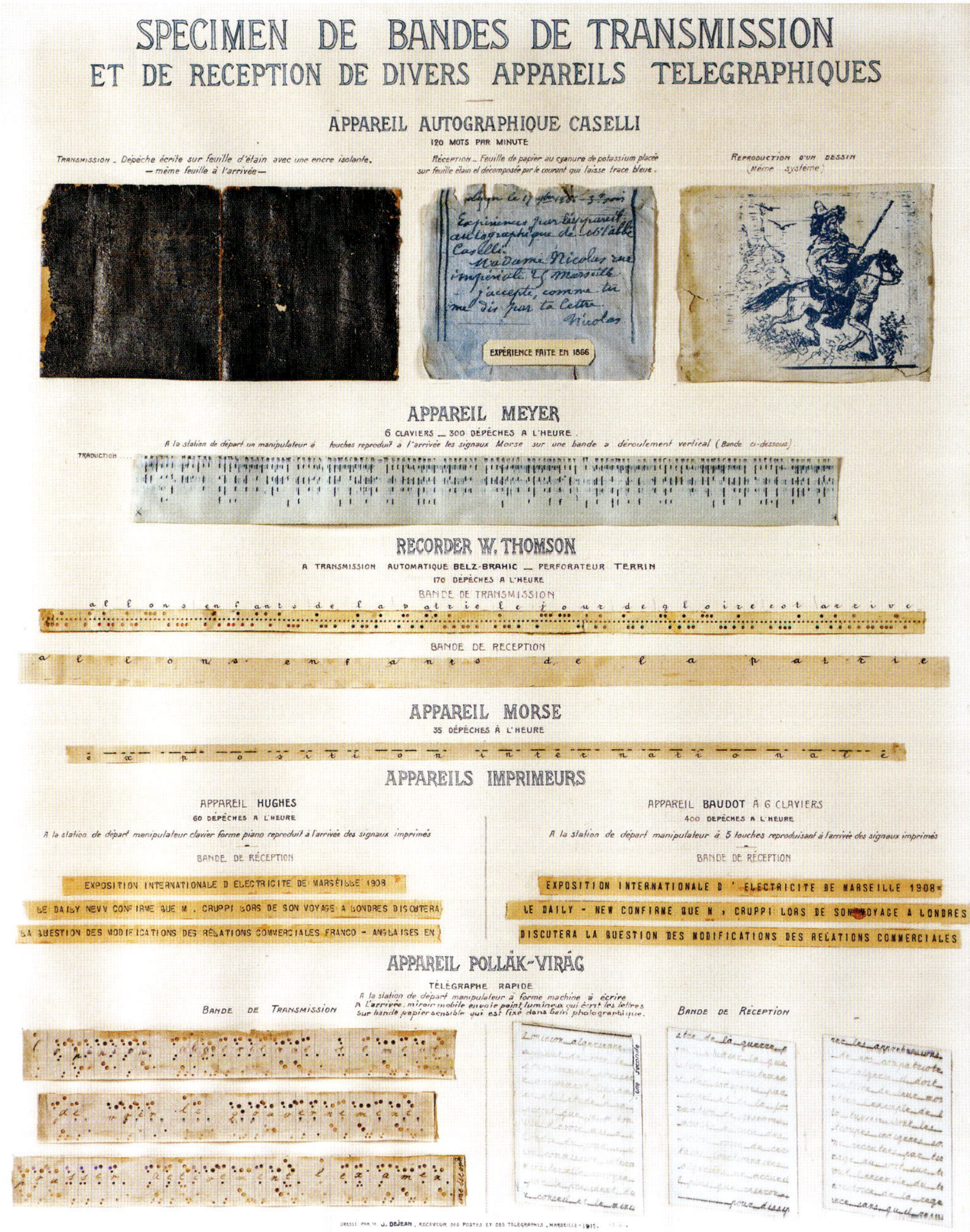

Figure 5.10

Specimen of outputs of different printing telegraphs: Caselli, Meyer, Thomson, Morse, Hughes, Baudot, Pollak-Virag (pantelegraph images at top, from 1866), 1911, paper, cardboard, glass, and wood, 79.5 × 65.7 × 2.2 cm. Musée des Arts et Métiers, Paris.

The pantelegraph is often described as a forerunner of the fax machine.[30] Yet, now that the many forms of image transmission that preceded the digital—CRT television and faxes, for instance—have themselves become extinct, visible only as fossilized forms or as vehicles for the nostalgia industry, the association between the pantelegraph and these "modern" forms of communication can no longer be assimilated as assuredly into a narrative of technological progress. The pantelegraph might as usefully be connected backward in time to the telegraphic techniques that preceded it. Furthermore, where it has been discussed as anything other than a quaint forerunner to a technology whose potential was not realized at the time, the pantelegraph has tended to be incorporated into a history of telegraphic communication that has focused on the transmission of written characters. How could this all appear differently if we shift the focus to the imagistic potential of the pantelegraph, which was in fact its primary innovation in the minds of many contemporary observers?

Pantelegraphy imparted a powerful visual message, but rather than maintaining a strict division between image and text, we might better consider the system in light of Lisa Gitelman's persuasive case for the representative function of technological objects and technological knowledge, as embedded in and produced by a rhetorical, discursive field of inscriptions.[31] Gitelman observes that Edison's phonograph (her main object of study) emerged "amid a cluster of mutually defining literary practices, texts, and technologies, among them shorthand reporting, typescripts, printing telegraphs, and silent motion pictures."[32] The inscriptive functions of Caselli's pantelegraph were likewise enfolded within a broad range of objects, ideas, and practices and had numerous conceptual and material affinities with other "new media" in the visual ecology of nineteenth-century France, as well as with more established forms of representation (among other inventions, Caselli himself designed a cinemograph for measuring the speed of trains). While some connections were clear—other forms of telegraphy, for instance, especially optical techniques—others were more oblique. Indeed, the happenstance coincidence of Caselli's invention with that transitional moment in narratives of modern painting, the Salon des refusés, held separately from the official display at the Palais de l'Industrie in 1863, throws both events into a new kind of relief. While one was met with Imperial approbation and the other official veto, in both cases the dislocation of images was invoked as a marker of both novelty and, in time, legitimacy, grounded in a shared faith that the image apart might offer a superior truth. Displacement could dilute an image's authority, but it might equally provide alternative forms of validation. Such comparisons cannot be sustained, for avant-garde art's endorsement of individual artistic creativity and the originality of its productions was at odds with the pantelegraph's radical replication and diffusion of images. However, in their different ways both offered forms of image making and display that were indubitably modern.

It is hard to escape the sense that sending an image so that it might appear in identical form elsewhere would have registered as a crucial moment in the history of mechanical reproduction, as well as announcing a paradigm shift in the history of how we encounter images more generally. Surely it *must* have changed everything? Yet the pantelegraph did not, in and of itself, transform inexorably the temporality and authenticity of visual images. In commercial terms, at least, it failed. Following his arrival in Paris in 1857, Caselli had benefited from the assistance of Paul Gustave Froment, to whom he had been recommended by the renowned physicist Léon Foucault, and the two collaborated on the design of the pantelegraph. (As a boy, Foucault had invented an optical telegraph, in imitation of the nearby Chappe machine at Saint-Sulpice.)[33] Caselli and Froment's invention was the subject of a great deal of interest in the French scientific community and was initially a great success, attracting the support of Napoleon III, who suggested that Caselli use the Parisian telegraph lines to conduct his experiments. The composer Gioachino Rossini wrote a short piece for piano, the "Allegretto del pantelegrafo," inspired by his use of the machine in 1860 to send a message between Paris and Amiens. In 1863, Caselli received authorization for the commercial exploitation of a line from Paris to Marseille, and he also experimented with a pantelegraph line between London and Liverpool. A US patent was received in 1863.[34] However, within a decade the pantelegraph had ceased to operate. It never attracted enough users, and it did not survive the traumatic events of 1870–1871.[35]

Poorly supported by the "Société anonyme du télégraphe pantographique Caselli," set up to defend and market the invention, the pantelegraph was introduced at a difficult moment in the aftermath of the shift from the Chappe system and the abolition of a state monopoly on telegraphic transmission. Telegraphic companies refused to charge lower prices for handwritten messages (in fact, they charged more), and the system was compromised by laws introduced to appease the anxieties of powerful interests threatened by the transition away from state control. These laws were particularly hostile to the visual dimension of pantelegraphy. They required that all messages be sent in an intelligible language and include the signature of the sender, and in apprehension of the possible seditious uses to which telegraphy might be put, they did not respond favorably to a device that could send maps, drawings, or coded messages. Finally, although one of the pantelegraph's key uses was the transmission of signatures for banking purposes, Morse code already provided an effective means of sending figures, and the expense ultimately proved prohibitive.[36] The device provoked interest in China as it could be used to transmit ideographic characters (two Chinese emissaries visited Froment's lab as early as 1863), although an attempt in 1884 to export the pantelegraph from Italy to China broke down.[37] While Jules Verne included a pantelegraph in his vision of a future Paris in his novel *Paris au XXe siècle*, crediting Caselli with its invention, no unbroken chain links the pantelegraph to the devices that succeeded it.[38] There is no forgotten history of technological triumph to be recuperated here, and any effects we might attribute to pantelegraphy ultimately transcend the technology itself.

Despite the legal restrictions that subordinated the telegraphic image to the text that accompanied it, the blue drawings transmitted by the pantelegraph remained firmly within the realm of the visual. A sheet of examples demonstrates its range (fig. 5.11). The fragile little portraits in blue ink materialize the machine's promise of bilocation, while a Rosetta Stone of different texts asserts the authenticity of the transmissions: "There is no possible error in the transmission of autographic telegrams. The copies of dispatches, reproduced by means of an electrochemical process, and not by the work of the employees, are always consistent with the originals," reads one.

This disavowal of the handmade is disingenuous, for the pantelegraph required a significant degree of intervention by the person making the drawings or writing the text. Occupying a conceptual, if not technical, middle ground between the image drawn by the human hand and new forms of mechanical objectivity, the

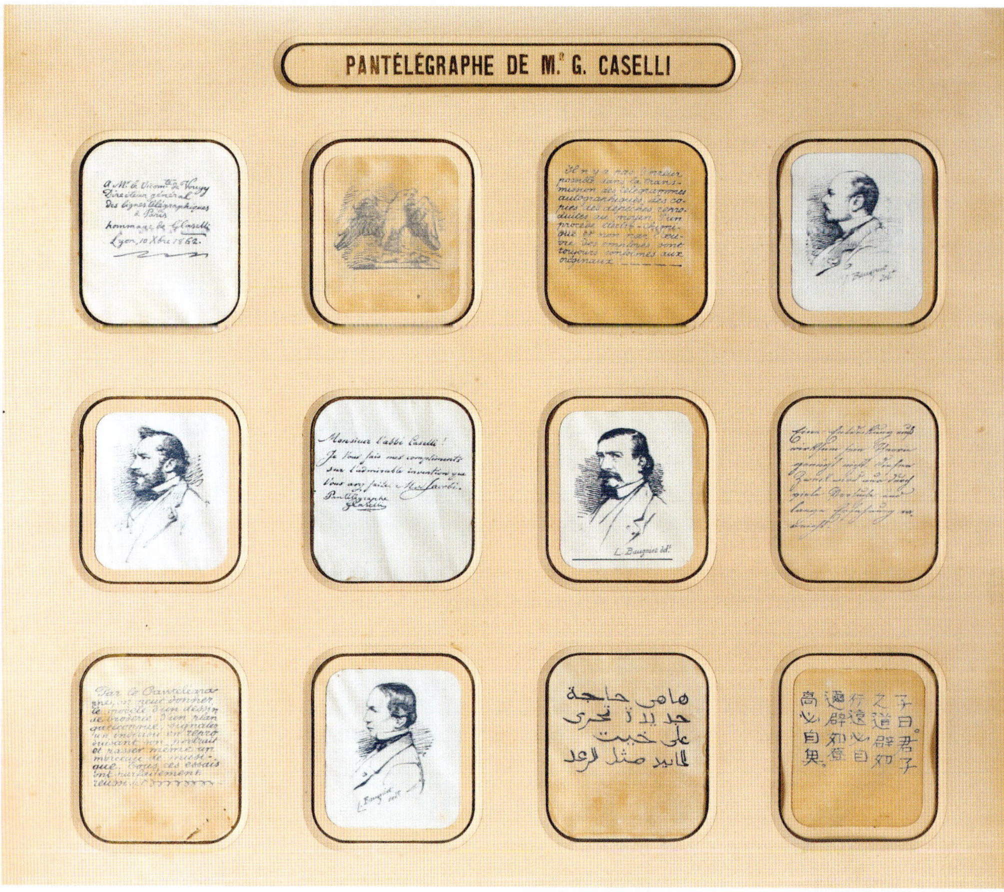

Figure 5.11
Giovanni Caselli and Paul Gustave Froment, sample of dispatches and drawings obtained with Caselli's pantelegraph, 1861, cardboard, paper, wood, glass, and ferrous alloy, 74 × 78 cm. Musée des Arts et Métiers, Paris.

relationship between pantelegraphy and established and evolving signs of authorship was unstable. In this sense, it was a fully telegraphic successor to Conté's engraving machine, which had at an earlier stage produced images that inhabited such ambiguous territory. Indeed, although their functions were quite different, there are some technical affinities between the device and Conté's machine, as a mechanized stylus, moving in lines, described an image for reproduction that challenged time and space. What, though, to make of the pantelegraph's little sketches and carefully delineated orthographies? In one, Caselli himself is the subject of the image (fig. 5.12). His portrait, drawn in Paris on January 1, 1867, was to be sent to a Monsieur Léon, 17 rue des Céléstins, Lyon. This image reveals little about its sitter, for the portrait of Caselli is standardized, and as a demonstration piece was likely not from life but appropriated from other representations.

Thanks perhaps to their combination of auratic presence and radical displacement, there is, undoubtedly, something rather ghostly about these images, these magical inscriptions from elsewhere.[39] It should come as no surprise that "spiritual" or "celestial" telegraphy was, by midcentury, regularly invoked globally to describe communication with the afterlife.[40] This characteristic of the pantelegraph did not go unnoticed by contemporary observers, exacerbated perhaps by the fact that Caselli, like Chappe before him, was a priest. In Gabriel Delanne's 1909 *Apparitions matérialisées des vivants et des morts*, the ghost of pantelegraphy was harnessed to the study of apparitions from beyond the grave, as well as to the manipulation of the living. Delanne described how a "community of sensation"—the same term Sensier had earlier used to describe the context for Georges Michel's work—might allow for the transmission of images and impressions between an operator (an interestingly telegraphic locution) and their somnambulist subject.[41] Via a curious process of magnetism, and by way of autosuggestion, a wound made to a photographic image might manifest as a stigmata on the skin of the person represented in the photograph. By the early twentieth century, the pantelegraph as functioning apparatus was already a distant memory, but for Delanne it continued to "offer an analogy for this phenomenon, because we know that thanks to an ingenious device, all trace left on the departure apparatus is reproduced automatically on the receiving apparatus located in the distance, electricity connecting each point of the two surfaces at a determined time."[42]

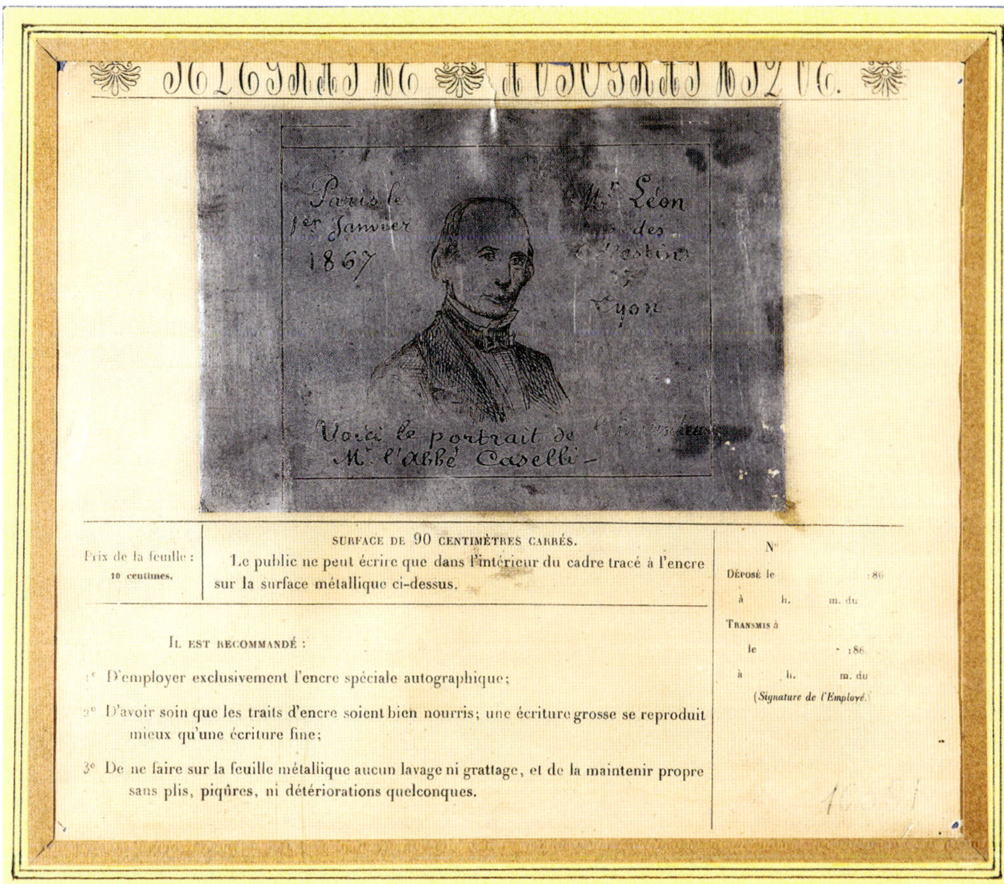

Figure 5.12
Dispatch on tin foil featuring a portrait of Caselli transmitted by the pantelegraph on January 1, 1867, attached to a printed sheet giving rates and instructions, 1867, cardboard, paper, tin, glass, 10 × 15 cm. Musée des Arts et Métiers, Paris.

Delanne described an experiment that took place in the studio of the photographer Nadar in front of several medical practitioners, in which a certain Madame O., under hypnosis, was successfully put to sleep by exposure to a photographic portrait taken while under the influence—a photograph that, unbeknownst to her, had been superimposed with an identically sized photograph of the right hand of the operator. According to M. de Rochas, who recounted this story, the image of the operator's hand communicated the vibrations produced by the hypnosis to the image of Mme O., which, serving simply as a relay, transmitted them to Mme O. herself. When the time came, Mme O. could be woken up by blowing on the photograph.[43] In 1861, the year Caselli registered his pantelegraph, Nadar exhibited *La Main du banquier D.*, a photograph of the hand of an anonymous banker, shot in electric light, which anticipates the operator's superimposed hand in this tale. The semaphoric quality of Nadar's image, which draws together contemporary interests in handwriting, spectrality, and palmistry, is worth pausing over.[44] The lines on the banker's palm tell of its owner's past and prophesy his future, the hand's communicative possibilities reinforced by its adamant gesture. In Delanne's telling, the placement of such a hand assumed pantelegraphic qualities of mysterious transmission.

As early as 1858, a language of sorcery pervaded the announcement of the "Télégraphe photographique" in *La Lumière*. Announcing Caselli's invention, the critic La Gavinie claimed: "The day is near when we will be able to write to each other from one hemisphere to the other and communicate our feelings, just as if we were speaking into each other's ears.—At each end of the telegraph line we will be able to confide in each another or exchange portraits.—What sorcerer would dare to announce this marvel!"[45] La Gavinie predicted the future course the technology would take and anticipated that Caselli's machine would prove particularly damaging to French notaries. He suggested that they would lose the income provided by their intermediary function, as anyone might sign a deed or certificate, from Paris to New York, London to Peking, without needing to be present in person.[46] The journalist took from an earlier report in *Le Magasin pittoresque* some basic information about the operation of the pantelegraph, including the claim that "To transmit by means of electricity, in an instant to a great distance, one's own portrait, or that of people with whom one finds oneself, or the view of the property where one lives, would surely be one of the finest applications of the combined findings of the electric telegraph and photography."[47] Pantelegraphy's present was avowedly visual and commercial, fusing personal identity and land ownership, and its future was photographic.

Although its images were produced at the point of reception by means of a chemical reaction, Caselli's machine was not, of course, truly photographic. However, as these accounts demonstrate, from the beginning it was considered in tandem with photography, and it was later incorporated into accounts of its development. As Simone Natale has observed, in the context of the United States, Daguerre's innovation coincided with the opening of early electric lines and Samuel Morse's interest in photography.[48] Subsequent to the demise of the pantelegraph, Willoughby Smith's 1873 discovery of the photosensitive qualities of crystalline selenium made photographic transmission down telegraph lines—in discussion since Becquerel's experiments in the late 1830s—a near possibility. Yet the connection between photography and telegraphy was not formed by a concurrence between the "inventions" of either technology, for embedded within this relationship were memories of earlier developments and long-standing practices in both fields. By the 1896 edition of Frédéric Dillaye's *Les Nouveautés photographiques*, a lengthy section on the history of "phototélégraphie" paid particular attention to Caselli's machine, which was attached posthumously to a longer history of photographic transmission.[49] Such narratives, which were predominantly technical in character, were preoccupied with affirming the successful progress of the medium of photography. However, four years after the publication of Dillaye's account, a more oblique, historically nuanced response appeared, in a text now taken to be a foundational, if somewhat eccentric, contribution to the history of photography. For the author of this text, photography's history belonged as much to the history of telegraphy as the other way around.

In Transit in Our City

Nadar's *Quand j'étais photographe* was published in 1900, toward the end of his life. The past tense of the title is crucial, yet although Nadar had moved away from photography to focus on his other scientific pursuits, he was still active in the medium.[50] Noting the title's strangeness, Rosalind Krauss observes that "Nadar's past tense has less to do with his personal fortunes and the trajectory of his own career through time, than with his status as witness."[51] For Nadar, photography's unique transformation of the world was crucial, and he stressed the extent to which it had surpassed the revolutionary-era achievements of "Laplace and the Montgolfiers, Lavoisier, Chappe, Conté, all of them."[52] As previously discussed, Nadar had earlier eulogized "this marvelous Conté," recognizing a fellow inventor and lauding his work in ballooning and telegraphy.[53] In an early section of *Quand j'étais photographe*, Nadar presented a strange fantasy of image transmission that not only referenced explicitly the telegraphic transmission of images, but seemed to consider telegraphs more generally and, remarkably, pantelegraphy in particular as a form of historical comprehension.

Nadar begins his story by describing a bizarre correspondence that took place in the autumn of 1856. A café owner named Gazebon had written to Nadar, telling him of an encounter he had had recently with a M. Mauclerc, "an actor in transit in our city."[54] Mauclerc had convinced Gazebon that he had in his possession a daguerreotype portrait of himself that Nadar had allegedly taken "by the electric process" while the photographer was in Paris and Mauclerc in Eaux-Bonnes, near the border with Spain. Although Mauclerc was disbelieved by some, Gazebon remained certain of the actor's credibility, having "dabbled in the process" himself.[55] Gazebon thus requested that Nadar photograph him at Pau—preferably in color, while seated at a table in his *salle de billards*—with the promise that if this distance photography technique were successful, he would soon send some business his way. Gazebon's name rang a bell: Nadar recalled that Gazebon had in fact contacted him two years previously, also at Mauclerc's instigation, regarding a gilded copper engraving—"a masterpiece of Restoration bad taste"—that Mauclerc had assured him was highly valuable.[56] Mauclerc had persuaded Gazebon that this engraving was a collectors' item whose only other copy was, strangely enough, in Nadar's possession. Nadar, slightly freaked by their author's persistence, ignored both letters, writing off Mauclerc as a crook and Gazebon as a gullible fool. He hung on to the correspondence, though, declaring, "It is not unpleasant and it is legitimate, in the last days of a long and sufficiently fulfilling career, to have received and to reread letters such as this one."[57]

After Nadar recalls this encounter, the story jumps forward in time. Some twenty years later, while relaxing with his friend

Hérald de Pages, Nadar was visited unexpectedly by a young man, a nineteen-year-old electrician from Clignancourt whose mother had allegedly been in service for Nadar's mother in Lyon, and who had himself worked for one of Nadar's friends. Eduardo Cadava has noted the allegorical function of Hérald de Pages ("herald of pages to come"), asking whether this character is in fact a "double of Nadar who simply enters through the always open back door of his unconscious."[58] Perhaps so, although either way, the telegraphic nomenclature here is worth noting too. Pushing for an audience with the famous photographer, the youthful visitor eventually wheedled his way in. He began by relaying his career to date: having already worked in Breguet's workshop, the young man had subsequently apprenticed with Trouvé while he was working on his dual-motored electric velocipede, with Froment as he developed his electric chronometers, with Marcel Desprez on his generator, and with Ader on his telephone. Each of these scientists was duly acknowledged for the magnitude of their achievements. Pride of place in this glittering vita, however, went to an invention that by the time Nadar was writing had long fallen into disuse: "I was even lucky enough to be accepted by M. Caselli to work on his autographic telegraphy. That is where, especially . . ."[59]

That is where . . . what? Transmission interrupted! An assumedly heartfelt evocation of the excellence of the invention and of the young man's memory of his career as a pantelegrapher was curtailed, for at this point, cut short in the telling of his story, he moved, cautiously, to the subject of his visit, which was no less than the possibility of long-distance photography. Claiming to have developed a new technique, which, it is implied, Caselli's pantelegraphy inspired, the mysterious visitor asked Nadar to grant him the opportunity to demonstrate his invention, asking that he commission one of his technicians to take, "in the isolated conditions indicated or that you will suggest yourself, with whatever model you choose," a photograph to prove or disprove his claim.[60] Nadar, as if already anticipating being photographed from afar, froze stockstill: "I did not move a muscle."[61] De Pages, on the other hand, was more effusive. "Do you hope to be able to take photographs from all distances, and out of sight?" he demanded. "I do not hope to be able to do it, sir," responded the young man, "I already do it. But I don't know how else to explain it to you, and you will see the rest yourself: I am not an inventor, I haven't invented anything; I have only encountered something that was always there."[62]

Admitting to the two men that he had already demonstrated his invention, the visitor showed them, with a performative flourish, a cutting from "an ordinary *Courrier* or *Écho de la Banlieue*," which documented his successful attempt to photograph the town of Deuil, near Montmorency, from Montmartre.[63] The man's appeal found a receptive audience, for it so happened that the day before this encounter, Nadar and de Pages had visited the International Exposition of Electricity, where they had marveled at the exhibits, yet remained slightly troubled by the "diabolical servant" of contemporary technology, which Nadar recounted in a memorable passage:

> We had seen it invisibly discharge all duties and perform all functions, realizing all the dreams of the human imagination. Obedient and ready to execute our commands, this all-powerful and impeccable agent, this servant unrivalled in all its liveries and all its names: telegraph, polyscope, phonophone, phonograph, phonautograph, telelogue, telephone, topophone, spectrophone, microphone, sphygmograph, pyrophone, etc., etc. We have seen it lifting and transporting our burdens for us, propelling our ships, our carriages, carrying our voice from region to region and storing sound, *ne variatur*, up to its least perceptible modulations, writing, drawing far beyond the reach of the hand, at all distances . . . indicating to the surgeon the bullet lodged in our body, stopping our galloping horses or our locomotives dead in their tracks, and also arresting thieves, ploughing our soil, winnowing our wheat, improving and aging our wine, and shooting game for us, monitoring our cashiers while guarding our cashboxes. . . . A first-class worker, in all the arts and professions, and good at everything, one at a time or all at once as you wish, market porter, postman, lampman, engraver, farmer, doctor, artilleryman, bookkeeper, archivist, carpenter, stand-in soldier, tenor, and policeman. . . . In fact, why not photographer, this universal Jack-of-all-trades, and even long-distance photographer?[64]

As Stephen Bann has noted, in his sustained examination of the text, despite Nadar's overwhelming commitment to a cult of progress and his fascination with the social and ontological implications of technological novelty, his scientific approach was eclectic and he was averse to triumphalist narratives of photographic exceptionalism.[65] Indeed, Nadar's vision of technological supremacy bears a resemblance to du Camp's dystopian account of telegraphic workers, their minds and bodies bound in servitude to an incessant stream of everyday and official information that is ours, not theirs. This is technology as regulating device—"monitoring our cashiers while guarding our cashboxes"—in the service of capital and the law. Anticipating this section of the text, the *Journal de Toulouse* of October 26, 1863, had suggested, shortly before an account of one of Nadar's aerial voyages, that these technologies might also combat criminality, with the microscope, the darkroom, and the telegraph operating as "three new gendarmes with which murderers will, from now on, have to contend."[66] Hugo's warning "in the slave in favor see the master to come" returns with a vengeance. Nadar's description represents technology, with telegraphy in a lead role among the other inscriptive devices on his list, as both suprahuman and subhuman, operating above the level of our own capacities and at the same time, for now at least, beneath contempt.

The trope of machines as servants was hardly new at the turn of the twentieth century. As Markus Krajewski has outlined, such subalterns, from early modern automaton retainers to digital servers, function as "peripheral media," yet "in fact, it is their (supposedly) insignificant actions, routine operations, and local, gestural, and silent knowledge that often give the decisive impulse, the solution to an epistemic impasse."[67] Telegraphic workers had, since the earliest days of the Chappe system, been considered particularly worthy of pity, if not empathy, although such feeling was always qualified and strategic. And as we have seen, telegraph operators provided a shorthand to talk about other forms of subjection. Chappe's invisible stationnaires were often viewed as indistinguishable from the apparatuses they maneuvered, while the cold and cramped conditions the Chappe operators endured, and the strain on their eyesight produced by staring through a telescope for long hours from an isolated turret, were thought to sap mental and physical strength. We recall the "living chrysalises" described by Alexandre Dumas, their monotonous insect lives wasted watching a "white-bellied, black-clawed fellow insect, four or five leagues distant."[68] Du Camp's description of robotic and oppressed telegraphic workers in the 1870s took this to its logical conclusion, updating the message for the electrical age. However, partly because of the telegraph's ability to condense ideas about obedience and emancipation, the possibility that machines and their operators might somehow achieve autonomy or resist their conditions had long lurked in the background of descriptions of the system. As an 1851 caricature by Nadar for the *Journal pour rire* showed (fig. 5.13), the optical telegraph in particular contained the potential to strike back: it wasn't a good idea to get too close.

Diabolical Masks

The telegraph's zoomorphic affinities were firmly in place before the publication of Dumas's novel and had a long shelf life in the years that followed, granting the servant-machine a different kind of agency. Here is Barthélemy, writing in the May 29, 1831, issue of his peculiar anti–July Monarchy verse satire *Némésis*:

> The minister issues floods of circulars;
> Everywhere the telegraph, obscure vocabulary,
> Bringing a clear meaning to the attentive prefects,
> Like a cut snake, wriggling in the air.[69]

And here is Aby Warburg, at the end of a 1923 lecture published in English in 1939 as "A Lecture on Serpent Ritual," drawing a sharp comparison between modern communications technology and primitive thought, as typified by the Pueblo Indian cult of the serpent:

> And away above his top hat runs the electric wire. In this copper-snake, invented by Edison, he has wrested the lightning from nature. The American of to-day no longer worships the rattle-snake. . . . Electricity enslaved, the lightning held captive in the wire, has produced a civilization which has no use for heathen poetry.[70]

Warburg's remarks are about a photograph of an "Uncle Sam in his tall hat" he had captured in 1896 walking past an unidentified pseudo-classical rotunda on a San Francisco street (in fact, the former City Hall, completed in 1899 and destroyed in the 1906

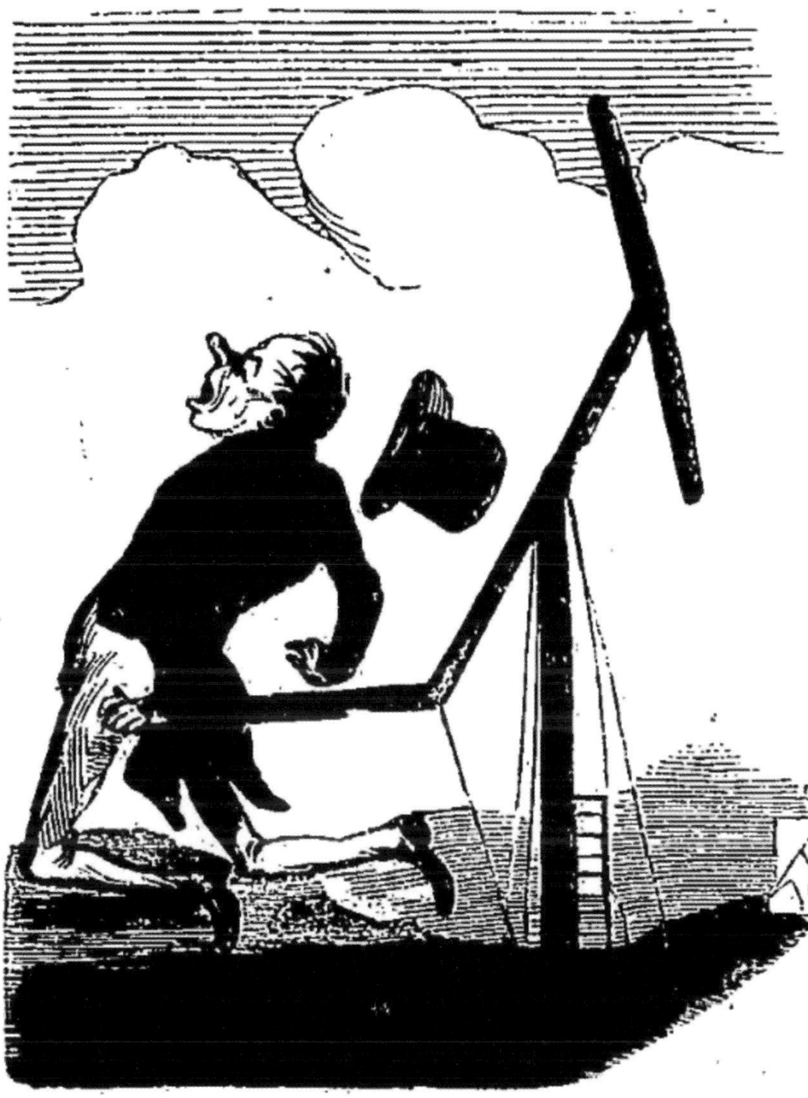

Mais il ne faut pas passer trop près du télégraphe.

Figure 5.13
Nadar, *Mais il ne faut pas passer trop près du télégraphe*, published in *Le Journal pour rire*, no. 13, December 26, 1851. Bibliothèque nationale de France, Paris.

earthquake) (fig. 5.14). This figure, shown in a city that hosts perhaps the best known of many Telegraph Hills worldwide (named for the semaphore that stood there from 1849), exemplified the type of man whose technological rationalism dispensed with the cult of the serpent and overcame an indigenous fear of lightning, allowing him to "enslave" electricity's power. Warburg's suspicion of the ability of Enlightenment "progress" to answer the fundamental questions of human existence found an eloquent metaphor in the electric telegraph, which, alongside the telephone, he derided for "destroying the cosmos."[71]

A final example: for Barthes, again, describing the Eiffel Tower, Paris's most prominent telecommunications mast, it was photography alone that was able to reveal the landmark's unique characteristics, although it did so in ways that breathed unexpected life into its iron struts:

> Photography, which often tells us the whole truth about an object, perhaps offers another metamorphosis: animal metamorphosis. Whether we think of it as an insect with a hard thorax whose legs have been ripped off, or we see it rising up into the sky like a wingless bird that is trying to push itself up higher, way above the clouds, or whether it appears, finally and more prosaically, like a huge giraffe put there for the sole purpose of inspiring awe among Parisians . . . there is a virtual animality to the tower.[72]

Barthélemy's, Warburg's, and Barthes's quite different turns to animal metaphor—their snakes, insects, birds, and giraffes—distinct in context and intention though they are, are notable for their invocation of an analogical language that had long attended telegraphy, even its pre-electric forms (Barthes's description recalls Caran d'Ache's giraffe telegraphs, discussed in chapter 3).[73] Equally, the line that connects Dumas's "black-clawed insect" to Warburg's "copper-snake" is one that also connects Dumas's wretched telegraphers and Warburg's diminished world. Nadar himself was implicated in this knotted technological imaginary. "An Icarus with replaceable wings" was how Jules Verne described him, rendering the photographer too a kind of hybrid—part-human, part-animal, part-machine—endlessly falling and self-generating, straddling the mythic past and the present. Nadar's response to the International Exposition was filtered through nearly a century's critique

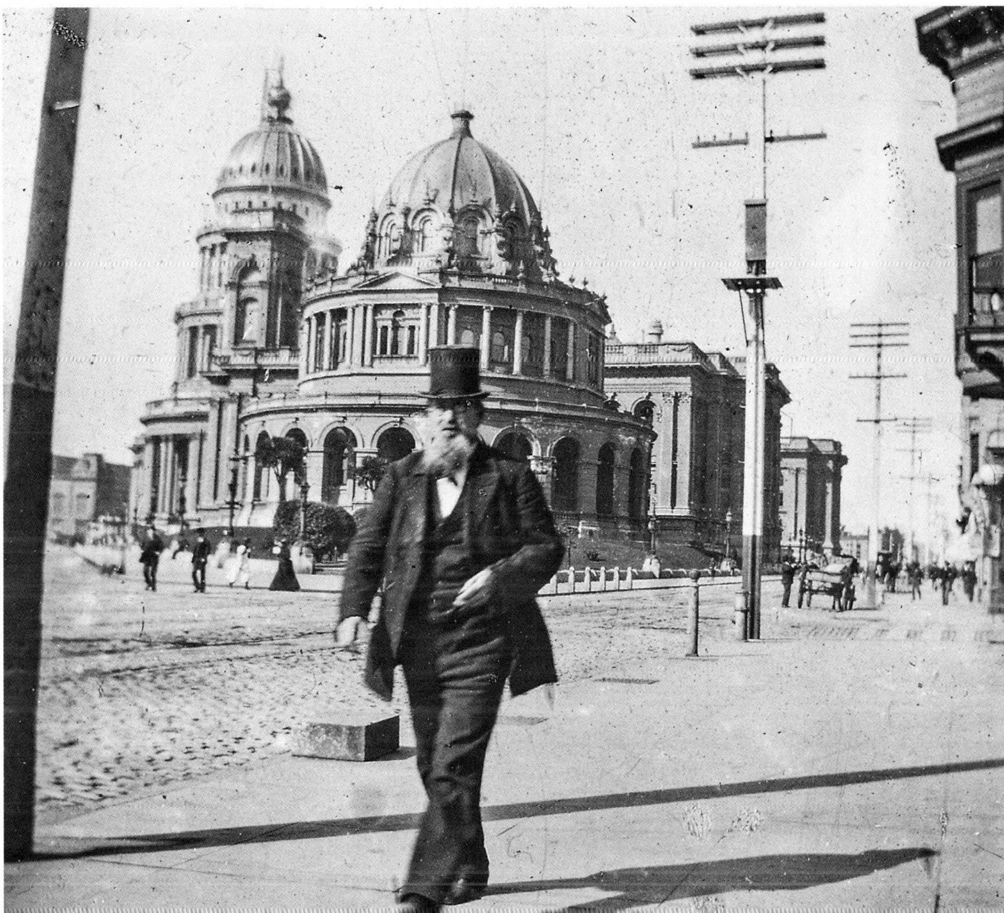

Figure 5.14
Aby Warburg, *Uncle Sam*, 1896, photograph, digital image from original glass negative. The Warburg Institute, London.

of telegraphy's complex dialectic between human, machine, and animal, stillness and movement, exploitation and progress, dead media and those that still pulsed.

The first International Exposition of Electricity, to which Nadar likely referred in *Quand j'étais photographe*, took place at the Palais de l'Industrie on the Champs-Elysées in autumn 1881 on the initiative of Adolphe Cochery, Minister of Posts and Telegraphs, following a smaller display at the Exposition Universelle of 1878. If Nadar's description seems functionalist in tone, his book, taken as a whole, reflects upon an earlier point in the nineteenth century, when the role of technology—contra Warburg—seemed neither so fixed nor so utilitarian. As John Tresch describes, the period until 1848 can be characterized by a confluence of seemingly opposed mechanist and romantic scientific worldviews.[74] Tresch recounts the role of spectacular, subjective, transformative machines in establishing truth, and notes the participation of scientists such as Arago and Foucault in the phantasmagoric operatic spectacles of Berlioz or Meyerbeer. These composers, conversely, demonstrated a consistent interest in technological advances; Berlioz proposed at one point the use of an electrical telegraph to keep players in time.[75] Honoré Daumier, who returned to the theme of the Chappe telegraph first seen in *Gargantua* several times, pointed to this meeting of art and science in a lithograph published in *Le Charivari* in April 1859, shortly before the introduction of the pantelegraph, in the hiatus after the withdrawal of the Chappe system (fig. 5.15). Showing three conductors raised above a massed crowd, Daumier's print refers to the "grand festival des orphéonistes" that took place at the Palais de l'Industrie in 1859. The text accompanying the image makes explicit that this was, indeed, an "alliance of telegraphy and music."

Such an alliance had, in fact, already been achieved, in the "universal musical language" of *téléphonie*, also known as Solrésol, developed from 1817 by musician François Sudre. Its early iterations were essentially an acoustic version of the Chappe system, with coded language transmitted over a distance by different musical notes; later the system was developed so that messages mapped onto the seven notes of the musical scale could also be represented by the seven colors of the rainbow (as well as by shapes or the numbers one to seven). Sound, color, numbers, or shapes could be used interchangeably to convey the same message: one might, in theory, make a speaking painting. Sudre demonstrated

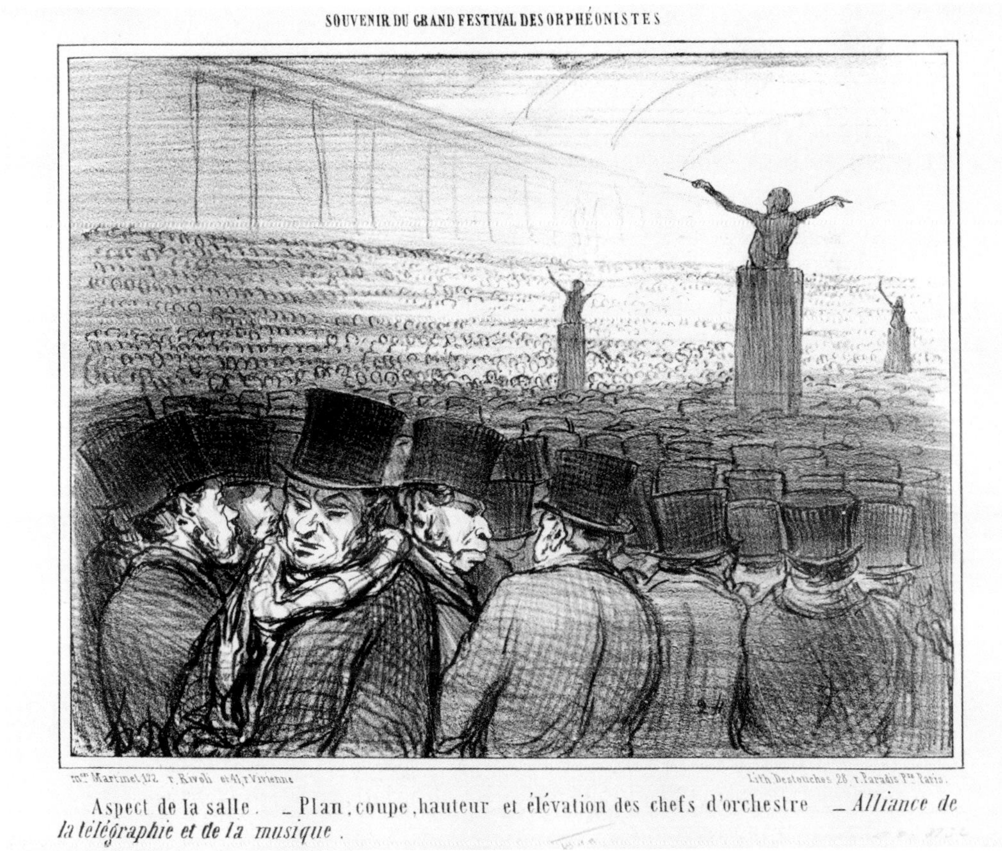

Figure 5.15
Honoré Daumier, *Souvenir du Grand Festival des Orphéonistes. Aspect de la salle.—Plan, coupe, hauteur et élévation des chefs d'orchestre—Alliance de la télégraphie et de la musique*, published in *Le Charivari*, April 2, 1859, lithograph, 20.8 × 26.5 cm. The Art Institute of Chicago.

téléphonie successfully on the Champ-de-Mars in 1829 and continued to pursue it over subsequent decades. He proposed special machines—telephones, he called them—that could be adapted for military use. Although it was not until four years after his death in 1862 that his wife Joséphine published his language in full, Sudre had been awarded 10,000 francs at the Paris Exposition Universelle in 1855 and a gold medal at the 1862 International Exhibition in London.[76] Such techniques were in the air in the early 1860s, contemporaraneous to the pantelegraph, expanding substantially the remit of Chappe's system but not relinquishing it fully.

We might also read an 1861 etching by Félix Bracquemond after Nadar's portrait of Meyerbeer in this context, for it too is expressive of such "alliances" (fig. 5.16). Bracquemond's print remediates Nadar's photograph but also ironizes it, with mock-heraldic grotesques and cartouches within the platemark sending up the composer and testing the limits of different media: print, drawing, and photograph. But Bracquemond's sketches and scribbles in graphite, black ink, and orange pencil also speak to the conditions of transmission—spectacular transmission at that—that characterized the careers of both Nadar and Meyerbeer. Like the slight fading of the image in a pantelegraphic message, the cartouches are repeated across the surface of the paper at varying densities, some of them split laterally between finished and unfinished sides of the same image, as if mid-transmission. This is less a matter of a direct causal relationship between telegraphy and the print than an instance of an image's animation by, and interrogation of, new possibilities for mechanical reproduction. Coincident with Caselli's invention, Bracquemond's defaced portrait of Meyerbeer is overlaid by an accretion of archaic decoration and whimsical masks that acknowledge the theatrical character of Meyerbeer's practice and its evolution in a climate of serially appearing, ghost-like reproductions.[77]

Returning to *Quand j'étais photographe*, at this point in his account, having listened to the young electrician's claims regarding long-distance photography, Nadar himself succumbed to a strange hallucination, an optical illusion in which his friend de Pages's features merged with those of the youthful visitor, revealing "a kind of diabolical mask which slowly took on the form of a face I had never seen before but that I recognized immediately: Mauclerc, Machiavellian Mauclerc, 'in transit in our city'; the electric image mockingly reared its head at me from the land of Henri IV."[78] As if

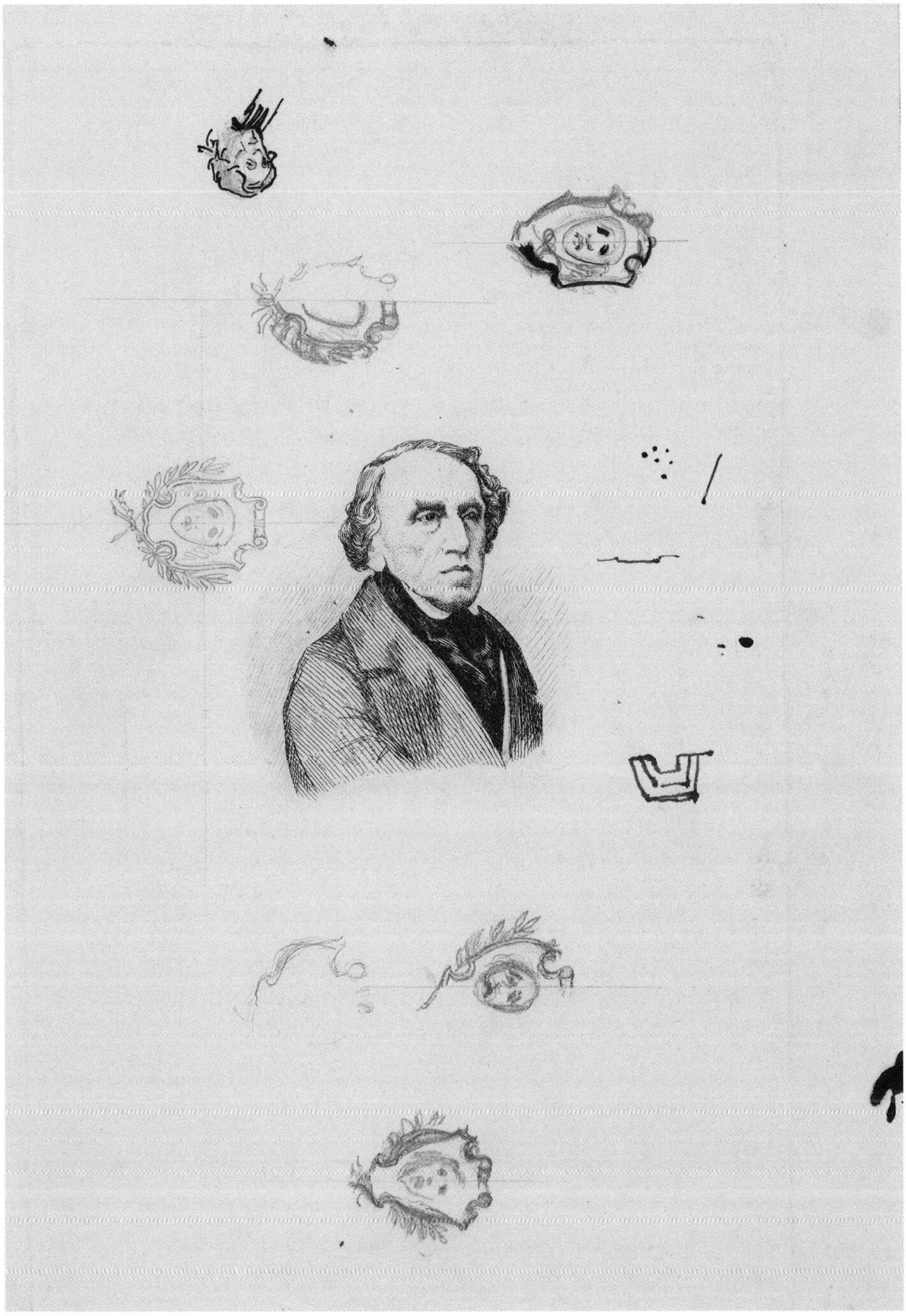

Figure 5.16
Félix Bracquemond, after Nadar, *Giacomo Meyerbeer*, c. 1861, etching touched with graphite, black ink, and orange chalk, 24.2 × 14.1 cm. British Museum, London.

in one of Francis Galton's composite photographs, superimposition of features revealed a criminal "type" that transcended location in a particular time and place.[79] It recalled Balzac's eidolic theory of specters, and perhaps also the autosuggestive images of Mme O. produced in Nadar's studio. However, it also pointed to ways of tracking information across time that were not unique to photography. Nadar's hallucination was a photographic effect, certainly, but rather than alluding solely to the temporal consequences of photography, it was grounded in the shared histories of photography and telegraphy.

When it came to it, the actual mechanics of the young man's proposed method were shady. The visitor stressed that no connecting wires were necessary, for the machine depended on the conducting properties of air alone. Having relieved an ironically amused Nadar of two *louis*, the young man left, swearing to return twelve days later. Needless to say (although the story is not resolved fully), we are left to assume that this never happened, that the two men had been scammed, albeit knowingly, by a consummate racketeer, and that the ghost of Mauclerc continued to stalk the streets. Pushed by de Pages as to whether he still denied the feasibility of long-distance photography, Nadar affirmed his agnosticism, refusing to deny or confirm the possibility. Two addenda to the section bring the story up to date. A first postscript notes recent work on precisely this technical question by Dr. Ed. Liesegang of Vienna, citing an article in the *British Journal of Photography* "which finally discredits Mauclerc to the greatest glory of Gazebon, who is rehabilitated." A final P.P.S. is even more adamant, asserting a contemporaneity that surpasses the long-ago story of Mauclerc and Gazebon and also eclipses its later recounting, making sure that readers are in no doubt that the text they are reading occupies the time of the now: "P.P.S. And from this very morning, with the definitive success of Marconi's wireless telegraphy, what can we not dream of?" Nadar signs off, finally, "Marseille, June 99."[80]

A Spider in Wet Weather

When Nadar was a photographer, he tells us, the young con man who called upon him was a telegrapher. Telegraphy, like photography, walked a fine line between truth and falsehood, fraud and sincerity, and like photography it confounded the stable identities it purported to fix. Nadar's visitor in *Quand j'étais photographe* was not an oppressed and exploited telegraphic worker nor even a sublimated form of one. Rather, in his humble work shirt, he carried the brio of the impetuous, creative inventor—even self-inventor—despite his contrary claim that he was just repeating preexisting innovations, including those of Nadar himself, that had been "always there." Caselli's device, after all, was (like photography) artistic in tone, if often bureaucratic in application (again, like photography), and as du Camp had observed, it occupied a different conceptual and physical space than the massed ranks of telegraph operators at the *bureau central*. Nadar encounters this young man, if not as an equal, then as a fellow traveler, although he also operates as a cipher for significant changes in both media, a transitional figure between pantelegraphy and Marconi's success. For Krauss, Nadar's skepticism in *Quand j'étais photographe* was another iteration of his commitment to photography's indexical qualities, his conviction that "photography can only operate with the directness of a physical graft."[81] Yet the text unfolds over a long time, bringing together a half-forgotten exchange from 1856 with a story from the 1880s and its telling in 1899. Nadar's "P.P.S." regarding Marconi challenges his own skepticism, affirming the subtly dialectical quality of his approach, for while pantelegraphy may have "failed" where photography transparently "succeeded," photography's future now looked likely to be realized by telegraphic means, reanimating Caselli's long-moribund project in the process.

As Bann has described, not only is *Quand j'étais photographe* one of the first attempts by a contemporary practitioner to document the history of photography's early industrial forms; it demonstrates too Nadar's particularly self-aware understanding of the relationship between images and history.[82] This is a relationship that extends into the future, for the conceptual attraction of the story is, Bann contends, an affinity between long-distance photography and "what we now banally term 'television.'"[83] Nonetheless, "the moment has not yet come," for Nadar seems to hint at something that had not yet transpired, and would not for some years.[84] But is it "Nadar," that curious confection of self and other in the photographer's memoir-story, who articulates this? Or is it rather the strange visitor, with his tall tale of previous work on Caselli's machine, who ventriloquizes for Nadar photography's displacement of bodies in space? Nadar's narrative betrays the extent to which the early history of photography was bound up with the ways in which it might be transmitted. The pantelegraph provided a language with which to understand something that had attended photography since its earliest days: the dream, and sometimes nightmare, of an image that might move seamlessly from one place to another. Collapsing the durational and spatial aspects of the new medium, this mobile image was tied to the mobility of Mauclerc ("in transit in our city") but also to the mobility of objects in time (du Camp's *sic transit*).

In *Charles Baudelaire intime: Le poète vierge*, published posthumously in 1911, Nadar describes a surprising meeting in the 1840s with "a strange, ghostly figure," whom he encountered on a walk across Paris with the writer and journalist Alexandre Privat d'Anglemont. When Nadar and Privat were able to identify this "apparition," they saw it was none other than Baudelaire himself.[85] Nadar gave the following description of his friend: "Assisted by the black of the costume, the restrained, meticulous, crushed gesture recalled the successive silhouettes of the optical telegraph which was then being taken apart on the towers of Saint-Sulpice or, better, the angular gymnastics of a spider in wet weather after her thread. The relationship with our new friend was already complete, despite his reserve, because things happened in this way then, long before the electricities of M. Edison."[86] Striking, in this account, is the attention Nadar pays to Baudelaire's physical similarity not only to a Chappe telegraph but to its destruction. Baudelaire's body is framed in terms of an interregnum between the dismantling of the optical telegraph on Saint-Sulpice and the new speed of human connection forged by Edison's electric marvels. This motif situates Nadar's recollection in a particular time and place, the Paris of his youth, a period to which he returned frequently in his writings. Baudelaire *intime*, Baudelaire in time. Death is always already encoded in this description, which appeared in print after the passing of both men, and which invoked, by association, photography's much commented upon ability to mediate past lives. Notably, however, Nadar used the time of telegraphy to calibrate and comprehend this past.

Ultimately, however, as he tells us, Nadar thought the spider's web provided a better analogy. Again, Nadar's description of the arachnoid "gymnastics" of Baudelaire's gestures prompts the question: who was Nadar really talking about here? After all, it was Nadar himself who, in 1852, at least a decade after his meeting with Baudelaire and nearly sixty years before the publication of his account of that event, described his own extended limbs and his exaggeratedly "spider-like manners" for humorous effect.[87] This coincidence in Nadar's and Baudelaire's anatomy and habits collapsed the temporal and physical difference between author and subject in much the same way that telegraphy's compression of time and space promised to turn sender and receiver into a single information system. The instantly recognizable scarlet *N* of Nadar's signature, a zigzag which Ségolène Le Men has described as "stamped sideways on the sky" in the form of the photographer's body in Daumier's famous 1862 caricature of *Nadar Elevating Photography to the Height of Art*, returns here as the rotatable *Z* or *N* shape of the Chappe telegraph.[88] A letter from Brunet to Chappe, sent on 5 Prairial Year II (May 24, 1794), had discussed painting telegraphs red to make them stand out, recommending vermillion in particular, while decades later Caselli experimented with pantelegraph messages in full color, sending from the Observatoire an image of "a magnificent rose" that "as if by magic was painted before the eyes of the elected officials" at the telegraphic administration.[89] Nadar's famous flaming red hair and robes served a similar purpose—both "magnificent" and immediately recognizable. The telegraphic convulsions that Nadar discerned in Baudelaire mirrored those he celebrated in himself.

And in fact, Nadar was not entirely novel in moving swiftly back and forth between the telegraph and the "spider in wet weather after her thread," nor were these things mutually exclusive: after the introduction of the electrical system, machine and arachnid became quite interchangeable. Popular entomologist Jean-Henri Fabre devoted an entire chapter of his celebrated *Souvenirs entomologiques* to the "telegraph wire" of the garden spider, a component of the web whose vibrations signaled the arrival of prey.[90] This filament required constant maintenance to ensure the survival of the connection. Earlier, Jules Michelet's *L'Insecte* described the spider's web as an "electric telegraph," sensitive to the lightest touch.[91] Nadar had photographed Michelet between 1855 and 1859 when the great historian was working on this topic, during which time electrical signals replaced the Chappe system; he also made a portrait of Fabre in 1880. When Nadar's preference for the spidery metaphor linked his own comportment with that of Baudelaire, it was also unavoidably telegraphic in its associations—the spider he seemed to have in mind, however, was Chappe's bygone machine, rather than the electrical webs that inspired Fabre and Michelet, even if one can sense the latter beginning to gain favor in his account.

Much of the discourse on telegraphy's relationship to both contemporaneous and "new" media has focused on its electric forms, particularly those that achieved some measure of endurance, aligning them either explicitly or implicitly with a future path often understood in overly deterministic terms. Yet the telegraphs with which Nadar punctuated his writing on photography had been mostly outmoded for some time—the Chappe system ceased to operate the year before Nadar's first correspondence from Gazebon in 1856—and even his references to technology from the ensuing twenty years until the electrician's visit, and from then until his narration of this story, recalled its former, obsolete iterations. At one level, telegraphy seemed to provide a useful framing device because of its longevity as a practice and its continuing relevance, a stable marker against which photography's progress might be measured. Yet, as Max Kozloff put it, apropos of Nadar (though he could have been talking about the pantelegraph too), "time, in 1860, was burning up."[92] If, for Nadar, the time of photography was informed by the many other devices that accompanied its introduction, such as the pantelegraph, through these associations he also reminded readers that telegraphy, like photography, offered a means to think the past in its complex relationship to the present and future.

Nadar's late writing exploits, in the process, a fault line between a vision of photography as determined by technology itself and one in which human subjectivity—including that of the artist—was primarily at stake. In other words, it expresses, but never really resolves, an uncertainty about the limits of human versus technological agency and about whether photography's role was to record the traces of human time or whether its key relation might in fact be with other machines. Nadar's image of an "all-powerful and impeccable agent" appropriating every aspect of human activity is undercut by the autobiographical tone of *Quand j'étais photographe*. Part of the force of telegraphy as it appeared to Nadar derived from the way it presented such ambiguities in concentrated form, for behind every telegraphic apparatus was a human operator, however subordinated to their device they might be.

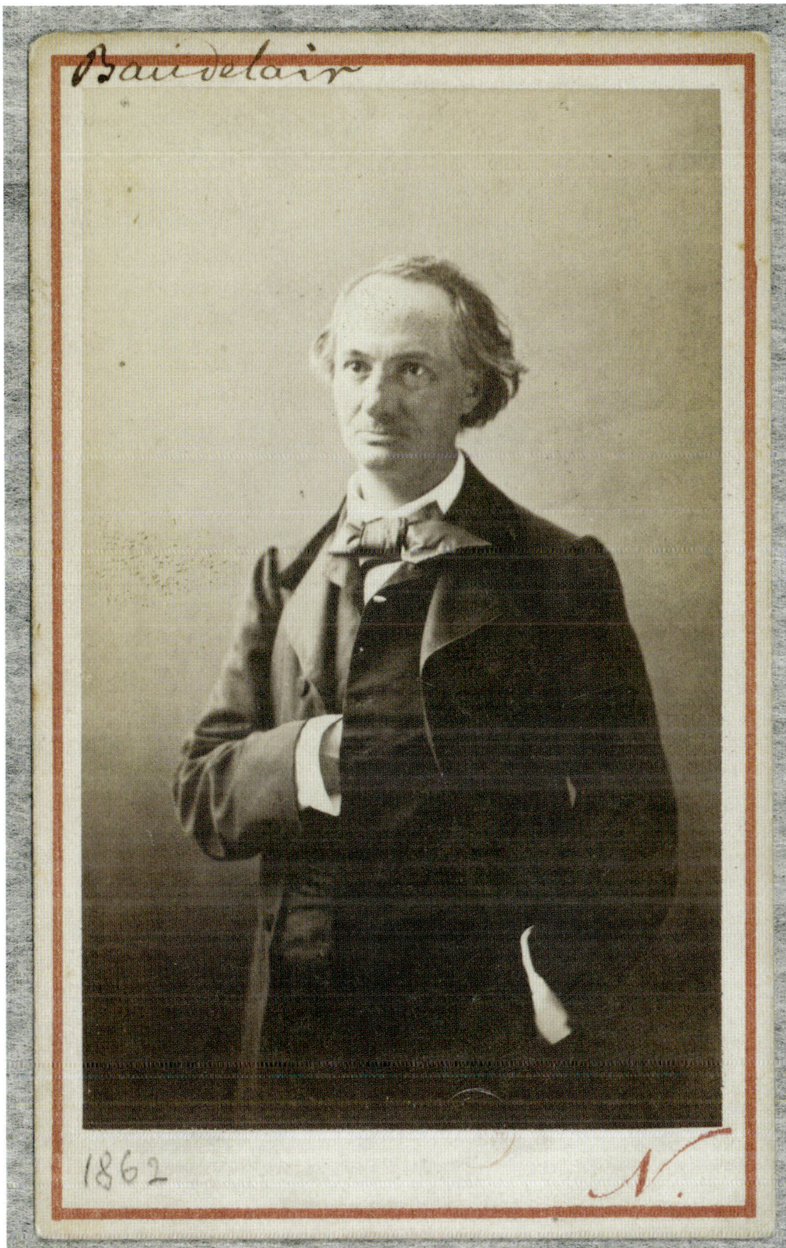

Figure 5.17
Nadar, *Charles Baudelaire en redingote, à mi-corps, de trois quarts*, 1862, albumen print on paper mounted on card, 8.5 × 9 cm. Bibliothèque nationale de France, Paris.

Yet Nadar also knew that to speak of current forms of telegraphic communication—and photography too, for that matter—was to invoke a shared technological genealogy that included the past time of telegraphy's visual world, from Caselli's little blue drawings further back to Chappe's network of semaphoric relays, materialized in the body of Baudelaire and captured repeatedly by Nadar's camera, his damp-spider-telegraph arm folded in jacket (fig. 5.17). In this way, Nadar outlined the continuing relevance of the optical telegraph to a world that had largely forgotten it. A cut snake—to paraphrase Barthélemy—but still wriggling in the air.

6 *The Machine Stops*

may hatch, may crawl,
may feed, may fly;
what survives is code.[1]

"We have arrived at the edge of our frame. The tower of the church of Saint-Michel appears in the distance with its telegraph and underground charnel house; sad emblems of life and death!"[2] These words, which appeared in *Excursions daguerriennes*, published in the early 1840s by Noël-Marie Paymal Lerebours, point us to the edge of this book's frame, too—and back to the beginning of the line I have drawn, starting from Chappe's tentative experiments at the site of the Belleville cemetery. In the text for *Excursions daguerriennes*, the author of this segment, de Contencin, narrated a brief tour around the Port des Quinconces in Bordeaux, culminating in a visit to the preserved mummies at the Basilica of Saint-Michel: "We shall descend into this dreadful tomb, whose walls are all lined with dried-up corpses; we shall cast a quick glance at these repulsive figures which have lived, and which have been torn from the worms to become the object of daily profanation, and we shall quickly climb the dark steps to shake off the dust of the sepulcher in the bright sunshine of the laughing city."[3] In the gruesome context of an encounter with Saint-Michel's bodies, exhumed from a nearby cemetery and on display until 1979, literally anything else might appear suffused with life.

Despite its "sadness," for de Contencin, the telegraph's somber animation offered the greatest contrast with the deathly display in the church below. "Serious as an epitaph, immovable as a rock," in composer Gustave Nadaud's description, the telegraph with its "waving and twisting" was, for *Le Figaro* in 1839, closer to a "Cirque-Olympique clown."[4] But de Contencin was not alone in drawing connections between the telegraph and Bordeaux's ghoulish spectacle. Victor Hugo, visiting Bordeaux in 1843, saw overlaps rather than contrasts, noting the eerie confluence of the "specters, exchanging in the night I don't know what communications" and the morose telegraph, this "poor wooden machine, controlled by a string, waving in the clouds," sending back and forth imperceptible information destined to fill tomorrow's newspapers.[5] The twin telegraphs at Saint-Michel, on the Bayonne line, received messages from Paris and retransmitted them onward toward the southwest.[6] They can be seen atop the church tower in the far-left background of one of the accompanying illustrations from the series (fig. 6.1), a collection of lithographs, etchings, aquatints, and photogravures made after daguerreotypes that presented Bordeaux's port as one more node in a global network linking monuments across France, as well as those in other European countries, North Africa, North America, Russia, and the Middle East. The remediation of the daguerreotype in this way, a decade before the emergence of paper photography in the 1850s, ensured for a while the life of the medium by piggybacking on print techniques that preceded it.

Not everyone felt as positively about the optical telegraph as de Contencin. For many more observers, it was the telegraph that was dying, in the process of being replaced by electromagnetic successors. By 1893 François Gautier felt able to write with confidence that "we are a long way today from the aerial telegraph, which, like oil street lighting, stagecoaches, and horse-drawn barges, has disappeared forever."[7] However, the Chappe telegraph was not fully replaced by electrical signals until the mid-1850s, lasting until 1860 in Algeria, and as has been described in the previous chapter, the repercussions of optical telegraphy extended for decades. The telegraph's close relative, naval semaphore, continued into the Second Empire and well beyond, with semaphore machines integrated into the electrical telegraph network at key ports and outposts.

All the same, Chappe's machine had its death rituals, some brisk, some protracted. As an example of the former, on October 31, 1858, a poster appeared in and around Gagny, a town in the eastern suburbs of Paris situated on the Paris-Strasbourg line. Pursuant to a law passed in May 1850, the *affiche* announced the sale of

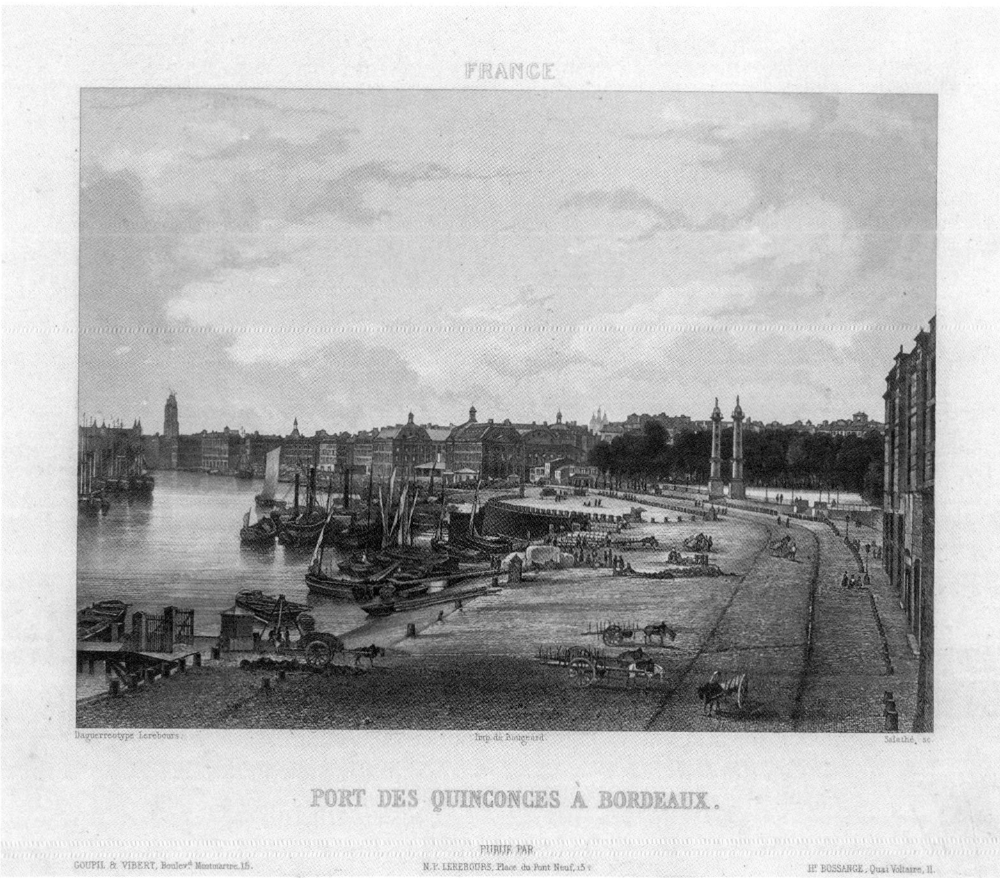

Figure 6.1
Noël-Marie Paymal Lerebours [daguerreotyper], Bougeard [printer], Friedrich Salathé [engraver], *Port des Quinconces à Bordeaux*, engraving after daguerreotype, 14.9 × 20.2 cm., from Noël-Marie Paymal Lerebours, *Excursions daguerriennes: vues et monuments les plus remarquables du globe*, vol. 2 (Paris: Goupil et Vibert; Lerebours; Bossange, 1842). Rijksmuseum, Amsterdam.

a former aerial telegraph station at the site of the Maison Guyot in Gagny.[8] For five hundred francs, one might purchase this five-floor stone tower and the land around it, built in 1797 for a telegraph post and renovated in 1839. Numerous such sales took place across France. The history of the optical telegraph was also a history of removal and transplantation, of architecture improvised, adapted, and repurposed as well as deliberate and systematic. Even at the height of their powers, telegraphs came and went, appearing and disappearing from the same sites throughout the time of the system's operation. Obsolescence and ephemerality were often a side effect of the desire to obtain and maintain the best possible view or to support updated technology, and they were built into the practice of telegraphy from its inception. These processes were intrinsic to telegraphy's success, rather than a sign of failure. Now, some of those telegraph towers have been lovingly restored as local heritage sites, at times with reconstructed telegraphs atop them. Commemorating the places to which the telegraph lent its name and recuperating its technology, they stand denuded of their haughty political symbolism and the once epochal power of their messages.

The Slowest Revolution

Susan Sontag's essay on Walter Benjamin, "Under the Sign of Saturn," first published in the *New York Review of Books* in 1978, took its title from Benjamin himself, who wrote: "I came into the world under the sign of Saturn—the star of the slowest revolution, the planet of detours and delays."[9] The influence of Saturn on Benjamin's own life, Sontag suggested, permeated his work, rendering him slow and blundering. Drawn to deciphering "compact codes," Benjamin was melancholic by nature.[10] As Sontag put it, "for the character born under the sign of Saturn, time is the medium of constraint, inadequacy, repetition, mere fulfillment. In time, one is only what one is: what one has always been. In space, one can be another person."[11]

The telegraph was the invention of a melancholic. In January 1805, at the age of forty-two, depressed and deeply distressed by claims from rival inventors that he had plagiarized his machine, Claude Chappe committed suicide, cutting his own throat and throwing himself down a well in the garden of the Hôtel Villeroy, which had been the administrative center of the network since 1794. Time, it seemed, had won out over space, and Chappe's self-silencing was eloquent. Exchanges with Breguet and Spanish engineer Agustín de Betancourt demonstrate changes in Chappe's personality from 1797, as he became troubled and irritable, not to mention hypochondriacal, possessed of what Ernest Jacquez called "this disease of great and tired minds."[12] A week later, Abraham Chappe published a letter in the *Journal de Paris* recounting his brother's death and claiming that, like his famous uncle, he had died a victim of zeal for science and his country. Abraham blamed a poisoning that purportedly took place during attempts to establish the line from Paris to Lyon for exacerbating his brother's nervous condition, an illness that "increased his melancholy and led him to the grave." The letter concluded by pleading "may their nephews learn to imitate their example and not to fear the fate of their uncles!"[13] The same newspaper had already relayed Chappe's suicide note, written in pencil (presumably one of those produced by Conté): "I am not happy, life overwhelms me, I die without reproach."[14]

It would be unwise to make too much of this tragic event, for Chappe's mental health had no direct bearing on the form his invention took nor, for the most part, the meanings it accrued—at least not until the time came to commemorate the technology and its inventor in the final decade of the century. The telegraph serenely practiced its repertoire of positions for another fifty years after Chappe's death. Nonetheless, there is something here worth repeating: for many contemporary viewers—those who lived under the twitching wings of the Chappe telegraph—the device and its secret codes were not only a means to communicate for military advantage. To be born under the sign of telegraphy was not just to be made aware of speed and efficiency (the sign of Mercury, presumably, a faster affair), but rather, in the melancholic terms outlined by Sontag, it was to be left behind by time, condemned to repetition, constraint, and suspicion while the important business of the world was transacted elsewhere, above one's head. Telegraphy was, for some who encountered it, maddening.

The pathological qualities of telegraphy, as du Camp and others noted, increasingly appeared intrinsic to its practice. Often, this was expressed most dramatically by residual connections to the optical system. There are, certainly, connections to be drawn between the Chappe telegraph's lugubrious presence and "the great, territorial, experimental, magical machine of hysteria," as it took shape within medical and juridical discourse in the nineteenth century.[15] For psychiatric pioneer Philippe Pinel, the upheavals of the revolutionary years were an obvious and widespread cause of mental distress. Of one hundred and thirteen people with a mental illness for which Pinel was able to discover a definitive cause, thirty were attributable to the events of the Revolution.[16] Taking their cue from Pinel's revolutionary liberations, the experiments of Jean-Martin Charcot later made hysteric possession and its gestures something female and sensational, produced within and by representation (especially photography). Charcot reported case studies of telegraph operators on at least two occasions, although hysteria was discounted as a diagnosis, and both operators were male.[17] Nonetheless, the staged performance of hysterical typologies documented in images from the Salpêtrière, encouraged by Charcot and put on display in his "Tuesday Lectures," frequently took telegraphic form, as patients such as Augustine (Louise Augustine Gleizes) arranged themselves into mournful tableaux vivants, their arms set to expressive angles. Interpreting these signals was the task the alienists set for themselves.

In the therapeutic environment, dramatic gestures—ineluctably telegraphic in appearance—could result from or be induced through hypnosis, as Delanne's experiments in Nadar's studio had intimated. Explosions, lights, deafening gongs, and other spectacular technologies summoned or tested the cataleptic subject.[18] Neurologist Jules Bernard Luys's 1890 *Leçons cliniques sur les principaux phénomènes de l'hypnotisme* recounts several such cases, and includes a photograph of eight people—six women and two men—sunk into a "somnabulent-cataleptic state" by means of a *miroir aux alouettes*, or lark mirror, positioned in front of them (fig. 6.2).[19] A lark mirror was a traditional hunting device, made of wood and often bird shaped, covered in small pieces of mirror and set atop a spindle. When spun, it enchanted small birds which, captivated by the shimmering kaleidoscope, could subsequently be trapped. The similarities between this entrancing machine and the photographic apparatus the group also face are unavoidable.[20] Once anesthetized, Luys reported that somnambulist subjects could be pinched or manipulated with no response, and they were also suggestible to verbal instruction. The human semaphores shown here—subjected to Luys's control and mesmerized by the flickering wings of his mirror-machine—return us to

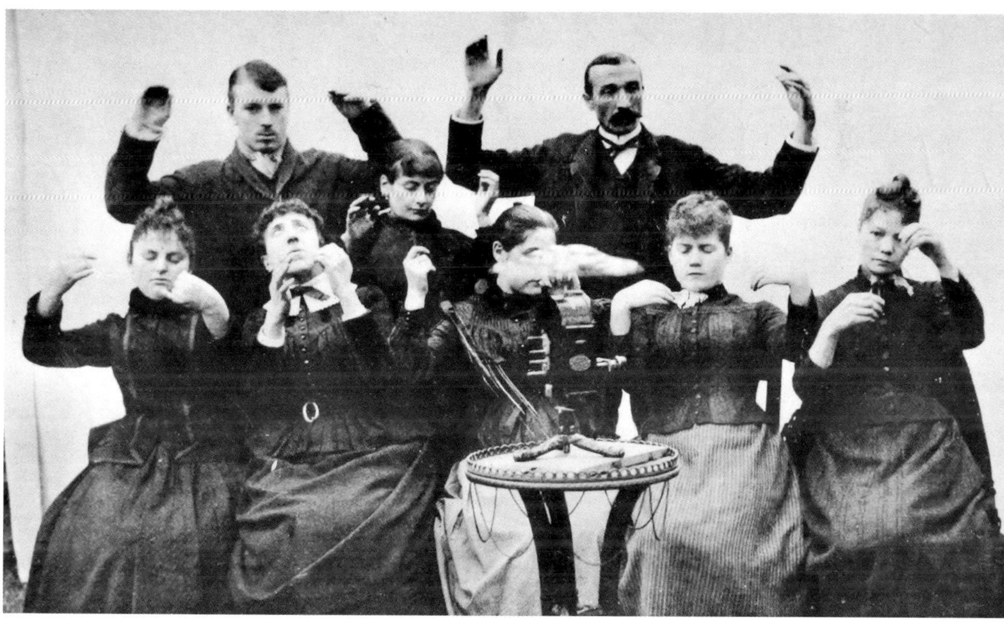

Figure 6.2
Georges Luys, "Planche XI: Groupe de huit personnes actionnées par un miroir à alouettes place sur une petite table," from Jules Bernard Luys, *Leçons cliniques sur les principaux phénomènes de l'hypnotisme* (Paris: Carré, 1890). Wellcome Collection, London.

the scene of telegraphy as dumb signaler, responding to coded messages that exceed the apparatus's comprehension (and that of its operator), their stupefaction at the reflections of the lark mirror recalling the telescope's distant gleam or the glancing shards of light thrown by heliographs.

There was another, longer history here, preceding the telegraph and stretching to our world of online conspiracy and affective malaise. In the early nineteenth century, it was perhaps inevitable that telegraphy, as a prominent form of media manipulation, would assume this role (its place in such fantasies later replaced by the radio, television, and the internet). In the late 1790s, soon after the telegraph's introduction, James Tilly Matthews's description of a conspiracy to take over his mind by means of an "Air Loom" had suggested that the clandestine communication networks of revolutionary France might offer fertile material for fantasy. Matthews's account provided the basis for John Haslam's 1810 *Illustrations of Madness*, while his drawing of the Air Loom was the first published artwork by an asylum inmate.[21]

As Mike Jay reports, the Air Loom plot that Matthews described was a complex affair, the eponymous Loom "fueled by combinations of 'fetid effluvia,' including 'spermatic-animal-seminal rays,' 'putrid human breath,' and 'gaz from the anus of the horse.'"[22] Having fitted Matthews's brain with a magnet, the nefarious Air Loom gang controlled his thoughts, actions, and words, as well as physically assaulting his body via terrifying procedures of "knee-nailing," "fiber-ripping," "apoplexy-working with the nutmeg grater," "bomb-bursting," "lengthening the brain," and "lobster-cracking." Matthews identified the gang as undercover Jacobin terrorists, led by Bill the King and featuring the florid characters of the Glove Woman, Augusta, and Jack the Schoolmaster, whose conspiracy spread to the highest tiers of government. Matthews had spent a significant amount of time in Paris during the Revolution, where he had traveled to assemble support against the Jacobins. Imprisoned in France during the Terror, and upon release having become convinced that he had been betrayed by William Pitt and Lord Liverpool, Matthews accused the latter of treason, leading to his confinement at Bedlam.

In Matthews's depiction of the Air Loom, "levers, barrels, batteries, brass retorts, and cylinders are rendered with the cool conviction of an engineer's blueprint," while the influencing rays it emitted are visualized as dotted lines.[23] While his machine seemed to work through some combination of animal magnetism and chemistry, the windmill-like device and its control by a pair of large levers also evoke the mechanisms of the telegraph. The idiosyncrasy of Matthews's delusions thwarts attempts to trace exact equivalences between the Air Loom and contemporary machines, for it drew eclectically on Enlightenment scientific culture. But by the time Matthews was released from prison in France, he would have encountered Paris anew, as a telegraphic city. The network prompted widespread anxieties about control, conspiracy, and surveillance, and although Matthews did not mention the telegraph by name, it is not unreasonable to speculate that it was a latent source for his machine.

Nearly half a century later, in 1844, Cham's *Charges parisiennes* returned to the paranoid "signes télégraphiques" of the recently defunct Chappe system to delineate diverse Parisian types and social situations in the interest of mild satire.[24] Within a few years, the optical telegraph would lie in ruins, the land and buildings on which it stood sold and its hardware flogged for scrap. Yet the people of Paris continued to dance on its grave as if in eccentric tribute. In Cham's vignettes, the telegraphic gestures of varied modern city dwellers—not only those confined to the asylum or treatment room—were configured as so many tics and unconscious repetitions. Psychoanalysis, of course, pursued this path, finding in gestures, slips of the tongue, and other glitches the irruption of the unconscious into everyday life. Developed in subsequent media histories, the reference has usually been to electrical signals of one kind or another, but as Cham indicated, these often overlapped with the optical telegraph's unmistakable if by then superannuated sign language, which was more expressive for such purposes.

Within this tradition, telegraphy was not, however, solely a matter of gesture, for later the analogy was to other processes, including those closer in spirit to Caselli's pantelegraph. Jacques Lacan, reading Norbert Wiener, described the mythic case of a man who was transmitted whole from Paris to New York by way of a telegraph that sent detailed information on everything comprising his individuality. The miracle of reconstituting a man via this "point-by-point resynthesis," breaking him down to pure information, was, Lacan argued, a mirage, dissolving "as soon as one points out that it would be no greater a miracle to telegraph over two centimeters. And we do nothing less when we move ourselves through the same distance."[25] The distance between communicants was immaterial; rather, it was the fact of communication—fatally banalized in contemporary culture, rather than being seen, as Lacan argued it should be, as a complex characteristic of our physical and psychic lives—that was ultimately of greatest importance. Telegraphy, by this point, offered a language for understanding consciousness—even life itself—and the smallest transport of the body. And while de Contencin, in the *Excursions daguerriennes*, contrasted the cheerless vivacity and thought processes of the telegraph to the tombs beneath it, death could also be understood in telegraphic terms. It was in this context, too, that the hallucinatory potential of the defunct machine came fully into view.

Scarecrow with Sparrows

A tiny painting by Jean-François Lebelle—measuring only sixteen by twelve centimeters—contains an improbable amount of detail (fig. 6.3). The painting represents another graveyard, Père Lachaise cemetery, inaugurated in 1804, that by the time the image was made in 1835 was established as the location for a new model of bourgeois burial.[26] A funeral is taking place in the foreground, while families promenade through the city of the dead, taking in the elaborate tombs of illustrious citizens. To the right-hand side of the painting, the telegraph at Montmartre can be seen, illuminated atop its escarpment by the glow of the setting sun. It is more surprising to see its message apparently received inside the cemetery itself, via a telegraph apparatus nestled among the sepulchers and weeping willows, directly above the semicircular memorial in the center of the painting.

This metal device, mounted on a cairn, was not made for action. Rather, it marked—and still marks—Claude Chappe's tomb, his surname inscribed in a cursive hand on the uppermost rock (fig. 6.4). Louis Lassalle's lithograph shows it standing alone, looking out over Paris from its lofty perch, although as Lebelle's painting makes clear, it was in fact surrounded by a clutter of other tombs, just as the telegraph itself frequently worked amid a confusion of chimneys and spires. Chappe was initially buried at the Vaugirard cemetery, his grave there marked by a double headstone, but his remains were moved to Père Lachaise after Vaugirard closed (the cemetery that now goes by that name is in a different location). Following the death of his brother Ignace in 1829, Claude

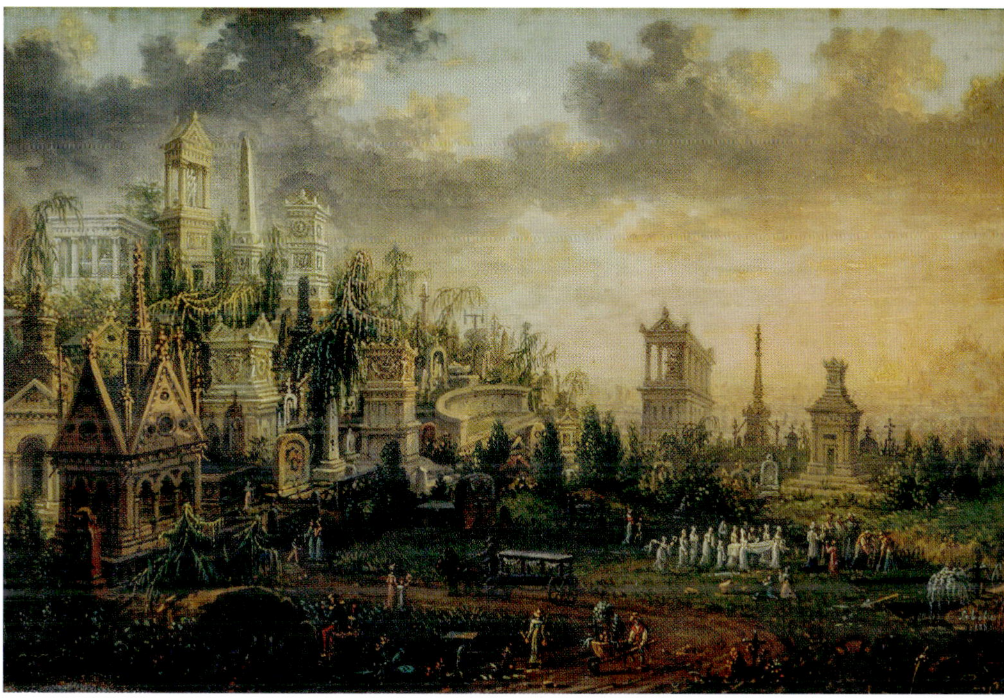

Figure 6.3
Jean-François Lebelle, *Le Cimetière du Père-Lachaise*, 1835, oil painting, 12 × 16 cm. Musée Carnavalet, Paris.

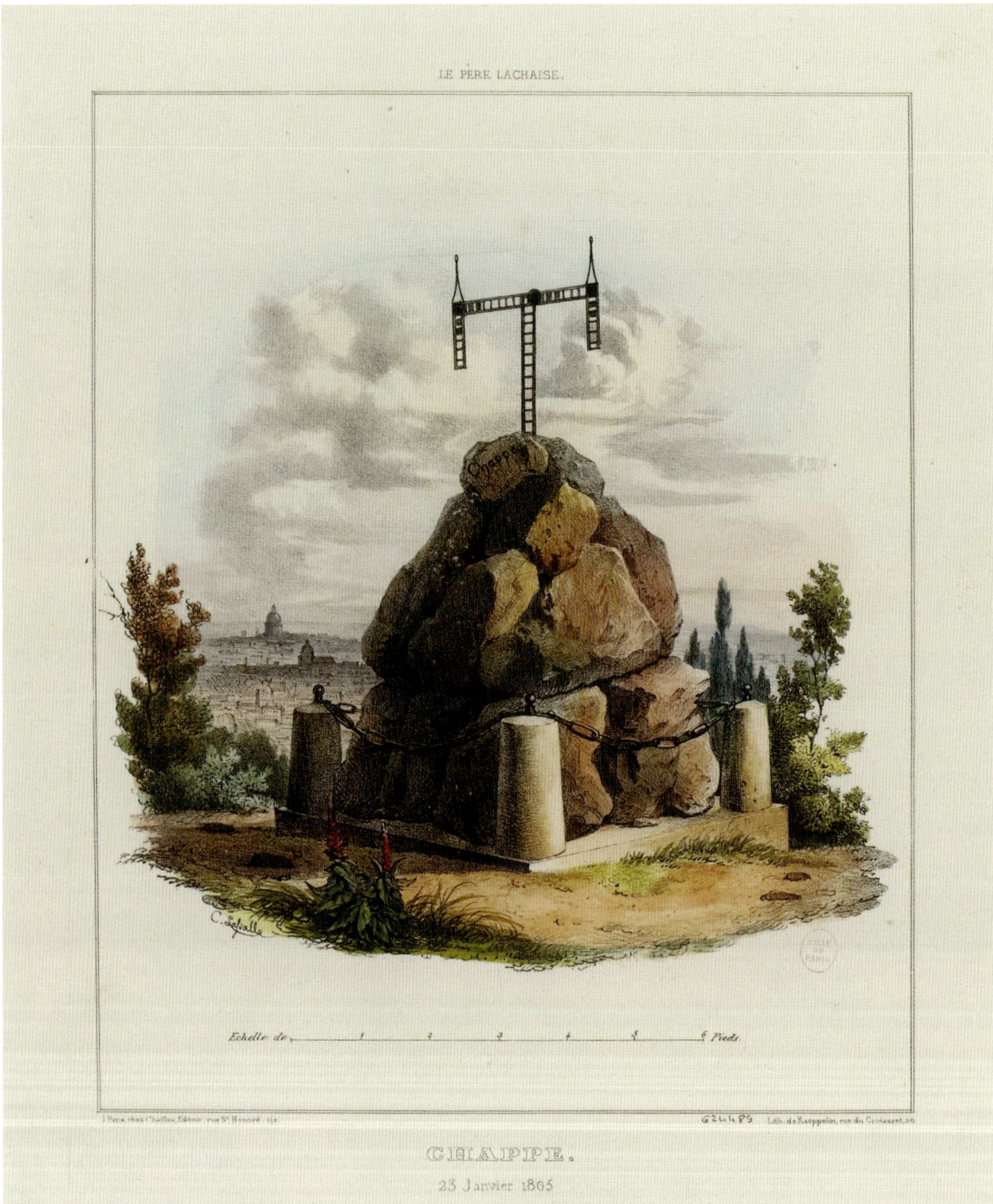

Figure 6.4
Louis Lassalle, *Monument funéraire de Chappe au Père-Lachaise*, n.d., colored lithograph, 36 × 27.2 cm. Musée Carnavalet, Paris.

was reburied alongside him that year at Père Lachaise. At that date a new monument to Chappe was installed. The Vaugirard headstone had been full of mistakes—Claude's birthdate was inscribed incorrectly (as 1765, instead of 1763), the date of the recapture of Condé was a year early (1793, not 1794), and there were errors in the telegraphic signs represented on the stone. In 1859, the stone was cut in half to display the designs on each side and was installed in the entrance hall to the central telegraph administration building on rue de Grenelle, where it would have watched over the arrival of the pantelegraph four years later.

The tomb at Père Lachaise was a more dramatic affair. It soon connected to other familial relays: Abraham Chappe, who died in July 1849, was also buried in a tomb with a convincing telegraph replica atop it, in their hometown of Brûlon. This, like Claude's grave marker at Père Lachaise, had its iron machine set in the "final signal" position. Yet by the last decade of the century, it became clear that Chappe's tomb at Père Lachaise required repair. Ernest Jacquez, librarian and archivist of the postal and telegraphic service, led a campaign in May 1891 to restore the monument, providing photographic evidence of its dilapidated state—one of its metal indicators had been blown clean off by the wind.[27] It was Jacquez who, as Bibliothécaire des postes et des télégraphes, acted as secretary for the committee of the Chappe centenary, and he was responsible for much important historical work to recuperate the history of the telegraph. With Adhémar Kermabon, Jacquez authored an atlas of the network from 1793 to 1852, each stop of which featured the outline of a tiny telegraph, printed in red (fig. 1.9). Kermabon was the son of a stationnaire, Auguste Kermabon, who later became inspector of the Paris-Brest optical telegraph line. In 1889 Adhémar made a series of elegant watercolors of postal and telegraphic operatives, in which uniforms and attributes were described in detail, historicizing the system in the year of the Revolution's centenary.

A couple of years later, the Chappe monument at Père Lachaise acquired company, this time on the streets of Paris. In the decades following the Franco-Prussian War, *statuomanie* in the French capital accelerated intensely, and Paris soon bristled with bodily correlates in bronze. In 1870 there were only around ten iconic statues:

those of Henri IV, Louis XIII, Lous XIV, Marshal Ney, and Marshal Moncey. By 1890, there were more than one hundred, and by 1940, two hundred.[28] Increasingly, these were geared toward Republican themes, national propaganda produced with a view to restoring French confidence after the traumatic events of 1870–1871.[29] Writers, artists, composers, poets, architects, and scientists were all represented, as well as the usual legislators and military leaders. Statue mania quickly gave way to concern over Paris's streets being blocked or uglified by these heroic avatars, which seemed to choke every major junction, as well as crowding parks and suburban corners. By the end of the century, excess statuary was considered an urban blight by many. The *Journal des débats*, assessing the excessive and dubiously accounted cost of Parisian statues in the "year of Panama," noted that 1893 had already seen the removal of several statues. However, "as if by an effect of the law of compensation," more were due to appear that same year. Sprouting like mushrooms, new likenesses of François Arago, Balzac, Beaumarchais, Condorcet, Émile Augier, Théophraste Renaudot, and Antoine-Louis Barye were to be installed. There was also to be a statue of Claude Chappe, which would additionally be a monument to his machine.[30]

In 1891, the administration of Postes et télégraphes, headed by its director Justin-Casimir de Selves, had decided that the centenary of the first telegraphic message should be marked by the erection of a monument honoring Chappe, whose history had receded in collective memory. Although requests to host the statue were made by representatives of Ménilmontant, Montmartre, and Belleville, it was agreed to place it in the 7th arrondissement, close to rue de l'Université, where Chappe had lived, and near the headquarters of the telegraph network on rue de Grenelle, at the junction of boulevard Raspail, boulevard Saint-Germain, and rue du Bac.[31] As was common, money for the statue was raised by subscription—a process frequently mired in accusations of corruption or political bias—and a national competition was launched, receiving twenty-six submissions.

The winning design, by Ernest Damé, showed Chappe standing beneath his telegraph, accompanied by a flag (fig. 6.5). In his left hand he grasped a telescope, his right hand at the telegraph's *manipulateur*. Reproducing with scrupulous detail the telegraph's mechanism, Damé's statue showed Chappe as if a stationnaire, frozen in the process of moving his machine. Unintentionally,

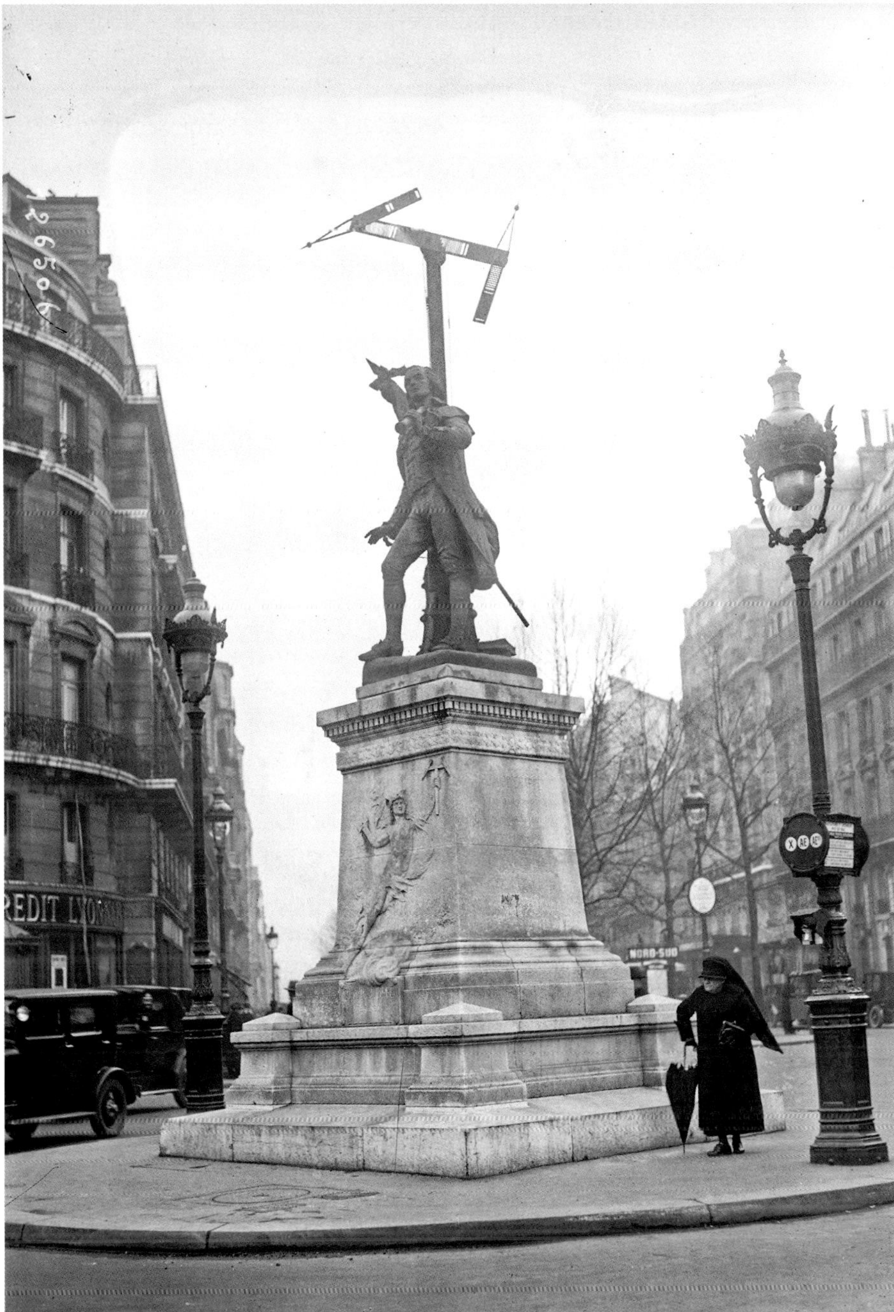

Figure 6.5
Agence Rol [photographer unknown], *Monument de Chappe*, 1928, photographic negative on glass, 18 × 13 cm. Bibliothèque nationale de France, Paris.

this pose had the effect of monumentalizing the telegraph's many unnamed and invisible operators, if only by proxy, or, depending on one's opinion, erasing them in favor of Chappe's charismatic presence. The plinth, designed by architect Georges Farcy, reprised the distinctive letterhead for the network. Featuring a relief sculpture of Mercury, swooping through the sky with a miniature telegraph held out before him, the names of Chappe's brothers, as well as his key achievements, were inscribed on the other side. Damé was reported to have been inspired in his image of Chappe by the only known portrait of the inventor, a print which in a strange coincidence had been engraved by his collaborator's grandfather, Ase Farcy.[32] Such connections to telegraphy's past were deliberate too. Inaugurated on a conspicuously freighted date, July 14, 1893, Chappe's statue was bonded to its revolutionary history. Minister of Commerce Louis Terrier presided over the ceremony, at which descendants of Chappe were present. Widely reported in the press, this ritual and the statue it recognized affirmed Chappe's place in the national republican patrimony. In its faithful reproduction of the telegraph apparatus, Damé's monument returned the machine's crooked shadow to the streets of Paris, nearly forty years after the last working apparatus had been dismantled.

During its residency in Paris, the statue found enthusiasts, if not exactly admirers. Surrealist poet Louis Aragon described the monument to Chappe at boulevard Saint-Germain as "the magnificent apotheosis of Chappe at the foot of a telegraphic scaffold"; as Simon Baker has recounted, the surrealists discovered in Paris's strange and prolific statues a powerful and absurd means of comprehending history.[33] In 1930, Robert Desnos's essay "Pygmalion and the Sphinx," published in *Documents*, included a photograph by Jacques-André Boiffard of Chappe's statue.[34] Dispensing with familiar complaints against statues, which were well known by then, Desnos and Boiffard's collaboration was a more complex interrogation of the failures of statuary's base materiality (tellingly, Boiffard's caption attributes the statue of Chappe to Farcy alone). As Baker describes, in Desnos's essay, "balloons of bronze, dust-clouds of stone, veins of marble, universally supported by the specious sophistry of the pedestal," are shown to be both marvelous and ludicrous.[35] Chappe's statue—and his telegraph—were an appropriate choice in this context, as from the outset the machine produced uneasy confluences and contradictions between weighty gestures and airy signs, between anthropomorphic metal and dusty clouds, and between the spindly device and its architectural support. Desnos's essay also reminds us of a key feature of the telegraph from its earliest days: its proximity to the monument (even as an antimonument), its status as a kind of quasi-sculpture, and its combination of surprising animation and stilled inexplicability. Both Galatea *and* Sphinx.

Shortly after Desnos's essay appeared, in *Le Surréalisme au service de la Révolution*, several correspondents—among them André Breton, Paul Éluard, and Tristan Tzara—responded to a questionnaire inquiring whether a selection of Paris's prominent statues should be retained, relocated, altered, transformed, or removed. Maurice Henry suggested that Chappe's statue should be left in place but hung with pieces of meat, ham, and sausages, which should be regularly replaced. Henry shared with Tzara the hope that the statue would be perfected by being painted in "brilliant colors imitating real colors." Georges Wenstein, channeling Nadar's preferred palette, claimed that with its telegraph painted in vermillion, the statue would "glorify pre-Romanticism."[36] These absurd plans for the monument recalled a satirical repurposing suggested five years earlier, in August 1928, by science fiction writer Gaston de Pawloski. Rejecting wholesale the recuperative enthusiasms of the Third Republic, de Pawlowski claimed that no one cared about Chappe any longer, and that his near-namesake Jean Chiappe, who had been appointed *préfet de police* the previous year, was far more relevant to contemporary concerns. De Pawlowski suggested that Chappe's head be replaced by that of Chiappe. The telegraph on the montaged statue would be animated by an electrical signal to direct traffic, which had become congested, with the arm of the Chappe-Chiappe machine rising every three minutes to press a button to change the direction of flow for cars coming from the rue du Bac, boulevard Raspail, or boulevard Saint-Germain.[37] Occupying its island at a busy junction, the Chappe statue had particular purchase in this context. In 1910, a pedestrian was run over when descending from a tram, the statue allegedly crowding them off the pavement into the oncoming traffic. As June Hargrove observes, "to all intents, the monument to Claude Chappe was accused of murder."[38]

De Pawlowski evidently enjoyed the similarity between the two men's names, but he may also have been slyly aware of the optical telegraph's reputation as a wing of the state, locked to outsiders and policing the futures of those who lived under it.

Perhaps, too, he considered the machine-human hybrids that merged telegraph operators with the devices they manipulated, which had populated the visual imagery of telegraphy for many years. However, while the history of Chappe's telegraph was briefly recuperated at the centenary of its first use and also survived in the rear-view mirror of enthusiasts such as Nadar, as the Third Republic drew to a close, with the Great Depression and war not far in the future, the optical telegraph seemed remote. The artwork, in the form of the statue, was a more pressing concern than the machine itself. Distant, ridiculous, and regressive, neither the telegraph nor the statue seemed sufficient to comprehend a tumultuous present. Except of course, as the responses of de Pawlowski and the surrealists showed, they became so precisely because of their quality of outmodedness, even if such a historical dynamic was not always admitted directly. These encounters were determined, as ever, by visual experience of the machine, which survived at scale in Damé's likeness.

Ultimately, the Chappe statue's tenure in Paris was short, for it was melted down by occupying forces in 1942. Sending the image of optical telegraphy back to the front, it seems appropriate, if depressing, that Chappe's telegraph was, in its monumental form, remade as military hardware, its mobilization for Vichy adding one more regime to those the machine had served. Louis Réau, following the removal of the statue, claimed that "no Parisian mourns the disappearance of Chappe, maneuvering at the crossroads of boulevard Raspail and boulevard Saint-Germain, like a scarecrow with sparrows, the dislocated arms of his aerial telegraph."[39] No fan of revolutionary *vandalisme*, Réau was perhaps serving cold revenge for Chappe's architectural interventions, as he had earlier described the telegraph "weirdly perched" on the chopped-down north tower of Notre-Dame-de-l'Épine.[40] But now the destruction of the monument can only feel like a loss that, as Raymond Queneau put it, is "hard to forgive."[41] What a different city we would see if the remaining statues to the great and good were accompanied by their archaic technical attributes—whether telegraphs or, as in the case of Bartholdi's extraordinary thirty-foot-high statue commemorating the aeronauts of the Siege of Paris, erected in 1906, a giant balloon with circling bronze pigeons, one attached by a wing tip.

A series of photographs by Pierre Jahan shows the remains of Parisian statues—including Chappe's—at a scrap metal dealer on

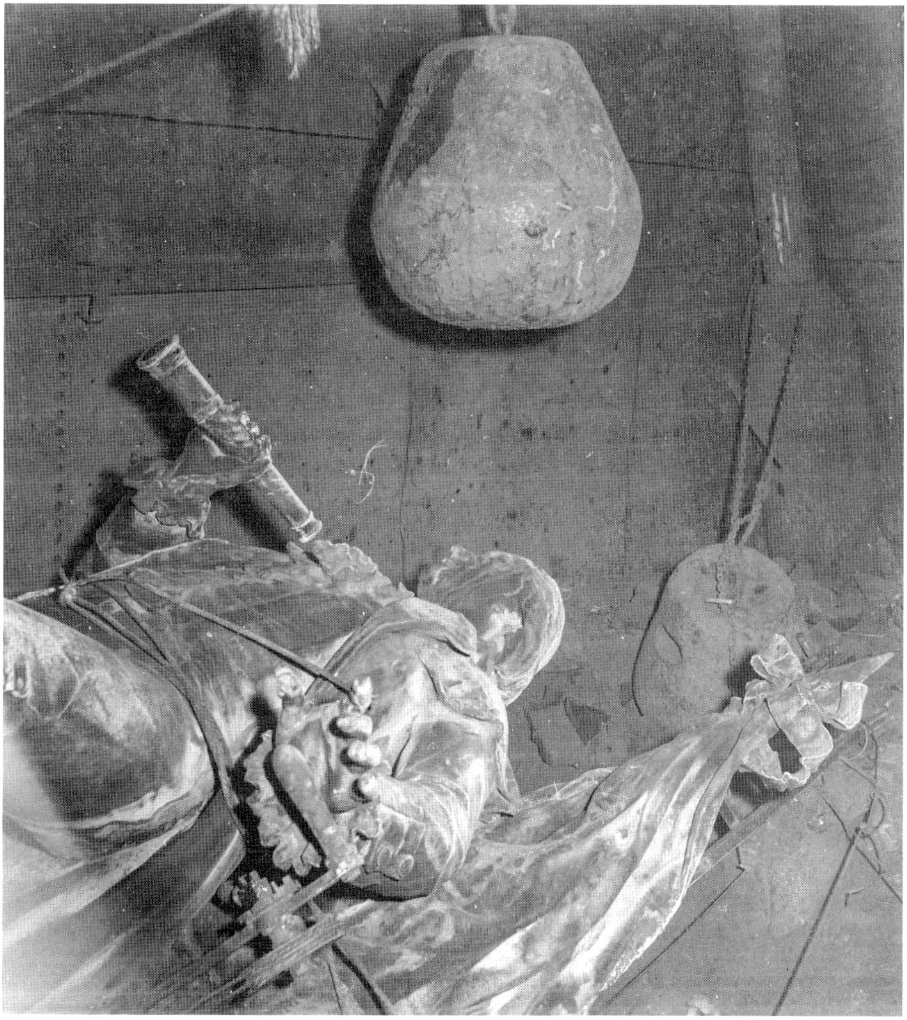

Figure 6.6
Pierre Jahan, *La Mort et les statues*, 1946, gelatin silver print, 29.8 × 23.9 cm. Centre Pompidou–Musée national d'Art moderne–Centre de création industrielle, Paris.

avenue du Général-Michel-Bizot, waiting to be repurposed (fig. 6.6). Seventeen thousand statues were removed in France during World War II; the revolutionary associations of Chappe and his machine did not endear him. Released of their heroic pretensions, these otherworldly fragments—fierce alligators from the place de la Nation; an armless Marat; the bear, eagle, and vulture from Square Montholon—appear strange and terrible in Jahan's photographs, their preposterous accouterments and sometimes jumbled iconography made to seem more fantastic and at the same time more material by abrupt dislocations, freakish couplings, and shifts of scale, as they were pulled horizontal, squashed, or dismembered.

In Jahan's image of Damé's Chappe statue, Claude is shown recumbent and still largely intact, his dusty body giving the image the appearance of a photographic negative. An enormous weight hangs above, Chappe reaching for the handle of his telegraph for the last time. As Jean Cocteau wrote in *La Mort et les statues*, the book containing Jahan's images, "Chappe awaits execution. Was he observing with his telescope the metal planet approaching from the depths of the centuries to annihilate him?"[42] Sontag's "planet of detours and delays" crash lands. Cocteau's tragic vision of history as a meteoroid strike-cum-militarized wrecking ball threatening imminent extinction would have been relatable in 1946, the year of its publication. But it also speaks to the way the Chappe telegraph had become a shorthand for the passing of time. Here, finally, the obsolescence of the machine—which had once been the very image of a vital modernity—is shown to be inseparable from that of the artworks representing it.

The Beating of Wings

E. M. Forster's 1909 novella *The Machine Stops* is a dystopian account of the consequences of advancing technologization. Published in response to H. G. Wells's *The Time Machine*, Forster's story describes a postapocalyptic world in which humans live below ground, their needs met by an all-powerful global "Machine." Communication happens via a kind of paratelegraphy, a premonition of instant messaging, by way of which people recirculate and reproduce "second-hand ideas." Leaving the Machine is a crime punishable by "homelessness." Ultimately, the Machine deteriorates, and since no one knows how to repair the failing technology, it stops, and the world with it.

In the time of the Machine, advanced thinkers warn against visiting the earth's surface, recommending viewing the "cinematophote," listening to the gramophone, or attending virtual lectures passed on by others, rather than seeking empirical knowledge. "'Beware of first-hand ideas!' exclaimed one of the most advanced of them," articulating the worldview of the Machine. Instead, "let your ideas be second-hand, and if possible tenth-hand, for then they will be far removed from that disturbing element—direct observation."[43] Thinking, here, is a matter of inscription and transmission, ideas passing from person to person, finessed by time and space, and stripped of danger or proximity. This satirical meditation on mediation segues into an example drawn from history:

> Do not learn anything about this subject of mine—the French Revolution. . . . You who listen to me are in a better position to judge about the French Revolution than I am. Your descendants will be even in a better position than you, for they will learn what you think. I think, and yet another intermediate will be added to the chain. And in time"—his voice rose—"there will come a generation that has got beyond facts, beyond impressions . . . which will see the French Revolution not as it happened, not as they would like it to have happened, but as it would have happened had it taken place in the days of the Machine.[44]

The telegraph may have gone, but in some ways we still think telegraphically. The past comes to meet the present, fooling us into believing that the present calls the shots. The French Revolution operates as an origin for knowledge in Forster's science fiction, but this information is ceaselessly deferred, passing through multiple relays, before being read back through contemporary concerns. And is this not, in truth, how we often encounter it? More than most political and social processes, the French Revolution has been mediated over 235 years to the point where its conflicting narratives are hard to disentangle from its source material or from its contemporary relevance, where "getting beyond facts" operates as a guise for celebrating history's diversions and distortions. "How this telegraph changes one's idea of life, and when the formalities and formulas are still more simplified, how full existence will be of facts and how free from uncertainties."[45] George Sand, writing to Flaubert in 1869, described well the optimistic view on such matters satirized by Forster. But it is stating the obvious to say that being "full of facts" doesn't make those facts more trustworthy.

Look, for instance, at a photograph by Paul Nadar, made around 1900, the same year his father published *Quand j'étais photographe* (fig. 6.7). It shows Juliette Beaumaine, an actress at the Théâtre des Variétés in Paris, set against a medley of newspapers. Arranged on the wall as if a trompe l'oeil, the papers spill onto the floor, polyphonous and apparently without limit. Among the titles *Le Figaro*, *La Justice*, *Gil Blas*, *La Nouvelle Presse*, and *L'Intransigeant* are what appear to be caricatures from illustrated weekly magazines. Beaumaine points with gloved hand to the front page of *Le Matin*, which carried a telegraph-wire motif on its masthead. Pens and an inkpot layer another means of inscription and relay. As if this were not plenty, Beaumaine's dress features, above the hem, the embroidered wires and poles of telegraph lines, one thread standing for another, while the Solomonic legs of the table recall the network's twisted cables, by then spanning the globe undersea and over land. Embroidered pigeons fly above the telegraph lines on her dress, and a syncopated diamond pattern evokes the stuttering rhythms of code or flag semaphore, the pattern detaching at the top of the dress and on Beaumaine's gloves.

To top things off, Beaumaine sports a memorable hat, composed of scrunched-up newspapers from which a taxidermied duck pokes its bill, mid-quack (a *canard*, slang for newspaper). It is a powerful look. A noisy photograph, the image squeezes in several competing informational systems. As Jonathan Sterne has pointed out, much of what we consider noise is in fact "simply a compression artifact," and indeed, this photograph seems to offer up compression as an object of aesthetic consumption.[46] The image is overlaid, in the bottom right-hand corner, by that familiar angular signature, inherited in this case.[47] Nadar's highly self-aware and overloaded photograph knowingly encapsulates its contemporary media world. But the language it uses to do so is based on artifice—this is a staging of a performance, after all—and the image prioritizes signs of transmission rather than the content of messages. In 1900 the optical telegraph seemed antiquated, a relic of a remote past, yet the overwhelming onslaught of news in Nadar's photograph and the difficulty of finding resolution to it in

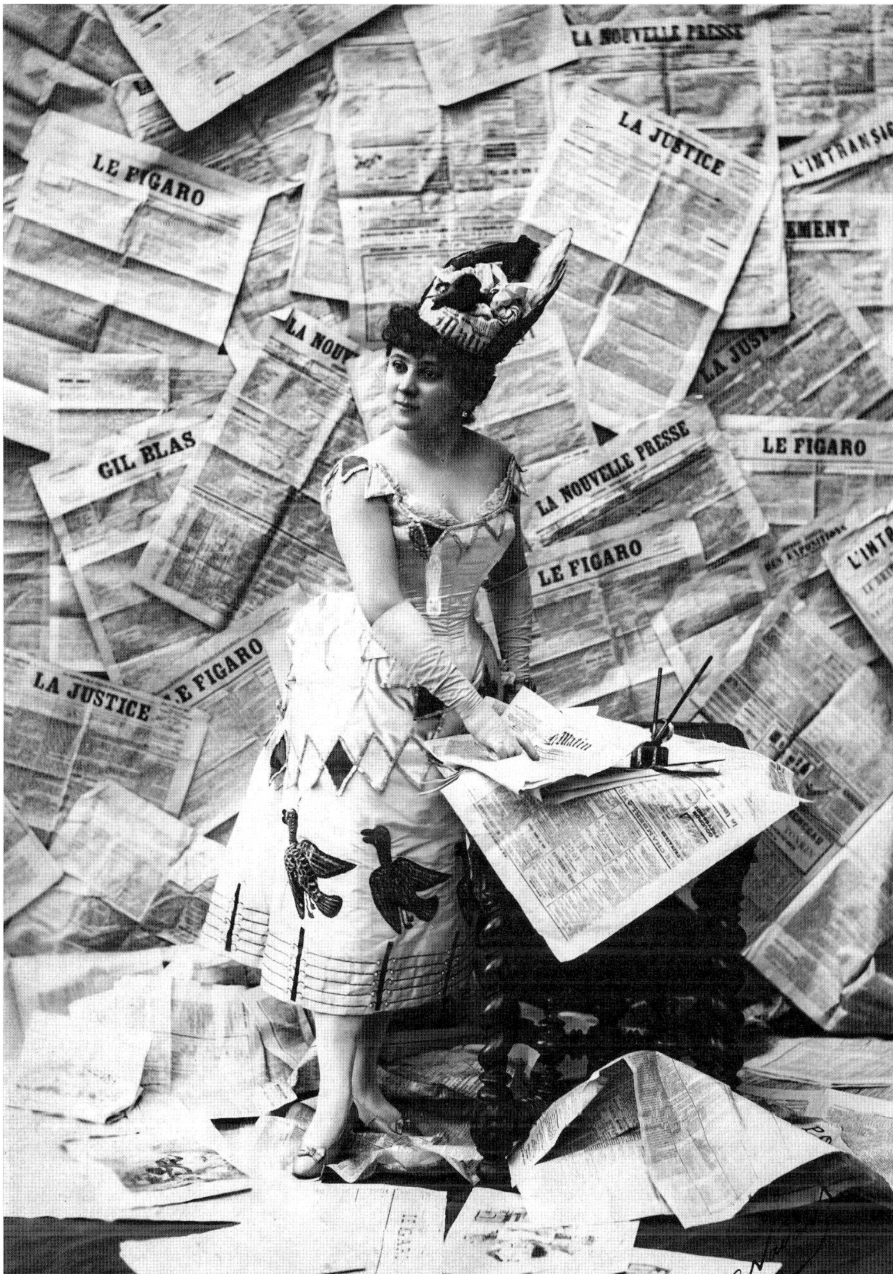

Figure 6.7
Paul Nadar, *Mlle Beaumaine, Théâtre des Variétés*, c. 1900, albumen silver print, 34.3 × 24.3 cm. The J. Paul Getty Museum, Los Angeles.

representation could be traced to the Chappe system's unreadable codes half a century earlier. While in some ways the optical telegraph's materiality and dependence on direct observation seem antithetical to the doomed virtual world described by Forster—a world that now chimes with our own—his book exaggerated for polemical effect tendencies that were first found in the telegraph's conspicuous yet sibylline acts.

Chappe's telegraph that appeared in the mid-1790s, at Belleville and later at Montmartre, animated a new social creature above the bones of the dead but did not leave the deceased behind. Reflecting on the dramatic times in which she had lived, anarchist, teacher, writer, and poet Louise Michel, best known for her role in the 1871 Paris Commune, remarked how "every era of history appears in spectral form. . . . This century is truly the red century, its purple shroud of blood covers the earth, but the air into which the harsh smell of mass graves rises is filled with the beating of wings."[48] Michel well understood that the past was never fully over, and that melancholy remainders of exploitation and tragic political failure necessarily coexisted with bright hope for a revolutionary future. Michel's "beating of wings" was at once ghostly and irreducibly material, insect-like, and alive: ephemeral, like a mayfly, dying to live again. Or perhaps like a telegraph: a more generative version of Victor Hugo's "big black insect."[49] Gesture, in her telling, became something historical, transformative, dynamic, and enduring.

A pair of paintings by Puvis de Chavannes, *Le Ballon* and *Le Pigeon*, both reproduced widely in lithographs by Emile Vernier, complement Michel's elegy. Although it put an end to Caselli's pantelegraph, the Franco-Prussian War and the Commune that followed witnessed versatile attempts to communicate in and out of the besieged city: by pigeon, by balloon, and underwater. Optical telegraphy reappeared briefly too—for example, the heliographic technique developed by aeronaut Eugène Godard—although it mainly retained its more predictable function as a means of government control.[50] While *Le Pigeon* more clearly reverberates with Michel's "beating of wings," Puvis's *Le Ballon*, completed during the siege, at the end of 1870, represents an armed woman on the ramparts of Paris, semaphoring to a balloon above Mont-Valérien, her flowing dress evoking the taffeta streamers proposed for

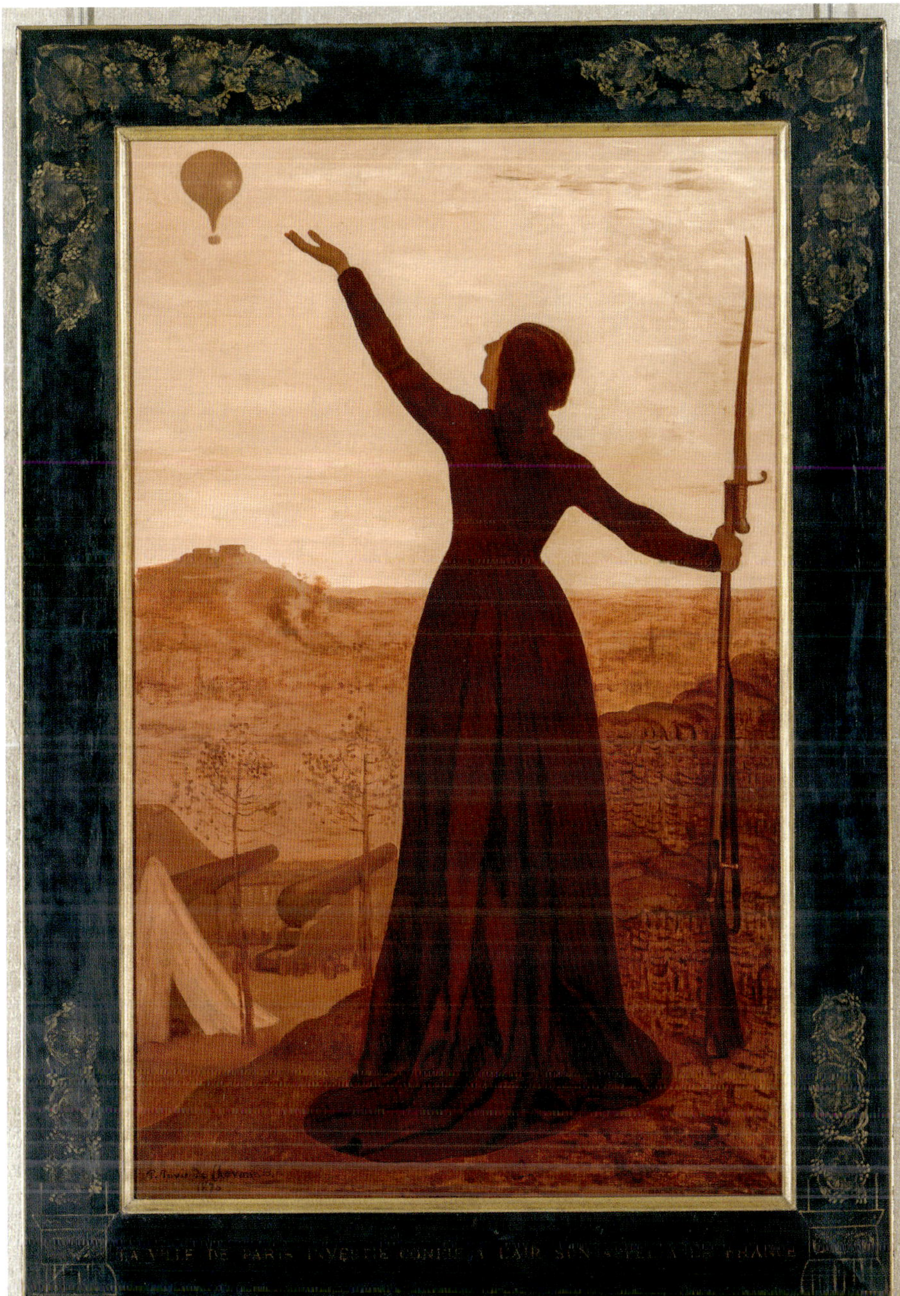

Figure 6.8
Pierre Puvis de Chavannes, *Le Ballon*, 1870, oil on canvas, 138.5 × 86 cm. Musée d'Orsay, Paris.

Conté's *aérographe* decades before (fig. 6.8). The site of an optical telegraph from the late 1790s, Mont-Valérien had been one of the first stops west out of Paris, located between Passy and Trou d'Enfer on the line to Brest. In 1841, with the construction of a fort at Mont-Valérien, the telegraph tower was moved inside the walls, where it stood long after the introduction of electrical signals. In 1893, the year Chappe's statue went up in Paris, a school of military telegraphy was installed there, and the year after the statue was melted down, in 1943, the optical telegraph tower at Mont-Valérien was destroyed by German troops, who used the fort as a prison and place of execution.

Puvis's paintings augured potential and hope, even in dark circumstances, but only for a brief window of time. A few months after they were painted, Versaillais forces used the strategic advantage of Mont-Valérien's key artillery position to brutal ends in the slaughter of the *semaine sanglante*, helping to crush the Commune. Belleville, and Père Lachaise, within sight of Chappe's telegraphic tomb, witnessed some of the last fighting, as well as cruel reprisals. Still, despite the somberness of *Le Ballon*, Puvis's figure signals to a future that remains luminous. It is an image that is telegraphic not just because it shows communication between points that were once linked by the optical telegraph, whose remains still occupied the space, or because its figure's stately gestures seem haunted by the machine (although both of those things are true of the painting). More importantly, in its evocation of temporal lags and accelerations, containment and liberty, it evinces a shared faith in communication, and in history. Writing or drawing at a distance was, years after the extinction of the optical system, connected to a long material, social, and political past and bound to forms of political potential, in ways that even its subsequent freezing on a Paris traffic island could not extinguish. Indeed, Michel's bracing allegorical interpretation of the nineteenth century had diagnosed a temporal and political dynamic shared by telegraphy, a historical process in which destruction, failure, or obsolescence might be countered by a dream image of possibility. Her image was vivid, visceral, and suffused with telegraphy's crooked insect metamorphoses. For optical telegraphy and its images had throughout their intertwined history described revolutionary survivals, optimism, and hope, as well as the wreckage of the past. As Michel concluded: "This is why our century, red with bloodshed, also has the glow of a new day, for the same entity contains at once both the monsters of our ancestors and the humanity of tomorrow; leaving scraps of its flesh in its chrysalis shell, it sets off with its wings torn but able to soar."[51]

In the days of the telegraphic machine in nineteenth-century France, visual culture was uniquely poised to appraise the telegraph's impassive gestures and to elucidate its reconfiguration of time. Equally, telegraphy gave new meaning to art's shifting temporality. The nineteenth century's kaleidoscopic world of images was open to the possibilities described by Michel, for it too was characterized by proliferation and endurance as well as disposability and waste. One of art's characteristics, and perhaps one of its enduring myths, is its survival. For the most part, we encounter art now, in the present, rather than in the time of its creation, a truism not unique to the works discussed in this book. But such survivals situate artworks in history, rather than setting them outside it. In the many images and objects that engage with the optical telegraph—which halt or reperform its signs; which explore its revelations as well as its forbidding inscrutability; which activate other aspects of its environment; which place it in a landscape or exclude it from one; which mobilize it to rethink ideas about repression, freedom, and subjectivity; which modify categories of obsolescence and durability; or which use it to expose afresh the characteristics of a medium—we are ultimately, like their initial audiences, seeing old news. Images, even those contemporary to the network, retreat to this past of speedups and slowdowns, then return to tell us about it, passing on the message "as it would have happened." It is art itself that is the time machine.

Acknowledgments

As I expect often happens, the origins of this book lie in a previous one. Years ago, while sifting through documents in the Archives nationales, I came across many applications for passports, signed "Chappe," requesting permission for workers to travel for the construction and administration of a new telegraph system. Who knew that there was a telegraph in the 1790s? I didn't. But it soon made sense, and my earlier preoccupation with how ephemeral things become the stuff of history carries into this book. Now, I see telegraphs everywhere I look. Maybe you will too, after reading this.

I couldn't have written this book without the insight, generosity, and encouragement of colleagues, family, and friends. I'm so grateful to colleagues in the History of Art Department at University College London for making it the uniquely exciting and exacting place it is, and for their friendship and solidarity. Thank you, especially, Robert Brown, Eleanor Day, Diana Dethloff, Natasha Eaton, Briony Fer, Emily Floyd, Tamar Garb, Jacopo Gnisci, Hanna Hölling, Cadence Kinsey, Aparna Kumar, Pip Laurenson, Hélia Marçal, Robert Mills, Rosemary Moore, Jenny Nachtigall, Mignon Nixon, Jacob Paskins, Naomi Quinn-Friedman, Emma Richardson, Frederic Schwartz, and Alison Wright. Mechthild Fend, Charles Ford, Maria Loh, Nicholas Robbins, Rose Marie San Juan, Stephanie Schwartz, and Allison Stielau have been inspiring interlocutors, and their input on this project has been particularly valuable. My students at UCL have endured my telegraphic obsessions with astonishing fortitude, and I'm tremendously thankful to them for their eagerness to engage with these and all ideas. This book is for them, too.

The award of a Philip Leverhulme Prize in 2012 was transformative, and I thank the Leverhulme Trust for granting me the time and resources to get this project moving. A Mid-Career Fellowship from the British Academy made its completion possible. Also stimulating was a period in 2014 as a Visiting Fellow at the Bard Graduate Center. Thank you, Jeffrey Collins, Marc LeBlanc, Daniel Lee, Peter Miller, and Paul Stirton for being such generous hosts, as well as Meredith Martin and Sina Najafi for welcoming me in New York.

In the spirit of the topic, communications with scholars and friends spread far and wide enriched this book and my experience of writing it. I am beyond grateful to Katie Hornstein, Jennifer L. Roberts, and John Tresch for their breathtakingly wise readings of parts or all of the manuscript. Colin Jones was, as ever, a model of collegiality and intellectual rigor, taking the time to send me telegraphic tidbits while conducting his own research (all of which, of course, found their way into the book). I value deeply Katie Scott's and Tom Gretton's exemplary encouragement to interrogate the relationship between the material, the technical, and the political. Stephen Bann's expertise and support at several crucial junctures has been indispensable, as has that of David Bindman. Inspiring discussions with Susan Siegfried have motivated me to think about the stakes of this project, while Iris Moon gave remarkable feedback and has been a brilliant and imaginative collaborator. I hope they see traces of our conversations in this text; any errors that remain are, of course, all mine.

Discussions with telegraph scholars Caroline Arscott and Clare Pettitt have been generative, and I'm grateful to the Scrambled Messages research group led by Caroline and Clare for being such ingenious and fun allies. Collaboration with Natalie Adamson, and earlier with the Multigraph Collective in Montreal, has been hugely energizing and a great source of camaraderie. In London, I've been lucky to talk about telegraphic matters (and also relieved to talk about anything but) with Steven Adams, Alexandra Ault, Nicholas Babbington, Emma Barker, Sam Bibby, Esther Chadwick, Allison Deutsch, Kirsty Sinclair Dootson, Steve Edwards, Seb Franklin, Andrea Fredericksen, Marie Giraud, Allison Goudie, Mark Hallett, Rosalind Hayes, Simon Macdonald, Eckart Marchand, Jann Matlock, Lynda Nead, Satish Padiyar, Nina Pearlman, Caspar Pearson, Barbara Penner, Sanja Perovic, Domenico Pino, Marcia Pointon, Alex Potts, Dorothy Price, Kate Retford, Adrian Rifkin, Isabelle Sagraves, David Solkin, Tom Stammers, Glynnis Stevenson, Sarah Victoria Turner, Francesca Whitlum-Cooper, Hannah Williams, Joanna Woodall, and Richard Wrigley. Equally, I have benefited greatly from the wisdom, advice, and friendship of revolutionary thinkers and *dix-neuvièmistes extraordinaires* Juliet Bellow, Daniella Berman, Francesca Berry, Sarah Betzer, Keith Bresnahan, Ting Chang, Holly Clayson, Malcolm Crook, Kathryn Desplanque, André Dombrowski, Emily Doucet, Julia Douthwaite, Nina Dubin, Alex Fairfax-Cholmeley, Jennifer Ferng, Alan Forrest, Linda Goddard, Lela Graybill, Daniel Harkett, Lynn Hunt, Mary Hunter, Matthew Hunter, Laura Kalba, Marika Takanishi Knowles, Jillian Lerner, Camille Mathieu, Sophie Matthiesson, Neil McWilliam, Jeremy Melius, Sarah Monks, Erika Naginski, Stephanie O'Rourke, Kelly Presutti, Jordan Rose, Christina Smylitopoulos, and Charles Walton.

For essential conversations, collaborations, and kindnesses—either fleeting yet formative or sustained over many years—I thank too James Baker, Joshua Bauchner, Julia Bryan-Wilson, Thomas Dodman, Michael Gaudio, Anthony Geraghty, Diana Greenwald, Hanneke Grootenboer, Hans Christian Hönes, Anne Hultzsch, Mari

Hvattum, Richard Johns, Evgenii Kozlov, Karen Lang, Petra Lange-Berndt, Mari Lending, Hélène Lipstadt, Isabelle Masse, Guillaume Mazeau, Richard Menke, Jennifer Milam, Asher Ethan Miller, David Mitchell, Simone Natale, Alison O'Byrne, Melissa Percival, John Durham Peters, Amalia Pica, Jeff Ravel, John David Rhodes, Alistair Rider, Dietmar Rübel, Kristel Smentek, Alice Spawls, Jonathan Sterne, Angela Vanhaelen, and Genevieve Warwick. I was fortunate to have had the opportunity to discuss parts of this book, in various states, at a number of institutions. I'm enormously grateful to organizers, participants, and audience members at those events—some named above—whose responses shaped this book decisively.

Several archives and collections were essential to my research. The Musée de La Poste in Paris holds an unparalleled collection on the history of communication in France. Patrick Bruneau, Julie Carbonnier, Elodie Goëssant, and Chantal Reynaud were helpful in granting me access, even during the museum's renovation, and provided important assistance. The singular Musée des Arts et Métiers has in many ways provided an intellectual frame for this book, a space to think with as well as in. I thank Anne Chanteux, Agnès Meynard, and Isabelle Taillebourg, as well as staff of the CNAM Archives and Library, especially Lise Cloitre and Emmanuelle Minault-Richomme. The exceptional collections of the Musée Carnavalet and the Bibliothèque nationale de France have been vital to this project and deserve special acknowledgment.

I am also indebted to the staffs of the Archives de Paris; Archives départementales des Yvelines; Archives nationales; Bibliothèque historique de la Ville de Paris; Bibliothèque historique des postes et des télécommunications; British Museum Department of Prints and Drawings; Fogg Art Museum and Houghton Library, Harvard University; Fondation Custodia; Getty Research Institute; Metropolitan Museum of Art; Morgan Library and Museum; Musée de l'Armée; Musée de la Révolution française, Vizille; Musée des Beaux-Arts, Rouen; Musée Fabre, Montpellier; Museum für Kommunikation, Frankfurt; New York Public Library; UCL Art Museum; UCL Library; and the Warburg Institute Library. Last, a special—un-libraryishly loud—shout-out to the staff of the Rare Books and Music Reading Room at the British Library, where much of this book was written.

It has been a delight to work with Thomas Weaver, my editor at MIT Press. I'm so grateful to Tom for seeing something in this project, and for his wit, wisdom, and patience as the book has progressed. Thank you, also, Gabriela Bueno Gibbs and the team at MIT, especially Matthew Abbate and Paula Woolley, for expertly steering the book to completion. Parts of *Time Machines* have appeared in *nonsite.org* (2014); *Art History* (2016); *Oxford*

Art Journal (2016); *Photography and Other Media in the Nineteenth Century*, ed. Simone Natale and Nicoletta Leonardi (Penn State University Press, 2018); *A History of the European Restorations, 1815–1848*, vol. 2, ed. Michael Broers, Ambrogio A. Caiani, and Stephen Bann (Bloomsbury, 2019); and *Time, Media, and Visuality in Post-Revolutionary France*, ed. with Iris Moon (Bloomsbury, 2021). I'm grateful to the editors and anonymous readers of those publications, in particular Bridget Alsdorf, Stephen Bann, Sam Bibby, Nicoletta Leonardi, Margaret Michniewicz, Iris Moon, Simone Natale, Katie Scott, Genevieve Warwick, and Richard Wrigley. For generous assistance with images, I'm obliged to Scott Krafft and Olivier Lacroix. My great thanks to Fiona Benson for graciously allowing me to borrow a few lines from her stunning poem.

Delays and disjunctures, but also powerful moments of connection and love, occasioned by lives lived apart, then coming together again, shaped this book at the start, and later, when the pandemic arrived, in ways none of us could anticipate—nothing has made me more sensitive to the affective dimensions of media than relationships separated by distance or clouded by time. The loss of three people marked the writing of this book: Gareth Jones, my oldest friend and art-adventurer; Helen Weston, an inspirational scholar and teacher; and Lisa Chandler. They are missed terribly.

My family has been an unwavering source of encouragement and cheer, and I dedicate this book to them with all my love and boundless gratitude. A few years ago, Marina de Alarcón responded to my tentative, belated semaphore with an immediacy and warmth that exceeded all expectations. Thank you, Marina and Ollie, and my wonderful nephews, Solomon, Salvador, and Santiago. My thanks also to Marion, Karen, Darren, Lizzie, Toby, and Paige. James, much missed, delighted everyone he met. My mother Elizabeth Taws is my greatest champion, and I owe her everything. Jo Applin—who travels through time with me—is both the signal and the code.

Illustration Credits

Fig. 1.1 Musée du Domaine royal de Marly, Marly-le-Roi.
Fig. 1.2 CCO Paris Musées/Musée Carnavalet–Histoire de Paris.
Fig. 1.3 Museumsstiftung Post und Telekommunikation.
Fig. 1.4 Musée de La Poste, Paris–La Poste, 10923. © Musée de La Poste, Paris/La Poste, 2023/Photo TD.
Fig. 1.5 Museumsstiftung Post und Telekommunikation.
Fig. 1.6 Musée des Arts et Métiers–Cnam/Michèle Favareille.
Fig. 1.7 Musée de La Poste, Paris–La Poste, 10929. © Musée de La Poste, Paris/La Poste, 2023/Photo TD.
Fig. 1.8 Musée de La Poste, Paris–La Poste, 2079. © Musée de La Poste, Paris/La Poste, 2023/All rights reserved.
Fig. 1.9 Musée de La Poste, Paris–La Poste, 13021. © Musée de La Poste, Paris/La Poste, 2023/Photo TD.
Fig. 1.10 Yale Center for British Art, Paul Mellon Collection.
Fig. 1.11 Musée de La Poste, Paris–La Poste, 4008. © Musée de La Poste, Paris/La Poste, 2023/Photo TD.
Fig. 1.12 BnF.
Fig. 1.13 Musée de La Poste, Paris–La Poste, 2008.143.1. © Musée de La Poste, Paris/La Poste, 2023/All rights reserved.
Fig. 1.14 Musée de La Poste, Paris–La Poste, 2012.104.1. © Musée de La Poste, Paris/La Poste, 2023/Photo MF.
Fig. 1.15 BnF.
Fig. 1.16 BnF.
Fig. 1.17 © RMN-Grand Palais (Musée du Louvre)/René-Gabriel Ojeda.
Fig. 1.18 Museumsstiftung Post und Telekommunikation.
Fig. 1.19 Harvard Art Museums/Fogg Museum, Museum Collection. © President and Fellows of Harvard College, M22040.
Fig. 1.20 Harvard Art Museums/Fogg Museum, Bequest of Grenville L. Winthrop. © President and Fellows of Harvard College, 1943.245.
Fig. 1.21 CCO Paris Musées/Musée Carnavalet–Histoire de Paris.
Fig. 1.22 National Gallery, London, Presented by Emilie Yznaga 1945. © The National Gallery, London.
Fig. 1.23 CCO Paris Musées/Musée Carnavalet–Histoire de Paris.
Fig. 1.24 © Northwestern University. Peter Moore Photography Archive, Charles Deering McCormick Library of Special Collections, Northwestern University Libraries.
Fig. 2.1 BnF.
Fig. 2.2 © RMN-Grand Palais (Château de Versailles)/Gérard Blot.
Fig. 2.3 CCO Paris Musées/Musée Carnavalet–Histoire de Paris.
Fig. 2.4 Musée de La Poste, Paris–La Poste, D11038. © Musée de La Poste, Paris/La Poste, 2023/All rights reserved.

Fig. 2.5 Musée de La Poste, Paris–La Poste, 22411. © Musée de La Poste, Paris/La Poste, 2023/All rights reserved.

Fig. 2.6 bpk/Hamburger Kunsthalle/Hanne Moschkowitz.

Fig. 2.7 BnF.

Fig. 2.8 BnF.

Fig. 2.9 Musée de La Poste, Paris–La Poste, 7491. © Musée de La Poste, Paris/La Poste, 2023/Photo MF.

Fig. 2.10 BnF.

Fig. 2.11 CC0 Paris Musées/Musée Carnavalet–Histoire de Paris.

Fig. 2.12 CC0 Paris Musées/Maisons de Victor Hugo Paris–Guernesey.

Fig. 2.13 Musée de La Poste, Paris–La Poste, 2083. © Musée de La Poste, Paris/La Poste, 2023/Photo MF.

Fig. 2.14 BnF.

Fig. 2.15 © Musée Fabre de Montpellier Méditerranée Métropole/ photographie Frédéric Jaulmes–Reproduction interdite sans autorisation.

Fig. 2.16 BnF.

Fig. 2.17 BnF.

Fig. 2.18 © Nancy, Musée des Beaux-Arts.

Fig. 2.19 © Coll. Musée de la Révolution française–Département de l'Isère.

Fig. 2.20 © Agence La Belle Vie/Réunion des Musées Métropolitains Rouen Normandie.

Fig. 2.21 Photograph by author.

Fig. 2.22 © Coll. Musée de la Révolution française–Département de l'Isère.

Fig. 2.23 CC0 Paris Musées/Musée Carnavalet–Histoire de Paris.

Fig. 2.24 Courtesy of the Smithsonian Libraries and Archives, Washington, D.C.

Fig. 2.25 CC0 Paris Musées/Musée Carnavalet–Histoire de Paris.

Fig. 2.26 © RMN-Grand Palais (Musée du Louvre)/Michel Urtado.

Fig. 2.27 © Musée Fabre de Montpellier Méditerranée Métropole/ photographie Frédéric Jaulmes–Reproduction interdite sans autorisation.

Fig. 3.1 Courtesy of the Getty Open Content Program.

Fig. 3.2 BnF.

Fig. 3.3 © The Board of Trustees of the Science Museum.

Fig. 3.4 Courtesy of the Library of Congress, Washington, D.C.

Fig. 3.5 © Musée des Arts et Métiers–Cnam/Pascal Faligot.

Fig. 3.6 New York Public Library Digital Collections.

Fig. 3.7 New York Public Library Digital Collections.

Fig. 3.8 © Musée des Arts et Métiers–Cnam/Sylvain Pelly.

Fig. 3.9 Heidelberg University Library.

Fig. 3.10 Heidelberg University Library.

Fig. 3.11 © Musée des Arts et Métiers–Cnam/Philippe Hurlin.
Fig. 3.12 Photograph by author.
Fig. 3.13 © Musée des Arts et Métiers–Cnam/Jean-Claude Wetzel.
Fig. 3.14 Photograph by author.
Fig. 3.15 Musée de La Poste, Paris–La Poste, 2012.81.19. © Musée de La Poste, Paris/La Poste, 2023/Photo MF.
Fig. 3.16 Musée de La Poste, Paris–La Poste, 2012.81.20. © Musée de La Poste, Paris/La Poste, 2023/Photo MF.
Fig. 3.17 Photograph by author.
Fig. 3.18 © Paris–Musée de l'Armée, Dist. RMN-Grand Palais/Pascal Segrette.
Fig. 3.19 New York Public Library Digital Collections.
Fig. 3.20 New York Public Library Digital Collections.
Fig. 3.21 Museum of Modern Art (MoMA), New York, Ruth Vollmer Bequest, 318.1983. Digital Image © 2004, The Museum of Modern Art/Scala, Florence.
Fig. 4.1 CC0 Paris Musées/Musée Carnavalet–Histoire de Paris.
Fig. 4.2 Brown Digital Repository.
Fig. 4.3 CC0 Paris Musées/Musée Carnavalet–Histoire de Paris.
Fig. 4.4 CC0 Paris Musées/Musée Carnavalet–Histoire de Paris.
Fig. 4.5 © Musée Fabre de Montpellier Méditerranée Métropole/photographie Frédéric Jaulmes–Reproduction interdite sans autorisation.
Fig. 4.6 Library and Archives of the Natural History Museum, London.
Fig. 4.7 CC0 Paris Musées/Musée Carnavalet–Histoire de Paris.
Fig. 4.8 CC0 Paris Musées/Musée Carnavalet–Histoire de Paris.
Fig. 4.9 CC0 Paris Musées/Musée Carnavalet–Histoire de Paris.
Fig. 4.10 BnF.
Fig. 4.11 © Musée Fabre de Montpellier Méditerranée Métropole/photographie Frédéric Jaulmes–Reproduction interdite sans autorisation.
Fig. 4.12 CC0 Paris Musées/Musée Carnavalet–Histoire de Paris.
Fig. 4.13 CC0 Paris Musées/Musée Carnavalet–Histoire de Paris.
Fig. 4.14 CC0 Paris Musées/Musée Carnavalet–Histoire de Paris.
Fig. 4.15 BnF.
Fig. 4.16 BnF.
Fig. 4.17 CC0 Paris Musées/Musée Carnavalet–Histoire de Paris.
Fig. 4.18 CC0 Paris Musées/Musée Carnavalet–Histoire de Paris.
Fig. 4.19 © RMN-Grand Palais (Musée du Louvre)/Michel Urtado.
Fig. 4.20 Metropolitan Museum of Art, The Horace W. Goldsmith Foundation Gift, through Joyce and Robert Menschel, 1987.
Fig. 4.21 Belvedere, Vienna, inv. 7946.
Fig. 4.22 National Galleries of Scotland, Presented by George R. MacDougall 1911.

Fig. 4.23 Musées de Strasbourg, M. Bertola.
Fig. 4.24 Yale University Art Gallery, Stephen Carlton Clark, BA 1903, Fund.
Fig. 4.25 Fondation Custodia, Collection Frits Lugt, Paris.
Fig. 4.26 © RMN-Grand Palais (Musée du Louvre)/Michel Urtado.
Fig. 4.27 The Morgan Library & Museum, New York, 2004.6, Gift of John M. Thayer.
Fig. 4.28 The Metropolitan Museum of Art, New York, Harris Brisbane Dick Fund, 1932.
Fig. 4.29 Canadian Centre for Architecture, Montreal.
Fig. 5.1 © Musée des Arts et Métiers–Cnam/Sylvain Pelly.
Fig. 5.2 Wikimedia Commons/Moonik/CC BY-SA 3.0.
Fig. 5.3 Photograph by author.
Fig. 5.4 CC0 Paris Musées/Musée Carnavalet–Histoire de Paris.
Fig. 5.5 Harvard Art Museums/Fogg Museum, Gift of Horace M. Swope. © President and Fellows of Harvard College, M4827.
Fig. 5.6 Museumsstiftung Post und Telekommunikation.
Fig. 5.7 Courtesy of the Getty Open Content Program.
Fig. 5.8 © Musée des Arts et Métiers–Cnam/Pascal Faligot.
Fig. 5.9 © Musée des Arts et Métiers–Cnam/Pascal Faligot.
Fig. 5.10 © Musée des Arts et Métiers–Cnam/Studio Cnam.
Fig. 5.11 © Musée des Arts et Métiers–Cnam/Pascal Faligot.
Fig. 5.12 © Musée des Arts et Métiers–Cnam/Pascal Faligot.
Fig. 5.13 BnF.
Fig. 5.14 The Warburg Institute, London.
Fig. 5.15 The Art Institute of Chicago, William McCallin McKee Memorial Endowment.
Fig. 5.16 © The Trustees of the British Museum, London.
Fig. 5.17 BnF.
Fig. 6.1 Rijksmuseum, Amsterdam.
Fig. 6.2 Wellcome Collection, London.
Fig. 6.3 CC0 Paris Musées/Musée Carnavalet–Histoire de Paris.
Fig. 6.4 CC0 Paris Musées/Musée Carnavalet–Histoire de Paris.
Fig. 6.5 BnF.
Fig. 6.6 © Pierre Jahan/Roger-Viollet; © Centre Pompidou, MNAM-CCI, Dist. RMN-Grand Palais/Jean-Claude Planchet.
Fig. 6.7 Courtesy of the Getty Open Content Program.
Fig. 6.8 © RMN-Grand Palais (Musée d'Orsay)/Hervé Lewandowski.

Objects from the Musée des Arts et Métiers are on display in the Communication gallery, Musée des Arts et Métiers, 60 rue Réaumur, Paris. http://www.arts-et-metiers.net/les-collections.

The collections of the Musée de La Poste include objects, works of art, documents, and archives related to the history of the French postal service and its professions, as well as the history of telecommunications, from the sixteenth century to the present day. The history of the French stamp (created in 1849) is also kept at the museum, as well as an extensive collection of production archives and letters illustrating the evolution of postmarks. The collection contains various items related to Claude Chappe's telegraphic system, including images, archives, and pieces of technical equipment. https://www.museedelaposte.fr/.

Notes

All translations are mine unless stated otherwise.

Epigraphs

1. Jean Paul, quoted in Harry Franz, "La nécessité rend inventif! Le télégraphe Chappe de Strasbourg vu par des voyageurs allemands (1798–1816)," *Les Cahiers de la* FNARH 96 (2005). Founded in 1984, the FNARH (Fédération National des Associations de personnel de La Poste et d'Orange pour la Recherche Historique) has generated much important research on the optical telegraph, to which I am indebted.

2. Roland Barthes, *Roland Barthes by Roland Barthes*, trans. Richard Howard (New York: Hill and Wang, 2010), 90–91.

Chapter 1

1. Formerly chemin de la Glacière, the street was renamed rue du Télégraphe in 1863. See Jacques Hillairet, *Dictionnaire historique des rues de Paris* (Paris: Éditions de Minuit, 1972), 540.

2. The cimetière de Belleville was inaugurated in 1804. When the area was used for telegraphy, it comprised a château and park belonging to the Lepeletier de Saint-Fargeau family. The commune of Belleville was not incorporated into Paris until 1859.

3. Archives Nationales, DLII/1.

4. On such processes of naturalization, see Lisa Gitelman, *Always Already New: Media, History, and the Data of Culture* (Cambridge, MA: MIT Press, 2006), 5–7.

5. This device predated the final version of the machine and was based on a system of encoded shutters. Claude Chappe reported the attack to the Legislative Assembly on September 12, 1792, as follows: "I have, consequently, had several machines necessary for this operation built in great numbers; I had one established in Belleville, two others were about to be finished and placed when I learned that a gathering of some inhabitants of the commune of Belleville and its surroundings had broken up and destroyed all these preparations, believing that they were destined to serve the projects of our enemies; they are currently threatening my life, as well as that of a citizen living in Belleville whom they suspect of having conspired with me in the placing of this machine." Archives Nationales, F/90/1427–1428 A 3.

6. The description of the telegraph as a "terrifying logogriph" appears in an anonymous poem "Le Reveil," published in *La Caricature*, December 8, 1831, 462.

7. Gustave Nadaud, "Le Vieux Télégraphe," in *Chansons de Gustave Nadaud* (Paris: Henri Plon, 1870), 255.

8. In Stendhal's *Lucien Leuwen*, written between 1830 and 1840 and published posthumously in 1855, Leuwen describes the telegraph as "cette machine diabolique." The telegraph plays a central role in Stendhal's novel, which was originally intended to be titled *Le Télégraphe*. Stendhal, *Lucien Leuwen* (Paris: Livre de Poche, 2007), 375.

9. The Chappe brothers replaced *tachygraphe*, meaning "that which writes fast" with *télégraphe* (that which writes at a distance) on the suggestion of diplomat André-François Miot de Mélito, in conversation with Ignace Chappe in April 1793. "Sténographie numérique" was also an early name. *Tachygraphe* was used regularly before the Revolution to designate professional shorthand-takers transcribing official proceedings; the rise of postrevolutionary parliamentary democracies led to increasing professionalization, and by the time of the July Monarchy, they were replaced by stenographers. See, for instance, Igonel et Breton, *A la Convention Nationale. Établissement d'un Journal Sténographique des travaux de la Convention* (Paris: Lycée des Arts, Year III). On the historiographical role of *logographes*, *tachygraphes*, and *stenographes*, see Linda Orr, "The Blind Spot of History: Logography," *Yale French Studies* 73 (1987): 190–214. For Orr, although the roots of the stenographic turn lay in an eighteenth-century obsession with economical speech and specifically with the Revolution's desire to couple writing and speech, it was only with the 1848 Revolution that the stenographer entered fully onto the historical stage.

10. Victor Adam, *Nouvel abécédaire en enigmes* (Paris: Imprimerie Lemercier, 1833).

11. *Le Télégraphe politique, ou, Journal des fondateurs de la République* (Paris: Imprimerie du Télégraphe, 1794). On the Haitian example, see Chelsea Stieber, *Haiti's Paper War: Post-Independence Writing, Civil War, and the Making of the Republic, 1804–1954* (New York: New York University Press, 2020). At the other end of the value spectrum, and well after the machine's demise, a stylized Chappe telegraph was adopted

from 1880 as the hallmark for silversmith Désiré Thorel fils.

12. See, importantly, Sophia Rosenfeld, *A Revolution in Language: The Problem of Signs in Late Eighteenth-Century France* (Stanford: Stanford University Press, 2001).

13. Yakup Bektas, "The Crimean War as a Technological Enterprise," *Notes and Records of the Royal Society* 71, no. 3 (September 2017): 233–262.

14. "Le télégraphe," *Vert-Vert, Gazette de Paris* 310 (November 6, 1835): 1.

15. On this issue, see Kirsty Sinclair Dootson, *The Rainbow's Gravity: Colour, Materiality and British Modernity* (New Haven: Yale University Press; London: Paul Mellon Centre for Studies in British Art, 2023), 8.

16. Baron to Claude Chappe, 10 Pluviôse Year II [January 29, 1794], Archives Nationales, F/90/1427–1428 A 20. Cited in Gérard Contant and Marie-Dominique Leclère, eds., *Correspondance des frères Chappe, 1792–1798* (place of publication and publisher unknown, 1993), unpaginated. Bibliothèque historique des postes et télégraphes, TA 00180. See also Alexis Belloc, *La Télégraphie historique: depuis les temps les plus reculés jusqu'à nos jours* (Paris: Firmin-Didot, 1894), 93–94. Belloc reports that Baron was one of Abraham Chappe's collaborators. A manuscript addition to this letter in Claude's handwriting records his response, which recommended to Baron that the machines should not be demonstrated at the present time.

17. "Chappe, ingénieur télégraphe, au Directoire Exécutif," Archives Nationales, AF/II/220, folder 8, document 24.

18. Thomas Carlyle, *The French Revolution, A History*, vol. 3 (London: Chapman and Hall, 1848), 290; *Instruction à l'usage des stationnaires télégraphiques concernant le mécanisme et autres parties du matériel* (Paris: Imprimerie de Béthune, 1827), 24.

19. Helmut Müller-Sievers, *The Cylinder: Kinematics of the Nineteenth Century* (Berkeley: University of California Press, 2012).

20. Müller-Sievers, *The Cylinder*, 42.

21. Carlyle, *The French Revolution*, 292. On the visual dynamics of the guillotine, see Daniel Arasse, *The Guillotine and the Terror*, trans. Christopher Miller (London: Allen Lane, 1989).

22. See the detailed technical analysis in Alexander J. Field, "French Optical Telegraphy, 1793–1855: Hardware, Software, Administration," *Technology and Culture* 35, no. 2 (April 1994): 315–347. See also Guy de Saint-Denis, ed., *La Télégraphie Chappe* (Nancy: FNARH/Éditions de l'Est, 1993); *La Télégraphie aérienne et optique: bicentenaire du télégraphe Chappe: colloque international, 8 octobre 1993, Paris* (Nancy: FNARH, 1995); and Gerard J. Holzmann and Björn Pehrson, *The Early History of Data Networks* (Los Alamitos: Wiley-IEEE Computer Society Press, 1995).

23. Jonathan Sterne makes this point regarding the optical telegraph in "Compression: A Loose History," in *Signal Traffic: Critical Studies of Media Infrastructures*, ed. Lisa Parks and Nicole Starosielski (Urbana: University of Illinois Press, 2015), 41–43.

24. *Lettre de l'ingénieur Fréminville au comité de salut public a.s. de la fabrication des lunettes*, 24 Vendémiare Year III [October 15, 1794], Musée de La Poste, Chappe Archives, 621.1, carton 7 (dossier central télégraphiques Paris I), D7544. Fréminville was subsequently named *commissaire* in charge of a workshop to manufacture lenses. See *Arrêté ouverture atelier de fabrication lunettes*, 25 Vendémiaire Year III [October 16, 1794], Musée de La Poste, Chappe Archives, 621.1, carton 7 (dossier central télégraphiques Paris I), D6269.

25. Patrice Bret, "Instruments of Knowledge and Power in a Colonial Context: Scientific Instruments during the French Occupation of Egypt, 1798–1801," in *Scientific Instruments between East and West*, ed. Neil Brown, Silke Ackermann, and Feza Günergun (Leiden: Brill, 2019), 209–210.

26. Field, "French Optical Telegraphy," 342.

27. See Eva Horn, "Air as Medium," *Grey Room* 73 (Fall 2018): 6–25.

28. Louis Figuier, *Les Merveilles de la science, ou description populaire des inventions modernes*, vol. 2 (Paris: Furne, Jouvet, et Cie, 1868), 1, 68.

29. Lewis Mumford, *Technics and Civilization* (1934; repr., Chicago: University of Chicago Press, 2010), 89.

30. Ernest Jacquez and Adhémar Kermabon, *Atlas des lignes télégraphiques aériennes, construites en France de 1793 à 1853* (Paris, 1892). Eric Wertheimer sees the map of the French network as closer in appearance to a "networked humanoid," articulating an "early infographic of modern power." Eric Wertheimer, "The Telegraphed Republic: Semaphores and the New Executive," *b2o: an online journal*, August 18, 2018, https://www.boundary2.org/2018/08/eric-wertheimer-the-telegraphed-republic-semaphores-and-the-new-executive/.

31. On this image, see Pascal Dupuy, “Guillotine et télégraphe, ou l’invasion ne tient qu’à un fil . . . caricatures anglaises et innovations techniques, 1780–1802,” in *L’Image satirique face à l’innovation: Actes du colloque de Nanterre d’octobre 1996*, ed. Hélène Duccini and Jean-Claude Gardes (Paris: Université Paris X-Nanterre, 1998): 41–51.

32. Victor Hugo, *Le Rhin. Lettres a un ami*, vol. 1 (Paris: Hachette, 1858), 37.

33. David Charles, *La Pensée technique dans l’oeuvre de Victor Hugo* (Paris: Presses Universitaires de France, 1997), 220.

34. Victor Hugo, *Notre Dame de Paris*, trans. Alban Krailsheimer (Oxford: Oxford University Press, 1993), 192–206. See too Wolfgang Ernst, “Let There Be Irony: Cultural History and Media Archaeology in Parallel Lines,” *Art History* 28, no. 5 (November 2005): 582–603. In a discussion of the work of Stephen Bann, Ernst mentions Hugo’s famous lines in connection to a print of the Chappe telegraph at Strasbourg Cathedral (585–586). Relatedly, see Stephen Bann, *Parallel Lines: Printmakers, Painters, and Photographers in Nineteenth-Century France* (New Haven: Yale University Press, 2001).

35. *Voyages pittoresques* was published between 1820 and 1878. The four volumes on Champagne included one volume of text and three atlases of plates, and they appeared in 1857.

36. See E. Ferret, *Les Breguet* (Paris: E. Mauchaussal, n.d.)

37. On such temporal acceleration see the influential argument in Reinhart Koselleck, *Futures Past: On the Semantics of Historical Time*, trans. Keith Tribe (New York: Columbia University Press, 2004), especially 43–57. Lynn Hunt, *Measuring Time, Making History* (Budapest: Central European University Press, 2008), offers a nuanced analysis of such debates.

38. Colin Jones, *The Fall of Robespierre: 24 Hours in Revolutionary Paris* (Oxford: Oxford University Press, 2021), 1.

39. See Hanneke Grootenboer, “A Clock Picture as a Philosophical Experiment: The Tableau Mécanique in the Physics Cabinet of Bonnier de la Mosson,” *Art History* 39, no. 2 (April 2016): 340–355.

40. See Peter Galison, *Einstein’s Clocks, Poincaré’s Maps: Empires of Time* (New York: W. W. Norton, 2003), and the classic account of rail travel in Wolfgang Schivelbusch, *The Railway Journey: The Industrialization of Time and Space in the Nineteenth Century* (Oakland: University of California Press, 2014). See also André Dombrowski, *Monet’s Minutes: Impressionism and the Industrialization of Time* (New Haven: Yale University Press, 2023).

41. See Robert Darnton and Daniel Roche, eds., *Revolution in Print: The Press in France, 1775–1800* (Berkeley: University of California Press, 1989); Jürgen Wilke, “History as a Communication Event: The Example of the French Revolution,” *European Journal of Communication* 4 (1989): 375–391.

42. On these phenomena, see Ken Alder, *The Measure of All Things: The Seven-Year Odyssey and Hidden Error That Transformed the World* (London: Abacus, 2004); Anne Marie Claire Godlewska, *Geography Unbound: French Geographic Science from Cassini to Humboldt* (Chicago: University of Chicago Press, 1999); Sanja Perovic, *The Calendar in Revolutionary France: Perceptions of Time in Literature, Culture, Politics* (Cambridge: Cambridge University Press, 2012); Matthew Shaw, *Time and the French Revolution: The Republican Calendar, 1789–Year XIV* (London: Boydell and Brewer, 2011). See also Edmond Soreau, “Les inventions françaises sous la Révolution,” *Annales historiques de la Révolution française* 157 (July–September 1959): 240–251.

43. Joseph Lakanal, *Rapport sur le télégraphe fait au nom du Comité d’instruction publique, réuni à la commission nommée par le décret du 27 avril dernier, vieux style; par Lakanal; réimpr. par ordre de la Convention nationale* (Paris: Imprimerie Nationale, 1794), 2.

44. For a helpful survey of these competing systems, see Holzmann and Pehrson, *Early History of Data Networks* (especially on Edelcrantz, 97–128).

45. *Institut National. Programme pour la continuation de la description des arts, séance publique du 15 vendémiaire an VII* (Paris: Baudouin, Year VII).

46. *Détail exact du Télégraphe, au moyen duquel la reprise du Quesnoy, de Valenciennes et de Condé a été connue à Paris long-temps avant l’arrivée des couriers* (Paris: Imprimerie Maudet, Year III), 6.

47. See, for instance, Archives du Conservatoire national des arts et métiers, P114. This pervasive idea was introduced at the telegraph’s inception, in Lakanal’s *Rapport*, 3.

48. Simone M. Müller, “From Cabling the Atlantic to Wiring the World: A Review Essay on the 150th Anniversary of the Atlantic Telegraph Cable of 1866,” *Technology and Culture* 57, no. 3 (July 2016): 507–526. See also Müller, *Wiring the World:*

The Social and Cultural Creation of Global Telegraph Networks (New York: Columbia University Press, 2016).

49. Wertheimer, "The Telegraphed Republic." On telegraphy's commercial potential, see Jean-Claude Bastian, "L'argent et la naissance des télécommunications françaises," *Les Cahiers de la FNARH* 107 (2008): 5–15.

50. "Article de haute politique, a propos du télégraphe," *Le Charivari* 143 (May 22, 1840): 1.

51. "Article de haute politique," 1.

52. The Blancs' scheme has been described as the "world's first cyber-attack." Tom Standage, "The Crooked Timber of Humanity," *The Economist*, October 5, 2017.

53. "Le télégraphe," *Vert-Vert, Gazette de Paris* 310 (November 6, 1835): 1.

54. On comparisons with the internet, see Müller, "From Cabling the Atlantic," 512. Also see Tom Standage, *The Victorian Internet: The Remarkable Story of the Telegraph and the Nineteenth Century's On-Line Pioneers* (London: Bloomsbury, 1998). For a sustained critique of this metaphor, and of presentism in histories of communication, see Richard R. John, "Debating New Media: Rewriting Communications History," *Technology and Culture* 64, no. 2 (April 2023): 308–358.

55. See, among others, Laurent Dubois, *A Colony of Citizens: Revolution and Slave Emancipation in the French Caribbean, 1787–1804* (Chapel Hill: University of North Carolina Press, published for the Omohundro Institute of Early American History and Culture, 2004), and Jan Goldstein, *The Post-Revolutionary Self: Politics and Psyche in France, 1750–1850* (Cambridge, MA: Harvard University Press, 2008).

56. Chaussard clarified in a footnote that the telegraph was the subject of this verse. This section of the text also contained reference to developments in French horology and a Montgolfier project to elevate water. Pierre-Jean-Baptiste Chaussard, *L'Industrie ou les arts: ode publiée à l'occasion de la fête du 1er vendémiaire an VII* (Paris: Imprimerie Didot l'aîné, 1798–1799), 16.

57. Walter Benjamin, "Theses on the Philosophy of History," in *Illuminations*, trans. Harry Zohn (New York: Schocken Books, 1969), 257.

58. Friedrich Kittler, *Optical Media: Berlin Lectures 1999*, trans. Anthony Elms (Cambridge: Polity, 2010), 129; Paul Virilio, *Bunker Archaeology* (Princeton: Princeton Architectural Press, 2009), 17. Virilio invokes the Chappe telegraph to describe "the necessity of *controlling* constantly expanding territory, of scanning it in all directions (and, as of now, in three dimensions) while running up against as few obstacles as possible."

59. Friedrich Kittler, "Code (Or, How You Can Write Something Differently)," in *Software Studies: A Lexicon*, ed. Matthew Fuller (Cambridge, MA: MIT Press, 2008), 43. See also Kittler, "The City Is a Medium," trans. Matthew Griffin, *New Literary History* 27, no. 4 (Autumn 1996): 717–729; Kittler, "Lakanal und Soemmerring. Von der optischen zur elektrischen Telegraphie," in *Wunschmaschine Welterfindung. Eine Geschichte der Technikvisionen seit dem 18. Jahrhundert*, ed. Brigitte Felderer (Vienna: Stadt Wien Kunst, 1996), 286–295; and Kittler, "Von der optischen Telegrafie zur Photonentechnik," in *The Moderns. Wie sich das 20. Jahrhundert in Kunst und Wissenschaft erfunden hat*, ed. Cathrin Pichler and Susanne Neuberger (Vienna: Springer, 2012), 161–174.

60. David F. Bell, *Real Time: Accelerating Narrative from Balzac to Zola* (Champaign: University of Illinois Press, 2004), 86. In a similar context, see the discussion of the gun as a complex political artifact in Ken Alder, *Engineering the Revolution: Arms and Enlightenment in France, 1763–1815* (Chicago: University of Chicago Press, 1997).

61. While the stationnaires of the optical system were male, the same was not true of those who operated its electrical successors. From the mid-nineteenth century, telegraph offices provided a space for the employment of women in a technologically emergent field, and they played a significant role in first-wave feminism as well as in key labor disputes. See Thomas C. Jepsen, *My Sisters Telegraphic: Women in the Telegraph Office, 1846–1950* (Athens: Ohio University Press, 2000).

62. James W. Carey, "Technology and Ideology: The Case of the Telegraph," *Prospects* 8 (1983): 303; republished in Carey, *Communication as Culture: Essays on Media and Society* (Boston: Unwin Hyman, 1989), 201–230.

63. Carey, "Technology and Ideology," 305.

64. Carey, "Technology and Ideology," 305.

65. A notable example is found in the characteristic dashes of Emily Dickinson's poems. On telegraphic literature in Dickinson and in other contexts, see Kay Yandell, *Telegraphies: Indigeneity, Identity, and Nation in America's*

Nineteenth-Century Virtual Realm (Oxford: Oxford University Press, 2019).

66. On the visual culture of the Atlantic telegraph and its aftermath, see Caroline Arscott and Clare Pettitt, eds., *Victorians Decoded: Art and Telegraphy* (London: Courtauld Books Online, 2016), and Anne Chapman and Natalie Hume, eds., *Coding and Representation from the Nineteenth Century to the Present: Scrambled Messages* (London: Routledge, 2021).

67. See, for instance, Carolyn Marvin, *When Old Technologies Were New: Thinking about Electric Communication in the Late Nineteenth Century* (Oxford: Oxford University Press, 1988), 3: "It starts with the invention of the telegraph, the first of the electrical communications machines, as significant a break with the past as printing before it."

68. Carey, "Technology and Ideology," 313–314.

69. See Siegfried Zielinski, *Deep Time of the Media: Toward an Archaeology of Hearing and Seeing by Technical Means*, trans. Gloria Custance (Cambridge, MA: MIT Press, 2006).

70. Ignace Chappe to Claude-Pierre Molard, 19 October 1807, Archives du Conservatoire national des arts et métiers, P116.

71. Important discussions of the telegraph in the context of European literature include Bell, *Real Time*, 76–130; Mary Fairclough, "*The Telegraph*: Radical Transmission in the 1790s," *Eighteenth-Century Life* 37, no. 2 (Spring 2013): 26–52; Valentina Fortunato, "Le télégraphe optique et la littérature au XIXe siècle," *Les Cahiers de la FNARH* 116 (2010): 150–157; Richard Menke, *Telegraphic Realism: Victorian Fiction and Other Information Systems* (Stanford: Stanford University Press, 2008); Deven Marie Parker, "Revolutions in Communication: The Politics of Mediation in Romantic Literature" (PhD diss., University of Colorado, 2012); Bernhard Siegert, *Relays: Literature as an Epoch of the Postal System*, trans. Kevin Repp (Stanford: Stanford University Press, 1999), 165–185; Matthew Tiews, "Arcana of Modern Communication: Telegraphy, Cryptography, and Artificial Languages" (PhD diss., Stanford University, 2004); and David Trotter, *The Literature of Connection: Signal, Medium, Interface, 1850–1950* (Oxford: Oxford University Press, 2020).

72. Yves Lecouturier, ed., *Le Patrimoine des télécommunications françaises* (Paris: Flohic, 2002), 86.

73. Jacques Rancière, *The Time of the Landscape: On the Origins of the Aesthetic Revolution*, trans. Emiliano Battista (London: Polity, 2023), 46.

74. Perhaps surprisingly, Walter Benjamin mentions the optical telegraph only fleetingly in his monumental work on the Paris arcades.

75. William Makepeace Thackeray, *The Paris Sketch Book: by Mr Titmarsh. With Numerous Designs by the Author, on Copper and Wood*, vol. 1 (London: John Macrone, 1840), 105–106. On Thackeray's response, see William Vaughan, "'David's Brickdust' and the Rise of the British School," in *Reflections of Revolution: Images of Romanticism*, ed. Alison Yarrington and Kelvin Everest (London: Routledge, 1993), 153–154. On satirical renderings of David's painting, see Peggy Davis, "Le *Serment des Horaces* face à la satire graphique," *RACAR: Revue d'art canadienne / Canadian Art Review* 37, no. 1 (2012): 26–40. Henry Angelo (1756–1835) was the proprietor of a well-known fencing academy on Bond Street, frequented by famous political figures of the day.

76. See the influential psychoanalytical reading of David's *Sabines* and the mirror in Ewa Lajer-Burcharth, *Necklines: The Art of Jacques-Louis David after the Terror* (New Haven: Yale University Press, 1999), 130–235.

77. Robert de La Sizeranne, "Ce que l'art doit à Napoléon," *Revue des Deux Mondes* 6, no. 3 (December 1921): 502.

78. André François Miot de Mélito, *Mémoires du comte Miot de Mélito, ancien ministre, ambassadeur, conseiller d'État et membre de l'Institut*, vol. 1 (Paris: Michel-Lévy frères, 1858), 38.

79. Rebecca Comay describes the "structural anachronism" that characterized the belated German (more particularly German philosophical) encounter with the French Revolution. Although the telegraph is not her subject, its temporal effects frequently align with her account of "traumatic dissonance," in which "the clocks are never synchronized, the schedules never coordinated, every epoch is a discordant mix of divergent rhythms, unequal durations, and variable speeds." Rebecca Comay, *Mourning Sickness: Hegel and the French Revolution* (Stanford: Stanford University Press, 2011), 4.

80. Frédéric Jean Laurent [Friedrich Johann Lorenz] Meyer, *Fragments sur Paris*, vol. 2, trans. Charles-François Dumouriez (Hamburg: 1798), 126–127.

81. Louis-Pierre Baltard and Amaury Duval [Charles-Alexandre-Amaury Pineux-Duval], *Paris et ses monumens, mesurés, dessinés et gravés par Baltard, architecte, Avec des*

Descriptions historiques par le cit. Amaury-Duval; ouvrage dédié à Napoléon Bonaparte (Paris: Imprimerie de Crapelet, 1803).

82. See John Harwood, "On Wires; or, Metals and Modernity Reconsidered," *Grey Room* 69 (Fall 2017): 108–136. Harwood discusses how Baltard's replacement as professor of theory at the École des Beaux-Arts, Guillaume-Abel Blouet, published at midcentury some designs for electrical telegraph installations (121). Also described is a telegraphic line developed by colonists in Jules Verne's 1874 *L'Île mystérieuse* (127–128).

83. Ignace Urbain Jean Chappe, *Histoire de la télégraphie* (Paris: chez l'auteur, 1824), 148. This is Chappe's translation from "the inventor of *synthématographie*," Johann Andreas Benignus Bergsträsser. See Bergsträsser, *Uebersichten und Erweiterungen der Signal-Order und Zielschreiberei in die Ferne oder neue Synthematographe und Telegraphe* (Leipzig: Breitkopf, 1795).

84. Ennemond Gonon, *Mémoire sur le système télégraphique nouveau, universel et perpétuel, pour le jour et pour la nuit* (Paris: Sirou, 1844). On Gonon, see Tiews, "Arcana of Modern Communication," 51.

85. See Thomas Crow, "The Oath of the Horatii in 1785: Painting and Pre-Revolutionary Radicalism in France," *Art History* 1, no. 4 (December 1978): 424–471.

86. Giorgio Agamben, "Notes on Gesture," in *Infancy and History: On the Destruction of Experience*, trans. Liz Heron (London: Verso, 2006), 135–140.

87. See Dylan Mulvin, *Proxies: The Cultural Work of Standing In* (Cambridge, MA: MIT Press, 2021).

88. Philippe-Alain Michaud, *Aby Warburg and the Image in Motion*, trans. Sophie Hawkes (New York: Zone Books, 2004), 268.

89. Amy Knight Powell, *Depositions: Scenes from the Late Medieval Church and the Modern Museum* (New York: Zone Books, 2012), 11.

90. Jennifer Roberts, *Transporting Visions: The Movement of Images in Early America* (Los Angeles: University of California Press, 2014), 142.

91. See, notably, Jonathan Crary, *Techniques of the Observer: On Vision and Modernity in the Nineteenth Century* (Cambridge, MA: MIT Press, 1990). On optical technologies at the other end of the nineteenth century, see Mary Ann Doane, *The Emergence of Cinematic Time: Modernity, Contingency, the Archive* (Cambridge, MA: Harvard University Press, 2002).

92. Louis-Sébastien Mercier, *Le Nouveau Paris*, ed. Jean-Claude Bonnet (Paris: Mercure de France, 1994), 878.

93. See Brian Seibert, "A Home Version of Trisha Brown's 'Roof Piece,' No Roof Required," *New York Times*, April 7, 2020, https://www.nytimes.com/2020/04/07/arts/dance/trisha-brown-roof-piece.html. On *Roof Piece*'s context and production, see Amanda Jane Graham, "Out of Site: Trisha Brown's *Roof Piece*," *Dance Chronicle* 36, no. 1 (2013): 59–76.

94. In "M. Asinard ou le Volcan de Montmartre" by Mario Coster and Ferdinand Oury, shown at Théâtre des variétés in 1809, the principal character was the stationnaire Longue-vue. The Bibliothèque nationale has a print of the actor Potier in the role of Longue-vue, made by A. Joly and published by Martinet, BnF département Arts du spectacle, ASP 4-ICO COS-1 (2,166). In 1834 a *vaudeville* titled "L'apprenti ou l'art de faire une maîtresse" by Cognard and Adolphe includes a telegraph employee named Gaillard. Jean-Claude Bastian, "Le Télégraphe Chappe à Montmartre," *Les Cahiers de la FNARH* 109 (2008): 27–28.

95. On forms of endurance in contemporary digital networks, see Wendy Hui Kyong Chun, "The Enduring Ephemeral, or the Future Is a Memory," *Critical Inquiry* 35, no. 1 (Autumn 2008): 148–171.

96. Rebecca Schneider, "Performance Remains," *Performance Research* 6, no. 2 (2001): 103.

97. See, for instance, Wolfgang Ernst, "Media Archaeography: Method and Machine versus History and Narrative of Media," in *Media Archaeology: Approaches, Applications, and Implications*, ed. Erkki Huhtamo and Jussi Parikka (Berkeley: University of California Press, 2011), 239–255.

98. For a nuanced critique of approaches that treat obsolescence in this way, see Shannon Mattern, *Code and Clay, Data and Dirt: Five Thousand Years of Urban Media* (Minneapolis: University of Minnesota Press, 2017), xv–xl. The description of media archaeology as a "traveling discipline" can be found in Erkki Huhtamo and Jussi Parikka, "Introduction: An Archaeology of Media Archaeology," in Huhtamo and Parikka, *Media Archaeology*, 3. The authors derive this term from the work of Mieke Bal.

99. Michelle Henning, "New Lamps for Old: Photography, Obsolescence, and Social Change," in *Residual Media*, ed. Charles R. Acland (Minneapolis:

University of Minnesota Press, 2006), 48–65.

Chapter 2

1. *Mémoire sur les moyens: 1. De communiquer sur-le-champ au Peuple, occupant les dehors du lieu où se tient l'Assemblée, les Délibérations qui y sont prises. 2. De se faire entendre dans une grande Assemblée, et notamment aux États-généraux, malgré le bruit des Délibérans* (Paris: Veuve Hérissant, 1789). The text is discussed briefly in Jacques Guilhaumou, "Zeitgenössische politische Lebesgeschichten aus der Französische Revolution (1793–1794). Autobiographischer Akt und diskursives Ereignis," in *Die Französische Revolution als Bruch des gesellschaftlichen Bewussteins*, ed. Reinhart Koselleck and Rolf Reichardt (Munich: R. Oldenbourg, 1988), 358–362. It is described as an "oeuvre facétieux" in Pierre M. Conlon, ed., *1789, ouvrages M–Z*, vol. 25 of *Le Siècle des Lumières: Bibliographie chronologique* (Geneva: Droz, 2007), 38.

2. *Mémoire sur les moyens*, 1.

3. Ignace Urbain Jean Chappe, *Histoire de la télégraphie* (Paris: Chez l'auteur, 1824), 64–65.

4. *Mémoire sur les moyens*, 3. See too Jacques Guillerme, "La 'parole à l'oeil'," *Revue de l'Art* 76 (1987): 94–95. Guillerme attributes the authorship of the *Tableau populaire* (and, by implication, the *Mémoire*) to Savart, in "Télégraphie et territoire: notes sur l'imagination des réseaux," in *L'Art du projet: Histoire, technique, architecture* (Wavre: Éditions Mardaga, 2008), 138.

5. See Erkki Huhtamo, *Illusions in Motion: Media Archaeology of the Moving Panorama and Related Spectacles* (Cambridge, MA: MIT Press, 2013).

6. *Mémoire sur les moyens*, planche II.

7. *Révolutions de Paris*, no. 21 (November 28–December 5, 1789), 20. Attributing the invention to "Sr. Gerard," the journal noted that the same image was available for sale at the cooper Mercier, of the rue Saint-Jacques next to the Collège du Plessis.

8. *Mémoire sur les moyens*, 7.

9. Jonathan Crary, "Géricault, the Panorama, and Sites of Reality in the Early Nineteenth Century," *Grey Room* 9 (Autumn 2002): 5–25 (especially 5–9). A thoughtful summary of sound as a means of historical comprehension, written with the French Revolution in mind, is Sophia Rosenfeld, "On Being Heard: A Case for Paying Attention to the Historical Ear," *American Historical Review* 116, no. 2 (April 2011): 316–334.

10. See especially Sophia Rosenfeld, *A Revolution in Language: The Problem of Signs in Late Eighteenth-Century France* (Stanford: Stanford University Press, 2001), and Nicholas Mirzoeff, *Silent Poetry: Deafness, Sign, and Visual Culture in Modern France* (Princeton: Princeton University Press, 1995). Telegraphy prompted responses relating it to blindness as well as deafness: Schwenger's *Mémoires sur les aveugles* concluded a synthesis of recent research into blindness with a proposal for a new kind of telegraph, a simplified variation on the Chappe model. Auguste-Guillaume Schwenger, *Mémoires sur les aveugles, sur la vue et la vision, suivis d'un déscription d'un télégraphe très simple* (Paris, 1800).

11. Belprey, *De l'optilogue, ou du cylindre parlant appliqué à la transmission des idées chez les sourds-muets* (Paris: Dabin, 1801).

12. *Détail exact du Télégraphe, au moyen duquel la reprise du Quesnoy, de Valenciennes et de Condé a été connue à Paris longtemps avant l'arrivée des couriers* (Paris: Imprimerie Maudet, Year III), 2. On telegraphy and the idea of a universal language, see Patrice Flichy, "The Birth of Long Distance Communication: Semaphore Telegraphs in Europe (1790–1840)," *Reseaux: Communication-Technologie-Société* 1, no. 1 (1993): 89–101.

13. On the hemispheric arrangement of the oath, see Wolfgang Kemp, "The Theater of Revolution: A New Interpretation of Jacques-Louis David's *Tennis Court Oath*," in *Visual Culture: Images and Interpretations*, ed. Norman Bryson, Michael Ann Holly, and Keith Moxey (Middletown, CT: Wesleyan University Press, 1994), 202–227.

14. It is tempting to speculate that the "Gérard" responsible for the crude drawings in the *Mémoire* was a knowing reference to Michel Gérard, a farmer and the only peasant member of the Assembly. Gérard, who was known for his direct manner of address, is represented in David's *Oath*, and was the model for Jean-Marie Collot d'Herbois's hugely popular *Almanach du Père Gérard*. Awarded a prize by the Paris Jacobin Club in 1791, this text narrates the return of Gérard to his home in Brittany, to communicate to his fellow villagers the benefits of the Constitution. See Rolf Reichardt, "The French Revolution as a European Media Event," in *European History Online (EGO)*, accessed December 30, 2023, http://www.ieg-ego.eu/reichardtr-2010-en; also Jürgen Wilke, "History as a Communication Event: The Example of the French

Revolution," *European Journal of Communication* 4 (1989): 375–391. On the curvilinear design of revolutionary legislative and judicial spaces and their relationship to theatre, see Katherine Fischer Taylor, "Geometries of Power: Royal, Revolutionary, and Postrevolutionary French Courtrooms," *Journal of the Society of Architectural Historians* 72, no. 4 (December 2013): 434–474.

15. Philippe Bordes, "Jacques-Louis David's 'Serment du Jeu de Paume': Propaganda without a Cause?," *Oxford Art Journal* 3, no. 2 (October 1980): 23.

16. See Richard N. Scarth, *Echoes from the Sky: A Story of Acoustic Defence* (Hythe: Hythe Civic Society, 1999). Also, Brian Dillon, "Listening for the Enemy," *Cabinet* 12 (Fall-Winter 2003).

17. See Hal Foster, "An Archival Impulse," *October* 110 (Fall 2004): 14–15. Foster discusses the Dungeness reflectors' "failed futuristic vision" in relation to the work of artist Tacita Dean, whose 1999 film *Sound Mirrors* focuses on them.

18. Of Marie-Marthe, and another sister Sophie-Françoise (b. 1767), we sadly hear little. It seems likely that they were well aware of their brothers' work on the telegraph, if not involved directly.

19. Jean Chappe d'Auteroche, *Voyage en Sibérie fait par ordre du roi en 1761, contenant les moeurs, les usages des Russes et l'état actuel de cette puissance, la description géographique et le nivellement de la route de Paris à Tobolsk, l'histoire naturelle de la même route* (Paris: Debure père, 1768). This claim is made in *Journal de Paris*, 13 Pluviôse Year XIII [February 12, 1805].

20. Jean Chappe d'Auteroche, *Voyage en Californie pour l'observation du passage de Vénus sur le disque du soleil, le 3 juin 1769* (Paris: Jombert, 1772).

21. Chappe's experiments with silk, including a new variety offering "the spectacle of a prism, where the primitive colors shine and, from their different mixtures, give rise to admirable shades," and which he compared to soap bubbles (which he had earlier attempted to electrify), are described in Claude Chappe, "Découverte d'un tissu transparent," November 1791, Musée de La Poste, Paris, dossier familial, Cote 621.1—Carton 1. D9229/1. See also "Expériences à toutes, sur la percussion des corps sonores," n.d., Musée de La Poste, Paris, dossier familial, Cote 621.1—Carton 1. D9229/4 ter.

22. Ernest Jacquez, *Claude Chappe: notice biographique* (Paris: Picard A. et fils, 1893), 7.

23. On early transnational experiments with electrical telegraphy—by Oersted, Galvani, Soemmering, Coxe, and Lesage, to name a few—see Edouard Ernest Blavier, *Cours théorique et pratique de télégraphie électrique* (Paris: Lacroix-Comon, 1857), 3–7.

24. Joseph Lakanal, *Rapport sur le télégraphe fait au nom du Comité d'instruction publique, réuni à la commission nommée par le décret du 27 avril dernier, vieux style; par Lakanal; réimpr. par ordre de la Convention nationale* (Paris: Imprimerie Nationale, 1794), 9.

25. Charles-Gilbert Romme, "Convention Nationale, présidence de Jean Debry, suite de la séance de lundi 1er avril," *Gazette nationale, ou Le Moniteur universel*, no. 94, April 4, 1793, 418.

26. M. Loiseau, "Le Télégraphe au Louvre," in *Le Télégraphe aérien: colloque de Nancy, mai 1981* (Nancy: FNARH, 1981), 2.

27. Loiseau, "Le Télégraphe au Louvre," 3.

28. Loiseau, "Le Télégraphe au Louvre," 4.

29. Yves Lecouturier, ed., *Le Patrimoine des télécommunications françaises* (Paris: Flohic, 2002), 32.

30. Bertrand Barère, "Convention nationale, présidence de Merlin de Douai, séance du 30 thermidor," *Gazette nationale, ou Le Moniteur universel*, no. 331, 1 Fructidor Year II [August 18, 1794], 1360.

31. Claude Chappe, quoted in Gerard J. Holzmann and Björn Pehrson, *Early History of Data Networks* (Los Alamitos: Wiley-IEEE Computer Society Press, 1995), 66.

32. Chappe, *Histoire de la télégraphie*, 200.

33. Jacquez, *Claude Chappe*, 73.

34. Jacques-Antoine Rabaut-Pommier, "Convention Nationale, présidence de Doulcet, suite de la séance du 29 messidor," *Gazette nationale, ou Le Moniteur universel*, no. 304, 4 Thermidor Year III [July 22, 1795], 1223.

35. On the administration of the optical telegraph network see Édouard Gerspach, *Histoire administrative de la télégraphie aérienne en France* (Paris: E. Lacroix, 1861).

36. These are now in the collection of the Musée de La Poste, Paris.

37. The association between Regnault's painting and vignettes is made in Philippe Bordes, "David's Contemporaries: A Generation of Artists Against the Revolution?," in

Culture and Revolution: Cultural Ramifications of the French Revolution, ed. George Levitine (College Park: University of Maryland at College Park, 1989), 67.

38. Charles Norry, *Relation de l'expédition d'Égypte; suivie de la description de plusieurs des monumens de cette contrée, et ornée de figures* (Paris: Magimel, Year VII).

39. On bells, see Alain Corbin, *Village Bells: Sound and Meaning in Nineteenth-Century France*, trans. Martin Thom (New York: Columbia University Press, 1998).

40. Schall's image was the pendant to a print representing *La Lanterne magique d'amour*. Similar themes to those in Schall's image were explored in Joseph Servières and Charles Henrion's *Le Télégraphe d'amour, comédie en un acte, mêlée de vaudevilles . . . Représentée à Paris, le 27 messidor, an 9* (Paris: Roux, 1801). In this play, the "mute eloquence" offered by signals transmitted between the lovers Valsain and Rose by way of a nearby windmill helps them avoid detection by the prudish widow Madame Robert. For discussion of the manuscript for this publication, see Bruno Bentz, "Le télégraphe d'amour," *Les Cahiers de la FNARH* 93 (2004): 28–31.

41. "Fanology, or Speaking Fan," published by William Cock after Charles Francis Badini, British Museum, Prints and Drawings, 1891,0713.508.

42. Mary Favret, *War at a Distance: Romanticism and the Making of Modern Wartime* (Princeton: Princeton University Press, 2010), 13.

43. See especially Katie Hornstein, *Picturing War in France, 1792–1856* (New Haven: Yale University Press, 2018).

44. "Que tour-à-tour j'accuse, en ma rage inutile, / Et ce siècle fécond et mon cerveau stérile; / Ce maudit Télégraphe enfin va-t-il cesser / D'importuner mes yeux, qu'il commence à lasser? / Là, devant ma lucarne! il est bien ridicule / Qu'on place un Télégraphe auprès de ma cellule! / Il s'élève, il s'abaisse; et mon esprit distrait, / Dans ces vains mouvemens, cherche quelque secret." Victor Hugo, *Le Télégraphe. Satire* (Paris: Imprimerie de Gille, 1819), 3. On Hugo and the telegraph, see Guy de Saint-Denis, "Victor Hugo et le télégraphe Chappe," *Revue des PTT de France* 6 (1981): 66–71; and David Charles, *La Pensée technique dans l'oeuvre de Victor Hugo* (Paris: Presses Universitaires de France, 1997), 197–238.

45. Yve-Alain Bois, "Thermometers Should Last Forever," *October* 111 (October 2005): 63. On noise and interference, see Jussi Parikka, "Mapping Noise: Techniques and Tactics of Irregularities, Interception, and Disturbance," in *Media Archaeology: Approaches, Applications, and Implications*, ed. Erkki Huhtamo and Jussi Parikka (Berkeley: University of California Press, 2011), 264–267. Also see Michel Serres, *The Parasite*, trans. Lawrence R. Schehr (Minneapolis: University of Minnesota Press, 2007). On electrical telegraphy and sound, see Jonathan Sterne, *The Audible Past: Cultural Origins of Sound Reproduction* (Durham: Duke University Press, 2003), 137–154.

46. Mathurin-Joseph Brisset, "Le Télégraphe," *La Patrie*, February 5, 1842, 1.

47. Civeton [draftsman], Couché fils [director], and Gossard [engraver], *Marché St Germain*, in Jacques-Antoine Dulaure, *Histoire civile, physique et morale de Paris*, vol. 9 (Paris: Baudouin, 1825), plate 78.

48. J. M. W. Turner, "Church of Saint-Sulpice, Paris," from *Paris and Environs Sketchbook [Finberg CCLVII]*, 1812, Tate Britain, London, D18575.

49. Hugo, *Le Télégraphe*, 4.

50. Victor Hugo, *Le Rhin. Lettres a un ami*, vol. 1 (Paris: Hachette, 1858), 42. David Charles, however, suggests that these three signs, when laid over one another, resemble a schematic image of a guillotine. Charles, *La Pensée technique dans l'oeuvre de Victor Hugo*, 219.

51. Victor Hugo, "Le Mont Saint-Michel," in *Oeuvres complètes. En voyage, tom. II: France et Belgique. Alpes et Pyrénées. Voyages et excursions* (Paris: Imprimerie Nationale and Editions Ollendorf, 1910), 54–56.

52. Victor Hugo, "Le Mont Saint-Michel," 56.

53. See, for instance, Adrien Dauzats, *Tour et flèche de la cathédrale de Strasbourg. Vue prise de la tourelle au dessus du télégraphe*, n.d., BnF Estampes et photographie, EST RESERVE VE-26 (R). Maupassant's aphorism appears in Roland Barthes, "The Eiffel Tower," trans. Richard Howard and Julie Rose, *AA Files* 64 (2012): 124.

54. Victor Hugo, "Le Mont Saint-Michel," 56.

55. The monkey and cat in *La Force de l'habitude* are the manipulative Bertrand and the duped Raton, respectively, characters from La Fontaine's fable "The Monkey and the Cat," and presumably a reference to General Bertrand, who remained loyal to Napoleon.

56. Alexandre Dumas, *The Count of Monte Cristo* (New York: Thomas Y. Crowell, 1894), 78. On motifs of transmission and

speed in Dumas's work, as well as in novels such as Stendhal's *Lucien Leuwen*, see David F. Bell, *Real Time: Accelerating Narrative from Balzac to Zola* (Champaign: University of Illinois Press, 2004), 76–130.

57. "Lettre de la préfecture de police à Chappe a.s. de la dénonciation d'un agent," 12 October 1815, Musée de La Poste, Paris, Cote 621.1—Carton 7 (dossier central télégraphiques Paris I), D6275.

58. Hugo, *Le Télégraphe*, 9.

59. This point is made in Armand Mattelart, "La communication et la promise de redemption," *Quaderni* 40 (Winter 1999): 70.

60. Walter Benjamin, *The Arcades Project*, trans. Howard Eiland and Kevin McLaughlin (Cambridge, MA: Belknap Press of Harvard University Press, 1999), 626. Fourier's ink sketch is in the Archives Nationales, 10 AS 23/dossier 18.

61. Thomas Bouchet, "Du pigeon courier au télégraphe électrique. Les systèmes de communication à distance vus par Fourier et par les phalanstériens," in *Dompter Prométhée: Technologies et socialismes à l'âge romantique (1820–1870)*, ed. François Jarrige (Besançon: Presses universitaires de Franche-Comté, 2016), 193–208.

62. Bouchet, "Du pigeon courier au télégraphe électrique," 5.

63. Ange-Marie d'Eymar and Claude Chappe, *Lettres sur un nouveau télégraphe* (Paris, 1797), 15.

64. Bouchet lists as some examples *La Phalange* (1836); François Cantagrel's *Le Fou du Palais-Royal* (1841); Victor Hennequin's *Théorie de Charles Fourier* (1847); and Henri Fugère's *Organisation du travail par la foundation d'une commune modèle* (1848). Bouchet, "Du pigeon courier au télégraphe électrique," 20.

65. Bouchet, "Du pigeon courier au télégraphe électrique," 15.

66. See Jules Allix, "Communication universelle de la pensée, à quelque distance que ce soit, à l'aide d'un appareil portatif appelé Boussole pasilalinique sympathique," *La Presse*, October 25 and 26, 1850; Justin E. H. Smith, "The Internet of Snails," *Cabinet* 58 (Summer 2015); Richard Taws, "Pasilalinic-Sympathetic Compass," in *Extinct: A Compendium of Obsolete Objects*, ed. Barbara Penner, Adrian Forty, Miranda Critchley, and Olivia Horsfall Turner (London: Reaktion, 2021), 240–243.

67. M. Clairville, *Les Escargots sympathiques; à-propos, mélé de couplets* (Paris: Beck, 1850), 4–5.

68. See Rosemary Lloyd, "Un peintre et ses critiques: Dominique Papety," *Romantisme* 50 (1985): 105–110.

69. Linda Nochlin, "The Invention of the Avant-Garde: France, 1830–1880," in *The Politics of Vision: Essays on Nineteenth-Century Art and Society* (London: Thames and Hudson, 1989), 7.

70. Nochlin, "Invention of the Avant-Garde," 8–9.

71. Louis Peisse in *Revue des Deux Mondes* (April 15, 1843), cited in Lloyd, "Un peintre et ses critiques," 106.

72. As far I have found, this detail of Daumier's *Gargantua* has not been discussed in the extensive literature on the print. The online catalogue of Daumier's works at the "Daumier Register," searchable thematically, lists several works with telegraphic themes, but not this one; see Daumier Register, accessed December 30, 2023, http://www.daumier-register.org.

73. Bell, *Real Time*, 89.

74. Benedict Anderson, *Imagined Communities: Reflections on the Origin and Spread of Nationalism* (London: Verso, 1991).

75. Lisa Gitelman, *Always Already New: Media, History, and the Data of Culture* (Cambridge, MA: MIT Press, 2006), 4.

76. These positions were encapsulated in the foreign policy debates in late 1791 and early 1792 between Brissot and Robespierre, the former in favor of war, the latter against.

77. Emma Rothschild, "Globalization and the Return of History," *Foreign Policy* 115 (Summer 1999): 112–113.

78. On the contradictions of French universalism, see Susan Buck-Morss, *Hegel, Haiti, and Universal History* (Pittsburgh: University of Pittsburgh Press, 2009).

79. Dumas himself was born in France in 1802, during the Haitian Revolution. His father, born in Saint-Domingue, was the son of a white French nobleman and an enslaved woman of African descent.

80. Barère, "Convention nationale, présidence de Merlin de Douai, séance du 30 thermidor," 1360; message cited in François Gautier, *L'Œuvre de Claude Chappe, créateur de l'administration française des télégraphes et inventeur du télégraphe aérien établi sous les auspices de la Convention Nationale* (Paris: Librairie de l'Armée, 1893), 3.

81. See, notably, Michel-Rolph Trouillot, *Silencing the Past: Power and the Production of History* (New York: Beacon Press, 1995). For wide-ranging discussion of the ways in which technology was allied to

imperialism in this period, see Daniel R. Headrick, *The Tools of Empire: Technology and European Imperialism in the Nineteenth Century* (Oxford: Oxford University Press, 1981). While Headrick takes account of telegraphy in this context, his point of departure is the implementation of a submarine cable between Britain and Ireland in 1852 (157–164).

82. On the communication of revolutionary ideas between the Caribbean, Europe, and across the Americas, see Julius S. Scott, *The Common Wind: Afro-American Currents in the Age of the Haitian Revolution* (London: Verso, 2020).

83. The painting was destroyed during the Allied bombing of Rouen on May 31, 1944. The most complete analysis of Lemonnier's work and career is given in Christine Le Bozec, *Lemonnier: un peintre en révolution* (Rouen: Publications de l'Université de Rouen, 2000).

84. *Musée de la Révolution française: Premières collections* (Vizille: Musée de la Révolution française, Conseil général de l'Isère, 1985), 39.

85. On Levasseur's contribution, see E. Delignières, *Catalogue raisonné de l'oeuvre de J. C. Le Vasseur, d'Abbeville* (Abbeville: Imprimerie Briez, 1865), 55.

86. The iconography of Franco-American cooperation is discussed in Le Bozec, *Lemonnier*, 51.

87. As Claudette Hould observes, the high finish of Levasseur's engraving set it apart from most printed images produced during the Revolution, as artists worked swiftly to adapt to changing political, social, and economic circumstances. Claudette Hould, *L'Image de la Révolution française* (Québec: Musée du Québec, 1989), 151.

88. Le Bozec, *Lemonnier*, 53.

89. *Mercure de France*, no. 26 (June 25, 1791), 152. Le Bozec suggests that the author was probably Lecouteulx de Canteleu, a Rouennais lawyer who likely played a key role in commissioning the painting, and whose family had organized the first regular service between France and America. Le Bozec, *Lemonnier*, 51.

90. For discussion of the combination of positive futurity and abolitionist paternalism in French art, with reference to Anne-Louis Girodet's portrait of Jean-Baptiste Belley, see Anne Lafont, *L'Art et la race: l'Africain (tout) contre l'œil des Lumières* (Dijon: Les presses du réel, 2019), 250.

91. Joan Dayan, *Haiti, History, and the Gods* (Berkeley: University of California Press, 1995), 46.

92. Robin Mitchell, *Vénus Noire: Black Women and Colonial Fantasies in Nineteenth-Century France* (Athens: University of Georgia Press, 2020), 3.

93. Anicet Charles Gabriel Lemonnier, *Séance du 1er juillet 1791, Délibérations de la Chambre du Commerce*, cited in Le Bozec, *Lemonnier*, 47.

94. See Charles-Louis Aubry, *Le Télégraphe du commerce et des finances de la République française, ou double moyen de ramener le commerce à ses véritables élémens, et de familiariser le people aux nouvelles Mesures* (place of publication and publisher unknown: n.d.), 1. Aubry started this pamphlet by connecting telegraphy, writing, and commerce: "Allow me to dedicate my TELEGRAPH to you; I believe it is fit to serve usefully commerce, finance, and the superb institution of Republican Measures, that is, three essential pivots of government."

95. Clare le Corbeiller, "Mercury, Messenger of Taste," *The Metropolitan Museum of Art Bulletin* 22, no. 1 (Summer 1963): 22–28.

96. Louis Figuier, *Les Merveilles de la science, ou description populaire des inventions modernes*, vol. 2 (Paris: Furne, Jouvet, et Cie, 1868), 13–16; Chappe, *Histoire de la télégraphie*, 66–71.

97. The Paris pneumatic post operated from 1866 to 1984 and was available for public use from 1879.

98. Sterne, *The Audible Past*.

99. Archives Nationales, F/17/1261, dossier 2, no. 18. See Richard Clay, *Iconoclasm in Revolutionary Paris: The Transformation of Signs* (Oxford: SVEC, 2013), 66.

100. Darcy Grimaldo Grigsby, *Extremities: Painting Empire in Post-Revolutionary France* (New Haven: Yale University Press, 2002), 15.

101. Anne Lafont, "How Skin Color Became a Racial Marker: Art Historical Perspectives on Race," *Eighteenth-Century Studies* 51, no. 1 (Fall 2017): 109.

102. See Jill H. Casid, *Sowing Empire: Landscape and Colonization* (Minneapolis: University of Minnesota Press, 2005).

103. Guy de Saint Denis, in *La Télégraphie Chappe*, ed. Guy de Saint-Denis (Nancy: FNARH/Éditions de l'Est, 1993), 290. On Toto, see also Guy de Saint-Denis, "Philippe Toto, stationnaire," *Les Cahiers de la FNARH* 137 (2018): 242–245.

104. In 1835, the correspondent for *Vert-Vert* noted that "the telegraph is the caduceus of modern Mercury." "Le

télégraphe," *Vert-Vert, Gazette de Paris* 310 (November 6, 1835): 1.

105. Chappe, *Histoire de la télégraphie*, 8–9.

106. Chappe, *Histoire de la télégraphie*, 11.

107. Pierre Puvis de Chavannes, *Deux figures volant vers la gauche, dans un paysage*, c. 1893, Département des Arts graphiques, Musée du Louvre, Paris, RF 15977, recto. For an insightful discussion of how art might register trans-Atlantic movement, see Jennifer Roberts, *Transporting Visions: The Movement of Images in Early America* (Los Angeles: University of California Press, 2014).

108. Alexis Belloc, *La Télégraphie historique: depuis les temps les plus reculés jusqu'à nos jours* (Paris: Firmin-Didot, 1894), 176.

109. Hellmut Wohl, "Beyond the Large Glass: Notes on a Landscape Drawing by Marcel Duchamp," *The Burlington Magazine* 119, no. 896 (November 1977): 763–772.

110. On Géricault's image in the context of contemporary media ecologies, including telegraphy, see Jordan Bear, "Adrift: The Time and Space of the News in Géricault's *Le Radeau de la Méduse*," in *Getting the Picture: The Visual Culture of the News*, ed. Jason E. Hill and Vanessa R. Schwartz (London: Bloomsbury, 2015), 182–189; and Tom Gretton, "From *La Méduse* to the *Titanic*: Géricault's *Raft* in Journalistic Illustration up to 1912," *19: Interdisciplinary Studies in the Long Nineteenth Century* 16 (2013): 1–38.

111. On *The Refusal of Time*, see Elfriede Dreyer, "Of Clocks Ticking: Heterotopic Space, Time and Motion in William Kentridge's *The Refusal of Time* (2012)," *Communicatio* 42, no. 3 (2016): 338–360. The Géricault-Chappe drawing can be seen in a notebook published for Documenta 13: William Kentridge and Peter Galison, *The Refusal of Time / Die Ablehnung der Zeit* (Ostfildern: Hatje Cantz, 2012).

112. Crary, "Géricault, the Panorama, and Sites of Reality," 23.

Chapter 3

1. Nadar, *Quand j'étais photographe* (Paris: Flammarion, 1900), 164. A section of Nadar's book was published in English as "My Life as a Photographer," trans. Thomas Repensek, *October* 5 (Summer 1978): 6–28, and a complete translation has appeared recently as *When I Was a Photographer*, trans. Eduardo Cadava and Liana Theodoratou (Cambridge, MA: MIT Press, 2015), 122.

2. Antoine-François Fourcroy, *Rapport fait par Fourcroy, au nom du Comité de salut public, sur les arts qui ont servi à la défense de la République, et sur le nouveau procédé de tannage découvert par le citoyen Armand Seguin* (Paris: Du Pont, 1795). A portrait of Fourcroy, painted by Anicet Charles Gabriel Lemonnier, is now in the Musée d'histoire de la Médicine, Paris.

3. Désiré van Monckhoven (1834–1882) wrote several early books on photography, which were translated widely, and he contributed to the development of enlargement techniques, the dry collodion process, and the carbon print process, among other innovations. See, for example, Désiré van Monckhoven, *Traité populaire de photographie sur collodion* (Paris: Leiber, 1862).

4. Nadar, *Quand j'étais photographe*, 159.

5. Stephen Bann, "'When I Was a Photographer': Nadar and History," *History and Theory* 48 (December 2009): 95–111; Eduardo Cadava, "Nadar's Photographopolis," *Grey Room* 48 (Summer 2012): 56–77.

6. Stephen Bann, *Distinguished Images: Prints in the Visual Economy of Nineteenth-Century France* (New Haven: Yale University Press, 2013), 87–120.

7. See Armand Seguin, *Rapport au Comité de Salut public, sur les nouveaux moyens de tanner les cuirs* (Paris, 1794).

8. Fourcroy, *Rapport*, 7–8.

9. Nadar, *Quand j'étais photographe*, 163.

10. Fourcroy, *Rapport*, 11–12. For a magisterial survey of such inventions, see Charles Coulston Gillispie, *Science and Polity in France: The Revolutionary and Napoleonic Years* (Princeton: Princeton University Press, 2004).

11. Nadar, *Quand j'étais photographe*, 160. On the development of weapons in revolutionary France, see especially Ken Alder, *Engineering the Revolution: Arms and Enlightenment in France, 1763–1815* (Chicago: University of Chicago Press, 1997). Also relevant in the context of this chapter is Gaspard Monge, *Description de l'art de fabriquer les canons* (Paris: l'Imprimerie du Comité de Salut Public, 1793).

12. See Nadar, *Mémoires du Géant* (Paris: E. Dentu, 1864). Also, Emily Doucet, "'The Idea Was in the Air': Nadar's Aerial Media," *Grey Room* 83 (2021): 112–137.

13. For a compelling reevaluation of this image, see Jillian Lerner, "Nadar's Signatures: Caricature, Self-Portrait, Publicity," *History of Photography* 41, no. 2 (2017): 108–125. Lerner

makes an insightful connection between this image and Nadar's caricature of Moisseu Réac "perched atop the telegraph to get the news from Paris," published in *La Revue comique* 22 (April 7, 1849): 322–323. Nadar's eight-image sequence, titled "Nouvelles télégraphiques," exploited the telegraph's anthropomorphism for comedic effect. See also Philippe Willems, "Between Panoramic and Sequential: Nadar and the Serial Image," *Nineteenth-Century Art Worldwide* 11, no. 3 (Autumn 2012), http://www.19thc-artworldwide.org/index.php/autumn12/willems-nadar-and-the-serial-image.

14. Marika Takanishi Knowles has described the performative qualities of work by Nadar and his brother Adrien at this time (particularly the Pierrot photographs), and the way in which they marked a new separation of figure and ground. These photographs, in which "Pierrot floats free, a supple and airy silhouette slipping out from between the sticky joints of figure and ground," bear comparison with Nadar's self-portrait photomontage. Marika Takanishi Knowles, "Lost Ground: The Performance of Pierrot in Nadar and Adrien Tournachon's Photographs of Charles Deburau," *Oxford Art Journal* 38, no. 3 (December 2015): 386.

15. Lerner, "Nadar's Signatures," 120.

16. George Kubler, *The Shape of Time: Remarks on the History of Things* (New Haven: Yale University Press, 1962), 17.

17. For a helpful list of representations of the *porte-crayon* in European portraiture, see the page prepared by Jacob Simon for the National Portrait Gallery, London, accessed February 23, 2024, http://www.npg.org.uk/research/programmes/artists-their-materials-and-suppliers/the-artists-porte-crayon.php.

18. Fourcroy, *Rapport*, 27.

19. Nadar, *Quand j'étais photographe*, 161.

20. Nadar, *Les Ballons en 1870. Ce qu'on aurait pu faire, ce qu'on a fait* (Paris: E. Chatelain, 1870).

21. Nadar, *Les Ballons en 1870*, 11. Nadar's copy of Edme-François Jomard, *Conté* (Paris: Imprimerie de Thunot, 1852), is in the Bibliothèque historique de la Ville de Paris, Collection Nadar 32486. Jomard's *Conté* remains influential and has informed many subsequent accounts. On Jomard, a leading member of the Egyptian expedition and editor of the *Description de l'Égypte*, see especially Yves Laissus, *Jomard, le dernier Égyptien, 1777–1862* (Paris: Fayard, 2004).

22. On Conté's life and career, see especially Patrice Bret, "Nicolas-Jacques Conté ou l'intelligence technique," *Revue du musée des Arts et Métiers* 25 (December 1998): 4–13; François Cornu-Thenard, *Nicolas-Jacques Conté: Curieux autodidacte du XVIIIe siècle. Peintre, chimiste, physician, aérostier et inventeur* (Dijon: Imprimerie Jobard, 1955); R. J. Maras, "Nicolas-Jacques Conté (1755–1805): Un savant et un inventeur sous la Révolution, le Directoire et l'Empire," *Revue d'histoire des sciences et de leurs applications* 14, no. 2 (1961): 155–168; and Alain Queruel, *Nicolas-Jacques Conté (1755–1805): Un inventeur de génie. Des crayons à l'Expédition d'Égypte en passant par l'aérostation militaire . . .* (Paris: l'Harmattan, 2004).

23. Conté worked as a *demonstrateur* at the Conservatoire, alongside Jean-Baptiste Le Roy and Alexandre-Théophile Vandermonde. In February 1803 Conté was charged with setting up a second Conservatoire des arts et métiers in Compiègne. See Archives Nationales, F/12/4867.

24. Jomard, *Conté*, 33.

25. Conté settled definitively in Paris in 1785. Conté's work as a painter is discussed in Cornu-Thenard, *Nicolas-Jacques Conté*, 14–24.

26. Bret, "Nicolas-Jacques Conté ou l'intelligence technique," 5.

27. Maras, "Nicolas-Jacques Conté," 157.

28. See Patrice Bret, "Recherche scientifique, innovation technique et conception tactique d'une arme nouvelle: l'aérostation militaire (1793–1799)," in *Lazare Carnot ou le Savant-Citoyen*, ed. Jean-Paul Charnay (Paris: Presses de l'Université de Paris-Sorbonne, 1990), 429–451; and Janis Langins, "Hydrogen Production for Ballooning during the French Revolution: An Early Example of Chemical Process Development," *Annals of Science* 40 (1983): 531–558. On the testing of balloons and weaponry at Meudon, see Archives Nationales, AF/II/220; also Gillispie, *Science and Polity in France*, 370–381.

29. Joseph Lakanal, *Rapport sur le télégraphe fait au nom du Comité d'instruction publique, réuni à la commission nommée par le décret du 27 avril dernier, vieux style; par Lakanal; réimpr. par ordre de la Convention nationale* (Paris: Imprimerie Nationale, 1794), 3.

30. Liza Oliver connects an epidemic of blindness in the Egyptian campaign to the encyclopedic gaze evident in the *Description de l'Égypte*. Oliver

does not mention Conté's portrait, but discusses his involvement in the project. See Liza Oliver, "Blindness Materialized: Disease, Decay, and Restoration in the Napoleonic *Description de l'Égypte* (1809–1828)," in *Seeing across Cultures in the Early Modern World*, ed. Dana Leibsohn and Jeanette Favrot Peterson (Farnham: Ashgate, 2012), 125–145.

31. Jomard, *Conté*, 72. Although Jomard does not specify which of the Montgolfier brothers he is referring to here, the antipathy toward paper he describes is striking given the Montgolfier family's status as prominent paper manufacturers.

32. Charles Dupin, *Essai historique sur les services et les travaux scientifiques de Gaspard Monge* (Paris: Bachelier, 1819), 89.

33. Deanna Petherbridge, *The Primacy of Drawing: Histories and Theories of Practice* (New Haven: Yale University Press, 2010), 121.

34. Joseph-Marie Degérando, *Éloge de M. Conté prononcé à la Séance de la Société d'Encouragement, 12 février 1806, par M. Degerando, Secrétaire de la Société, Secrétaire général du Ministère de l'Intérieur, Membre de l'Institut National, etc.* (Paris: l'Imprimerie de Madame Huzard, 1806), 5.

35. Erika Naginski, "Drawing at the Crossroads," *Representations* 72 (Autumn 2000): 67.

36. See Carol Armstrong, "Seurat's Media, or a Matrix of Materialities," *Grey Room* 58 (Winter 2015): 6–25. On the longer history of the Conté crayon, see Laurent Bigand, "De Nicolas-Jacques Conté à Conté SA: le parcours d'un groupe d'entreprises leader du marché français des instruments d'écriture, 1794–1979" (PhD diss., Lille 3—Université Charles de Gaulle, 2004). On the literary and cinematic afterlife of Conté's drawing materials, see Tadas Bugnevicius, "The Crayon and the Pencil: Balzacian Modernity in Jacques Rivette's *Ne touchez pas la hache*," *Revue Balzac* 6 (2023): 137–154.

37. *Exposition publique des produits de l'industrie française: catalogue des produits industriels qui ont été exposés au Champ-de-Mars pendant les trois derniers jours complémentaires de l'an VI* (Paris: l'Imprimerie de la République, 1798), 6; Bret, "Nicolas-Jacques Conté ou l'intelligence technique," 6.

38. Only the bust of Droz's statue now survives, the remainder having been melted down to make weapons during the Second World War. Upon Conté's death, the production of the crayons was taken over by his son-in-law Arnould Humblot-Conté, and the factory subsequently passed to Conté's granddaughter, who married a Desvernay. The business remained in the family until 1919.

39. Archives Nationales, F/17/1240B.

40. Archives Nationales, AF/II/220, folder 7 [1898].

41. *Instruction sur les qualités et propriétés des crayons de l'invention du citoyen Conté* (Paris: l'Imprimerie du Journal d'Indications, n.d.), Archives du Conservatoire national des arts et métiers, M494.

42. Tony Halliday, "Academic Outsiders at the Paris Salons of the Revolution: The Case of Drawings *à la manière noire*," *Oxford Art Journal* 21, no. 1 (1998): 69–86. On Isabey's use of Conté crayon, see too Cyril Lécosse, *Jean-Baptiste Isabey: petits portraits et grands desseins* (Paris: CTHS Editions and INHA, 2018). On the racialized dynamics of mezzotint and its role in the production of ideals of whiteness in the eighteenth century, see Jennifer Y. Chuong, "Engraving's 'Immovable Veil': Phillis Wheatley's Portrait and the Politics of Technique," *Art Bulletin* 104, no. 2 (2022): 63–88.

43. Halliday, "Academic Outsiders," 79–81.

44. Halliday, "Academic Outsiders," 82.

45. Seymour Howard, "The Steel Pen and the Modern Line of Beauty," *Technology and Culture* 26, no. 4 (October 1985): 789.

46. Bret, "Nicolas-Jacques Conté ou l'intelligence technique," 12.

47. Ernest Scheyer, "French Drawings of the Great Revolution and the Napoleonic Era," *Art Quarterly* 4 (Summer 1940): 187–204.

48. For the scientific report, see *Institut national des Sciences et arts. Classe des Sciences, physique, et mathématique. Extrait des Registres de la classe. Séance du 6 Prairial an 4 de la République française.* For the response of artists to Conté's pencils, see *Institut national des Sciences et arts. Classe de Littérature et Beaux-Arts. Extrait des Registres de la classe. Séance du 13 Prairial an 4 de la République française*; both Archives Nationales, AF/II/220, folder 7 [1898].

49. As Fourcroy described: "Even a fairly strong heat does not denature them; water does not soften them or dilute them in any way; the air does not alter them at all." Archives Nationales, AF/II/220, folder 7 [1898].

50. Jomard, *Conté*, 19, 31.

51. Jomard, *Conté*, 71.

52. Jomard, *Conté*, 31.

53. See, for instance, Fourcroy's report of 6 Prairial

Year IV, where the pencils are described in these terms. Archives Nationales, AF/II/220, folder 7 [1898].

54. Nicolas-Jacques Conté, letter of 18 Floréal Year VIII (May 8, 1805), Archives du Conservatoire national des arts et métiers, Bibl. 584.

55. On Conté's Egyptian inventions, see Patrice Bret, "Conté et 'les arts de la France au milieu des déserts': vicissitudes d'un transfert de technologie pendant l'occupation française," in *L'Égyptologie et les Champollion*, ed. Michel Dewachter and Alain Fouchard (Grenoble: Presses Universitaires de Grenoble, 1994), 331–345; and Bret, "Instruments of Knowledge and Power in a Colonial Context: Scientific Instruments during the French Occupation of Egypt, 1798–1801," in *Scientific Instruments between East and West*, ed. Neil Brown, Silke Ackermann, and Feza Günergun (Leiden: Brill, 2019), 206–228.

56. Jomard, *Conté*, frontispiece.

57. *Description de l'Égypte: ou, Recueil des observations et des recherches qui ont été fait en Égypte pendant l'expédition de l'armée française*, 20 vols. (Paris: Imprimerie Impériale, 1809–1828). The fullest account of Conté's engraving machine is given in Edme François Jomard, "Rapport fait par M. Jomard, au nom d'une Commission spéciale, sur les machines à graver en taille-douce," *Bulletin de la Société d'encouragement pour l'industrie nationale* 229 (July 1823): 169–183. For analysis of the plates of the *Description*, with attention to Conté's role in their engraving, see Melanie Ulz, *Auf dem Schlachtfeld des Empire: Männlichkeitskonzepte in der Bildproduktion zu Napoleons Ägyptenfeldzung* (Marburg: Jonas Verlag, 2008), 34–60. On Conté's machine in the context of other inscriptive technologies in the late eighteenth and early nineteenth centuries, see Peter Frieß, "Linea sub specie machinae: die mecahnische Linie—Ein Beitrag zur Druckgrafik 1750–1850" (master's thesis, Ludwig-Maximilian-Universität München, 1988). On Conté's work as an artist during this period, see Alain Mercier, "Portefeuille industriel: L'Égypte de Conté," *La Revue du Musée des Arts et Métiers* 19 (June 1997): 51–58.

58. On Conte's death, see Laissus, *Jomard*, 75. In addition to Roehn's portrait, Conté appears alongside his colleagues, including Napoleon, in an image of the interior of the house of Hasan Kâchef, appropriated for the Institut (*Description de l'Égypte*, vol. 1, *État Moderne*, pl. 55).

59. Conté was not the only scientist concerned with producing an engraving machine of this kind. See, for instance, the 1802 machine for drawing or engraving curved lines proposed in Archives du Conservatoire national des arts et métiers, C28. It is probable that Conté's invention was in fact preceded by an English invention, although his defenders have been at pains to state that he knew nothing of it, and that he had in fact already experimented with such a device himself before the Egyptian expedition. See Paul Vibert, *Mon Berceau: Histoire anecdotique, pittoresque et économique du 1er arrondissement* (Paris: Bellier, 1893), 280; and Jomard, "Rapport," 170. For a later English development of a comparable device, see the engraving machine proposed by C. L. Cooke in 1828, Archives du Conservatoire national des arts et métiers, V294.

60. On Gallet's contribution to the machine, and to "calcographie mécanico-linéaire," see F. E. Joubert père, *Manuel de l'amateur d'estampes, faisant suite au Manuel du libraire*, vol. 3 (Paris, 1821), 265–272.

61. Archives du Conservatoire national des arts et métiers, M211.

62. Rohberg-Herr de Vausenville, *Établissement en France, de l'art gammo-graphique, frère cadet de l'Imprimerie; Invention nouvelle et Françoise, pleinement executee, pour rayer toutes sortes de papiers, par une méthode plus prompte que l'impression et à l'instar de l'imprimerie typographique…* (Paris, 1792). See also *Proclamation du Roi sur plusieurs brevets d'invention. Du 2 Mai 1792* (Paris: l'Imprimerie Royale, 1792), 2.

63. Jomard, *Conté*, 62.

64. Edward Said, *Orientalism* (London: Penguin, 1995), 83.

65. Said, *Orientalism*, 87. On the European reception of the *Description*, see Andrew Bednarski, *Holding Egypt: Tracing the Reception of the Description de l'Égypte in Nineteenth-Century Britain* (London: Golden House Publications, 2005). On photography and Orientalism, see Ali Behdad and Luke Gartlan, eds., *Photography's Orientalism: New Essays on Colonial Representation* (Los Angeles: Getty Publications, 2013).

66. For discussion of the regulating ambitions of European colonialists in nineteenth-century Egypt, see Timothy Mitchell, *Colonising Egypt* (Cambridge: Cambridge University Press, 1988). See, too, Anne Godlewska, "Map, Text and Image. The Mentality

of Enlightened Conquerors: A New Look at the *Description de l'Égypte*," *Transactions of the Institute of British Geographers* 20, no. 1 (1995): 5-28; and Stephanie O'Rourke, "The Sediments of History in Napoleonic France," *Word and Image* 37, no. 1 (2021): 6-20.

67. Degérando, *Éloge de M. Conté*, 9–11. Conté made a similar point in his letter of 18 Floréal Year VIII (May 8, 1805), Archives du Conservatoire national des arts et métiers, Bibl. 584.

68. Darcy Grimaldo Grigsby, *Colossal: Engineering the Suez Canal, Statue of Liberty, Eiffel Tower, and Panama Canal* (Pittsburgh: Periscope Publishing, 2012), especially 20–41, which focuses on the *Description de l'Égypte*. Grigsby also discusses Conté's work briefly in *Extremities: Painting Empire in Post-Revolutionary France* (New Haven: Yale University Press, 2002), 66–67. On art and the Egyptian campaign, see also Todd Porterfield, *The Allure of Empire: Art in the Service of French Imperialism, 1798–1836* (Princeton: Princeton University Press, 1998).

69. Grigsby, *Colossal*, 38.

70. *Rapport des membres du conservatoire des arts et métiers au Ministre de l'Intérieur, sur l'utilité de l'enseignement théorique et pratique de l'art de filer le coton et autres matières par Machines dans tous le degrès de finesse*, 23 Prairial Year XII, Archives du Conservatoire national des arts et métiers, U95.

71. See Lynn Hunt, *Politics, Culture, and Class in the French Revolution* (Berkeley: University of California Press, 1984), 87–119.

72. Jomard, "Rapport," 172.

73. See the discussion of Monge's work on mirages in Camille Flammarion, *L'Atmosphère: Météorologie populaire* (Paris: Hachette, 1888). Conté's work on the *Description* undoubtedly owes something to Monge's work in descriptive geometry. See Gaspard Monge, *Géometrie déscriptive. Leçons données aux Écoles Normales, l'an 3 de la République* (Paris: Baudouin, 1798). Also see Grigsby, *Colossal*, 24–25.

74. Christopher Pinney, *The Waterless Sea: A Curious History of Mirages* (London: Reaktion, 2018).

75. Patrice Bret, "L'Homme, finalité de la technique," *Revue du musée des Arts et Métiers* 50 (December 2008): 36.

76. Bret, "Instruments of Knowledge and Power," 225–226. The use of telegraphs in Egypt was announced in a short notice "Etablissement des télégraphes en Égypte," *Courier de l'Égypte*, no. 74, 27 Messidor Year VIII [July 16, 1800], 4.

77. Jomard, *Conté*, 69.

78. Quoted in Queruel, *Nicolas-Jacques Conté*, 127.

79. Queruel, *Nicolas-Jacques Conté*, 60. Friedrich Meyer described the *aérographe*, with eight cylinders of fabric "stretched out on hoops like little paper lanterns," communicating successfully with the telegraph observatory at the Louvre; Frédéric Jean Laurent [Friedrich Johann Lorenz] Meyer, *Fragments sur Paris*, vol. 2, trans. Charles-François Dumouriez (Hamburg, 1798), 122–123.

80. Guy de Saint-Denis, "Nicolas-Jacques Conté et l'aérostation," *Les Cahiers de la FNARH* 115 (2010): 5–12.

81. [Claude-Joseph] Ferry, "Essais de télégraphie aérostatique," *Journal des sciences militaires des armées de la terre et de la mer* 5 (1826): 340–341.

82. See, for instance, Ignace Urbain Jean Chappe, *Dépêche télégraphique de Brest, 25 fructidor an XIII* [September 12, 1805], Pierpoint Morgan Library, 130450. This document reports to the Minister of the Marine the position of the British fleet and the current weather conditions, especially the "variable winds from the west to the north."

83. Charles-Gilbert Romme, "Convention Nationale, présidence de Jean Debry, suite de la séance de lundi 1er avril," *Gazette nationale, ou Le Moniteur universel*, no. 94, April 4, 1793, 2.

84. Paul Caraman, "Le télégraphe aérien de la tour Saint-Michel à Bordeaux, 1823–1853," *Revue historique de Bordeaux et du département de la Gironde* 4–6 (1911): 388–389.

85. John Durham Peters, *The Marvelous Clouds: Toward a Philosophy of Elemental Media* (Chicago: University of Chicago Press, 2015), 5.

86. Nina Athanassoglou-Kallmyer, "Blemished Physiologies: Delacroix, Paganini, and the Cholera Epidemic of 1832," *Art Bulletin* 83, no. 4 (December 2001): 689.

87. On Töpffer, see David Kunzle, *Father of the Comic Strip: Rodolphe Töpffer* (Jackson: University Press of Mississippi, 2007).

88. Claude Lévi-Strauss, *Tristes Tropiques*, trans. John and Doreen Weightman (London: Penguin, 1992), 262.

89. Lévi-Strauss, *Tristes Tropiques*, 272.

90. Jean-Baptiste Debret, *Voyage pittoresque et historique au Brésil, ou Séjour d'un artiste français au Brésil, depuis 1816*

jusqu'en 1831 inclusivement, vol. 3 (Paris: Firmin Didot, 1839), 233.

91. The association of giraffes with Egypt had been cemented by the 1827 gift from Muhammad Ali Pasha of Egypt to France's Charles X of a Nubian giraffe (then still widespread in North Africa). The creature, now known as Zarafa, lived in the Jardin des Plantes until 1845. The first giraffe to be seen in Europe for three hundred years, Zarafa inspired a giraffe craze in nineteenth-century Europe. See Daniel Harkett, "The Giraffe's Keepers and the (Dis)play of Difference," in *Of Elephants and Roses: French Natural History, 1790–1830*, ed. Sue Ann Price (Philadelphia: America Philosophical Society, 2013), 149–158.

92. On the history of this project, see *Napoléon au Chat Noir: l'Épopée vue par Caran d'Ache* (Paris: Adam Biro and Musée de l'Armée, 1999).

93. Roland Barthes, "The World as Object," in *Critical Essays*, trans. Richard Howard (Evanston: Northwestern University Press, 1972), 10.

94. Barthes, "The World as Object," 11.

95. Wolfgang Schivelbusch, *The Railway Journey: The Industrialization of Time and Space in the Nineteenth Century* (Oakland: University of California Press, 2014), xix.

96. On Barak, *On Time: Technology and Temporality in Modern Egypt* (Berkeley: University of California Press, 2013), especially 1–20.

97. Barak, *On Time*, 51.

98. Arthur Asseraf, "La mer immédiate: nouvelles, télégraphe et impérialisme en Méditerranée, 1798–1882," *Monde(s)* 2, no. 16 (2019): 59.

99. Barak, *On Time*, 51.

100. Françoise Viatte, "Weaving a Rope of Sand," *Yale French Studies* 89 (1996): 85–102.

101. A separate preliminary drawing for this plate from the *Description* focuses on this detail in isolation: *Description de l'Égypte. Dessins préparatoires*, Tome 2bis, Bibliothèque nationale de France, UB-181 (D)-FT 4.

102. Tim Ingold, *Lines: A Brief History* (London: Routledge, 2007), 152. See, too, Ina Blom, "'And Follow It': Straight Lines and Infrastructural Sensibilities," *Critical Inquiry* 45 (Summer 2019): 859–883.

103. Rosalind E. Krauss, "Grids," *October* 9 (Summer 1979): 59–60.

104. Rosalind E. Krauss, "Agnes Martin: The /Cloud/," in *Bachelors* (Cambridge, MA: MIT Press, 1999), 78.

105. Martin quoted in Krauss, "Agnes Martin," 78.

106. Krauss, "Agnes Martin," 82. See also Hubert Damisch, *A Theory of /Cloud/: Toward a History of Painting*, trans. Janet Lloyd (Stanford: Stanford University Press, 2002).

107. Jean-Baptiste Lamarck, *Annuaire météorologique pour l'an X*, vol. 3 (Paris, 1802), 149–164; Mary Jacobus, "Cloud Studies: The Visible Invisible," *Gramma* 14 (2006): 223. Later, John Ruskin developed this classificatory impulse, with his images diagramming the structure of cloud perspective. See John Ruskin, *Modern Painters*, 5 vols. (Boston: Estes and Laureat, 1894), vol. 5, 118–178. Also, Lorraine Daston, "Cloud Physiognomy," *Representations* 135 (Summer 2016): 45–71.

Chapter 4

1. Initiated in London by Barker in 1787, the panorama was opened by American Robert Fulton at the Passage des Panoramas in 1800.

2. Louis-Sébastien Mercier, "Paris vu dans son ensemble ou le Panorama," Pluviôse Year VIII (January-February 1800), ms 15079 (2), folios 110–111, in *Le Nouveau Paris*, ed. Jean-Claude Bonnet (Paris: Mercure de France, 1994), 1241. On this essay by Mercier, see Karlheinz Stierle, *La Capitale des signes: Paris et son discours*, trans. Marianne Rocher-Jacquin (Paris: Éditions de la Maison des sciences de l'homme, 2001), 108–109. Elsewhere in *Le Nouveau Paris*, Mercier compares the elevated position of a gesticulating coachman on one of the new *voitures* to a telegraph.

3. Mercier, "Paris vu dans son ensemble ou le Panorama," 1243.

4. Georges Perec, *An Attempt at Exhausting a Place in Paris*, trans. Marc Lowenthal (Cambridge, MA: Wakefield Press, 2010); Dolf Sternberger, *Panorama of the Nineteenth Century*, trans. Joachim Neugroschel (Oxford: Basil Blackwell, 1977), 27.

5. Sternberger, *Panorama of the Nineteenth Century*, 26. See too Laura Otis, *Networking: Communicating with Bodies and Machines in the Nineteenth Century* (Ann Arbor: University of Michigan Press, 2011).

6. Mercier, "Paris vu dans son ensemble ou le Panorama," 1244.

7. Stephen C. Pinson describes how specators were provided with opera glasses to inspect the detail of the diorama derived from this work. This preparatory study is painted in a much looser style than the finished diorama, and was used by Daguerre to experiment with composition and

coloristic effects. Pinson, *Speculating Daguerre: Art and Enterprise in the Work of L. J. M. Daguerre* (Chicago: University of Chicago Press, 2012), 65 and 192.

8. J. Montgolfier, "Au rédacteur," *Le Courrier des spectacles, ou Journal des théâtres*, 18 Floréal Year XI (May 8, 1803), 3.

9. On Texier and panoramic literature, see Jillian Taylor Lerner, "A Devil's Eye View of Paris: Gavarni's Portrait of the Editor," *Oxford Art Journal* 31, no. 2 (June 2008): 233–250.

10. Roland Barthes, "The Eiffel Tower," trans. Richard Howard and Julie Rose, *AA Files* 64 (2012): 126.

11. Perec, *An Attempt at Exhausting a Place in Paris*, 3.

12. Élie Berthet, *La Tour du télégraphe* (Paris: Degorce-Cadot, 1869), 8.

13. Theodor W. Adorno, *Minima Moralia: Reflections on a Damaged Life*, trans. E. F. N. Jephcott (London: Verso, 2005), 226.

14. In turn, the historic facade of the Samaritaine, which belongs to the LVMH luxury goods group—owned by Bernard Arnault, the wealthiest person in the world (as of 2024)—was at the start of this century threatened with demolition, before reopening in restored form in 2021.

15. Édouard André, *Swebach-Desfontaines: Peintre de la Révolution et de l'Empire* (Paris: Georges Rapilly, 1905), 11.

16. See *La Révolution par la gravure: Les tableaux historiques de la Révolution française* (Vizille: Musée de la Révolution française, 2002).

17. Jacques-François-Joseph Swebach, *Louis XVI Entering Paris, October 6, 1789*, Metropolitan Museum of Art, New York, 1986.309.2.

18. Susan L. Siegfried, *The Art of Louis-Léopold Boilly: Modern Life in Napoleonic France* (New Haven: Yale University Press, 1995), 96–103.

19. Jules Renouvier, *Histoire de l'art pendant la Révolution. Considéré principalement dans les estampes* (Paris: Jules Renouard, 1863), 146.

20. Georg Simmel, *The Philosophy of Money*, trans. Tom Bottomore and David Frisby, from first draft by Kaethe Mengelberg (London: Routledge, 2005), 122.

21. *Catalogue de tableaux, dessins, estampes, livres à figures . . . composant le cabinet et le mobilier de feu M. Swebach . . . qui seront vendus les 5, 6, 7 avril 1824, et jours suivans, etc.* (Paris, 1824), 55.

22. According to Étienne Gaspard Robertson, entrepreneur, inventor, and aeronaut, by the early nineteenth century, "It only took a slightly metaphorical imagination to transform the Seine into the river Lethe; because the phantasmagoria were principally located on its banks, there was not one *quai* . . . which did not offer you a little phantom at the end of a dark corridor or at the top of a tortuous staircase." Étienne Gaspard Robertson, *Mémoires: récréatifs, scientifiques, et anecdotiques d'un physicien-aeronaute* (Paris, 1830–1834), quoted in Terry Castle, "Phantasmagoria: Spectral Technologies and the Metaphorics of Modern Reverie," *Critical Inquiry* 15, no. 1 (Autumn 1988): 37. A one-act *vaudeville* titled *Une Matinée du Pont-Neuf* was described by the *Gazette Littéraire* in December 1805 as offering "a magic lantern on the Pont-Neuf . . . it reviews all the caricatures and cries of Paris." *Archives littéraires de l'Europe, ou Mélanges de littérature, d'histoire, et de philosophie*, vol. 8 (Paris: Henrichs, 1805), lxx.

23. The painting was cleaned in 2002.

24. The correspondent for the *Mercure de France* summarized his contributions: "A series of wash and gouache drawings caught my attention. I think they are all by the same artist. They represent halts, camps, and marches. I found in them wit, finesse, lightness of touch and a very rare art of bringing out the details without detracting from the main effect." *Mercure de France*, June 7, 1788, 35–46.

25. On Lebrun, see Darius A. Spieth, *Revolutionary Paris and the Market for Netherlandish Art* (Leiden: Brill, 2017), 128–245.

26. *Mercure de France*, July 4, 1789, 36.

27. See Philippe Havard de la Montagne, "Au Café Manoury avec Mercier: une lettre inconnue de Rétif," *Etudes rétiviennes* 12 (June 1990): 57–68.

28. David Lawday, *Danton* (London: Jonathan Cape, 2009), 27.

29. On Charpentier, see Valérie Mainz, "Charpentier, Constance" in *Dictionary of Women Artists*, vol. 1, ed. Delia Gaze (Chicago: Fitzroy Dearborn, 1997), 381–382. On Deseine, see Georges Le Chatelier, *Deseine le sourd-muet. Claude-André Deseine, statuaire (1740–1823), notice biographique* (Paris: Imprimerie de l'Institution nationale des sourds-muets, 1903) and Iris Moon, "Mirabeau in Biscuit: Political Reputations and the Changing Aesthetics of Porcelain during the French Revolution," *French Porcelain Society Journal* 7 (2018): 29–48.

30. An undated lithograph by Bellangé represents a water

carrier of the kind represented in the painting, in almost the same location, with the Pont-Neuf and its statue of Henri IV visible in the background. Hippolyte Bellangé, *Porteur d'eau. Que de 2 sous perdus*, n.d., lithograph by Villain for Gihaut frères, Musée Carnavalet, G.38385. See *Le Peuple de Paris au XIXe siècle* (Paris: Paris musées, 2011), 74.

31. According to Maxime du Camp, under Napoleon this device was made out of bronze from cannon captured at Austerlitz. Maxime du Camp, "L'Hôtel des Monnaies de Paris et la fabrication des espèces monétaires," *Revue des Deux Mondes* 78 (1868): 791–792.

32. Swebach exhibited and signed works under different names: often "Swebach dit Fontaine," and sometimes, more rarely, just "Swebach." In later years, the artist's signature became formalized as Swebach-Desfontaines. During the Consulate, he sometimes used the initials "S.W."

33. Iris Moon, *The Architecture of Percier and Fontaine and the Struggle for Sovereignty in Revolutionary France* (London: Routledge, 2016), 16.

34. Friedrich Kittler, "The City Is a Medium," trans. Mathew Griffin, *New Literary History* 27, no. 4 (Autumn 1996): 719, 726.

35. "Le pouvoir télégraphique," *La Renommée*, December 14, 1819, 718.

36. "Le pouvoir télégraphique," 718.

37. Christophe Morin, "Le 18e siècle: Les fontaines de pierre," in *Paris et ses fontaines: De la Renaissance a nos jours*, ed. Dominique Massounie, Pauline Prévost-Marcilhacy, and Daniel Rabreau (Paris: Délégation à l'action artistique de la ville de Paris, 1995), 82.

38. Michel Belloncle, *La Ville et son eau* (Paris: Éditions Serg, 1978), 42. On the construction and capabilities of the Machine de Marly and the pump at the Samaritaine, see Étienne Lécuyer de la Jonchère, *Nouvelle méthode de fortifier les plus grandes villes . . . Suivie de dissertations sur la machine de Marly, sur les pompes du Pont Nôtre-Dame & de la Samaritaine, avec des remarques très curieuses sur l'hydraulique, la mechanique, & la fortification* (Paris: Florentin Delaulne, 1718), 179–193.

39. *Henri IV et La Samaritaine. Entretien de six minutes* (Paris, 1790)

40. This measurement is given in Belloncle, *La Ville et son eau*, 138.

41. See Beatrice de Andia, "Autour des fontaines, l'art total," in Massounie, Prévost-Marcilhacy, and Rabreau, *Paris et ses fontaines*, 12; and Frédéric Graber, "Inventing Needs: Expertise and Water Supply in Late Eighteenth- and Early Nineteenth-Century Paris," *British Journal for the History of Science* 40, no. 3 (September 2007): 315–332. In 1794, the Samaritaine was criticized for obstructing the river's flow, and for its dilapidated state. See Capron, *Précis sur un moyen capable de fournir dans les places, dans toutes les grandes et petite rues, et jusqu'aux endroits les plus élevés de Paris, des fontaines, toujours coulantes, d'une eau salubre, plus pure, et d'une abondance, supérieure à tous nos besoins* (Paris: L. F. Longuet, 1794), 9–10.

42. Bouhot's painting is now in the Musée Carnavalet, Paris, P1286.

43. Amaury Duval, *Les Fontaines de Paris, anciennes et nouvelles; Ouvrage contenant soixante planches dessinées et gravées au trait, par M. Moisy; Accompagnées de descriptions historiques et de notes critiques et littéraires; par M. Amaury Duval* (Paris: Firmin Didot, 1812).

44. Duval, *Les Fontaines de Paris*, avertissement.

45. Duval, *Les Fontaines de Paris*, avertissement.

46. Matthew C. Hunter, "Curve, Line, Circle, Slash, Cross: A Diagram of Liquid Intelligence," *Grey Room* 69 (Fall 2017): 6–23.

47. Duval, *Les Fontaines de Paris*, 1.

48. On nineteenth century Parisian sewers, see Donald Reid, *Paris Sewers and Sewermen: Realities and Representations* (Cambridge, MA: Harvard University Press, 1993), and Matthew Gandy, "The Paris Sewers and the Rationalization of Urban Space," *Transactions of the Institute of British Geographers* 24, no. 1 (1999): 23–44.

49. Duval, *Les Fontaines de Paris*, 1–2.

50. Duval, *Les Fontaines de Paris*, 2.

51. Duval, *Les Fontaines de Paris*, 2.

52. Olivier Darrigol, *Worlds of Flow: A History of Hydrodynamics from the Bernoullis to Prandtl* (Oxford: Oxford University Press, 2005), v.

53. Darrigol, *Worlds of Flow*, vii. See also Julián Simón Calero, *The Genesis of Fluid Mechanics, 1640–1780*, trans. Veronica H. A. Watson (Dordrecht: Springer, 2008).

54. Victor Burgin, "Socialist Formalism," *Studio International* 191, no. 980 (March-April 1976): 148–154, reprinted in Burgin, *Parallel Texts: Interviews and Interventions about Art* (London: Reaktion, 2011), 25.

55. M. Loiseau, "Destruction des postes en 1815," in *Le Télégraphe aérien: colloque de Nancy, mai 1981* (Nancy: FNARH, 1981), 1–4.

56. Roland Barthes, *Michelet*, trans. Richard Howard (Oxford: Basil Blackwell, 1987).

57. Jacques Foucart, "Bouhot ou le triomphe d'une modestie hollandaise," in *Étienne Bouhot, 1780–1862*, ed. Sandrine Balan (Semur-en-Auxois: Musée de Semur-en-Auxois, 2001), 17. See also Jean-Luc Lasnier, ed., *Étienne Bouhot, 1780–1862: Un peintre bourguignon et le paysage du Chatillonais* (Châtillon-sur-Seine: G. Ramelet, 1984).

58. Barthes, *Michelet*, 63–64.

59. Barthes, *Michelet*, 27.

60. On urban construction in this period, see Allan Potofsky, *Constructing Paris in the Age of Revolution* (Basingstoke: Palgrave Macmillan, 2009).

61. See Marie-Claude Chaudonneret, *L'État et les artistes: De la Restauration à la Monarchie de Juillet (1815–1833)* (Paris: Flammarion, 1999).

62. Stephen G. Nichols, "Contingency and Post-Figural Historiography in the French Restoration," *MLN* 124, no. 4 (2009): 782.

63. Nichols, "Contingency and Post-Figural Historiography," 782.

64. Alexis de Tocqueville, *The Old Regime and the Revolution*, trans. John Bonner (New York: Harper and Brothers, 1856).

65. Hannah Arendt, "The Freedom to Be Free," *New England Review* 38, no. 2 (2017), accessed February 23, 2024, http://lithub.com/never-before-published-hannah-arendt-on-what-freedom-and-revolution-really-mean/

66. Arendt, "The Freedom to be Free"

67. Beth S. Wright, *Painting and History during the French Restoration: Abandoned by the Past* (Cambridge: Cambridge University Press, 1997), 19.

68. See *L'Invention du passé: tome 1, gothique, mon amour . . . 1802–1830* (Paris: Hazan, 2014) and Stephen Bann and Stéphane Paccoud, eds., *L'Invention du passé*, vol. 2: *Histoires de coeur et d'épée en Europe, 1802–1830* (Paris: Hazan, 2014). Also see Françoise Baudson, ed., *Le Style troubadour* (Bourg-en-Bresse: Musée de Brou, 1971).

69. Jeanne Magnin, "Un peintre bourguignon, Étienne Bouhot," *Les Annales de Bourgogne* 1, no. 4 (1929): 295.

70. Magnin, "Un peintre bourguignon," 295.

71. Sandrine Balan, "Étienne Bouhot, sa vie, son oeuvre," in Balan, *Étienne Bouhot*, 30.

72. *Affiches, annnoces et avis divers, ou Journal général de France* (Paris: Bureau des affiches, 1804), 2618.

73. Kittler, "The City Is a Medium," 717–729.

74. Balan, "Étienne Bouhot, sa vie, son oeuvre," 31. See John Zarobell, "Jean-Charles Langlois's *Panorama of Algiers* (1833) and the Prospective Colonial Landscape," *Art History* 26, no. 5 (November 2003): 638–668. Zarobell notes that in December 1833 Langlois closed his panorama to make additions, introducing more figure groups (659).

75. François Ligeret du Cloiseau, *Notice historique sur la vie et les ouvrages d'Étienne Bouhot* (Semur-en-Auxois: Verdot, 1854).

76. Magnin, "Un peintre bourguignon," 289.

77. Mantz, quoted in Ernest Chesneau, *La Peinture française au XIXe siècle. Les chefs d'école: L. David, Gros, Géricault, Decamps, Meissonnier, Ingres, H. Flandrin, E. Delacroix* (Paris: Didier, 1862), 399–400.

78. Balan, "Étienne Bouhot, sa vie, son oeuvre," 64–65.

79. Balan, "Étienne Bouhot, sa vie, son oeuvre," 43.

80. See Jean-Claude Bastian, "Le Télégraphe Chappe à Montmartre," *Les Cahiers de la FNARH* 109 (2008): 17–29.

81. Guy de Saint-Denis, ed., *La Télégraphie Chappe* (Nancy: FNARH/Éditions de l'Est, 1993), 40–47.

82. Ignace Urbain Jean Chappe, *Histoire de la télégraphie* (Paris: Chez l'auteur, 1824).

83. There is some doubt as to whether the work is by Bouhot—Balan cites its "cold and rather austere" composition, its unusual viewpoint, and its subject matter as evidence to the contrary—although the Musée Carnavalet claims it is by the artist; see Balan, "Étienne Bouhot, sa vie, son oeuvre," 41. On the contested authorship of Bouhot's paintings, see Foucart, "Bouhot ou le triomphe d'une modestie hollandaise," in Balan, *Étienne Bouhot*, 16–17.

84. Marc Gocel, *La Télégraphie aérienne de "A à Z"* (Florange: Les Presses du Tilleul, 2001), 12.

85. Pierre Pinon, "La Topographie parisienne d'Étienne Bouhot," in Balan, *Étienne Bouhot*, 79.

86. Alfred Sensier, *Étude sur Georges Michel* (Paris: Alphonse Lemerre, 1873), n.p.

87. On Sensier, see Étienne Charavay, "Alfred Sensier," *L'Amateur d'autographes, revue des collectionneurs, des archivists et des érudits* (January 1877): 12–15; Catherine Granger, "Sensier, Alfred," in *Dictionnaire critique des historiens de l'art* (Paris: INHA, 2009), https://www.inha.fr

/fr/ressources/publications /publications-numeriques /dictionnaire-critique-des -historiens-de-l-art/sensier -alfred.html; and Christopher Parsons and Neil McWilliam, "'Le Paysan de Paris': Alfred Sensier and the Myth of Rural France," *Oxford Art Journal* 6, no. 2 (1983): 38–58.

88. Sensier, *Étude*, n.p.

89. Sensier, *Étude*, 24. No evidence of Michel's restoration work can be found, although it is reasonable to assume that it might have formed part of an unofficial portfolio of activities undertaken by an artist of his kind. See Noémie Étienne, *The Restoration of Paintings in Paris, 1750–1815: Practice, Discourse, Materiality* (Los Angeles: Getty Publications, 2017).

90. Sensier, *Étude*, 1–2.

91. Asher Ethan Miller, *The Path of Nature: French Paintings from the Wheelock Whitney Collection, 1785–1850* (New York: Metropolitan Museum of Art, 2013), 17.

92. As Édouard André observed, Swebach-Desfontaines, unlike Michel, "cannot be considered forgotten." Swebach achieved professional success, and contemporary critics considered his paintings comparable to those of Boilly. André, *Swebach-Desfontaines*, 5–6.

93. Vincent Van Gogh, letter 432, 2 March 1884. Cited in Ger Luijten, "Maître Michel," in *Georges Michel: le paysage sublime*, ed. Magali Briat-Philippe and Ger Luijten (Paris: Fondation Custodia, Collection Fritz Lugt, 2017), 9.

94. Sensier, *Étude*, 70.

95. On landscape painting at this time, see Steven Adams, *Landscape Painting in Revolutionary France: Liberty's Embrace* (London: Routledge, 2019) and Kelly Presutti, *Land into Landscape: Art, Environment, and the Making of Modern France* (New Haven: Yale University Press, 2024).

96. See Jacques Foucart, "La Hollandomanie de Georges Michel, un défi de peintre"; Grégoire Hallé, "Georges Michel et le moulin de Rembrandt: L'artiste et les maîtres hollandais"; and Viktoria von der Brüggen, "Plaines mélancoliques d'Ile-de-France: Georges Michel et la peinture de paysage en Europe," all in Briat-Philippe and Luijten, *Georges Michel*, 24–31, 32–36, 52–58, respectively.

97. Sensier, *Étude*, 51.

98. John Durham Peters, *The Marvelous Clouds: Toward a Philosophy of Elemental Media* (Chicago: University of Chicago Press, 2015), 11.

99. Peters, *The Marvelous Clouds*, 11–12.

100. Georges Michel, *The Gathering Storm (Landscape with Quarry and Figures)*, n.d., The National Trust for Scotland, Brodie Castle, 73.296.

101. John Foster, "Extracts from Mr. Foster's Journal," in *Life and Correspondence of John Foster*, vol. 1, ed. J. E. Ryland (New York: Wiley and Putnam, 1846), 108; *Oeuvres de Barnave. Mises en ordre et précédées d'une Notice historique sur Barnave par M. Bérenger de la Drome*, vol. 1 (Paris: Jules Cherelle et Guiller, 1843), 210.

102. Iris Moon, "Stormy Weather in Revolutionary Paris: A Pair of Dihl et Guérhard Vases," *Metropolitan Museum Journal* 51 (2016): 112–127.

103. Moon, "Stormy Weather," 124.

104. Moon, "Stormy Weather," 124.

105. Rolf Reichardt, "Light against Darkness: The Visual Representations of a Central Enlightenment Concept," *Representations* 61 (1998): 95–148.

106. "Effets de l'électricité," *Supplément au journal des villes et des campagnes, des curés, des maires, etc.*, June 11, 1844, 126.

107. Charles Sellier, *Curiosités du vieux Montmartre: les moulins à vent, la porcelaine de Clignancourt, le Mont-Marat* (Paris: J. Kugelmann, 1893), 11.

108. *Vente d'une nombreuse et importante collection d'études peintes à l'huile et tableaux de M. Michel. Dessins et tableaux de l'École ancienne, quelques ustensiles d'atelier, et bordures dorées* (Paris: De Lacombe, 1842), 12, lot 137.

109. Lucien Lazard, "Les peintres de Montmartre," *Le Vieux Montmartre: Société d'histoire et d'archéologie du XVIIe arrondissement*, January 1899, 223.

110. On Montmartre's underground spaces and their condition during the July Monarchy, see Hippolyte Hageau, *Extrait du Rapport sur les carrières de Montmartre* (Paris: Dezauche, 1837).

111. On Cuvier's significance in postrevolutionary material culture, see Iris Moon, *Luxury after the Terror* (University Park: Penn State University Press, 2022), 135–169.

112. Johan Meerman, *Montmartre, poème hollandois, avec la traduction françoise* (Paris: De Bure, 1812), 34.

113. Meerman, *Montmartre*, 25–26.

114. Meerman, *Montmartre*, 26.

115. Sensier, *Étude*, 11.

116. Sensier, *Étude*, 20.

117. Dominique-Jacques-François Chéronnet, *Histoire de Montmartre: état physique de la butte, ses chroniques, son abbaye, sa*

chapelle du martyre, sa paroisse, son église et son calvaire, Clignancourt (Paris: Breteau, 1843), 60; Ernest Van Den Broeck, *Montmartre, son histoire religieuse et civile. Les moulins, les fontaines et les platriéres, les réservoirs, le sacré-coeur, les chansonniers* (Paris: Éditions de la Revue Lumineuse, n.d.), 16.

118. Ger Luijten, "Georges Michel—Dessinateur," in Briat-Philippe and Luijten, *Georges Michel*, 123.

119. "Effets de l'électricité," 126.

120. Jacques Payannet, "Moulins à vent et télégraphe aérienne," in *Histoire de la Poste et des Télécommunications. 7ème Colloque international, organisé par le* FNARH, *les 28 février, 1er et 2 mars 1991*, vol. 1 (Brûlon et Le Mans: FNARH, 1991), 451, 456.

121. A three-meter-high obelisk on the Impasse Girardon, the *mire du Nord*, was installed by Jacques Cassini in 1736 to mark the Paris Meridian at Montmartre.

122. See also the recto of this drawing, and Fondation Custodia, Collection Frits Lugt, Paris, inv. 2016-T.100, 2016-T.102, and 2016-T.111.

123. *Les Ouvriers de la montagne de Montmartre à messieurs les habitans de Paris* (Paris: Imprimerie de Moutard, n.d.).

124. Millet also included telegraph towers in paintings. See Jean-François Millet, *The Plain of Chailly with Harrow and Plough*, Belvedere, Vienna, 2450.

125. Louis Rhéal, "Tableau . . . de moeurs," *Le Tintamarre: critique de la réclame, satire des puffiste: Journal d'industrie, de littérature, de musique, de modes et de théâtres*, April 22, 1888, n.p.

126. Étienne Arago, *Notice des peintures, sculptures et dessins de l'École moderne, exposés dans les galeries du musée national du Luxembourg* (Paris: Imprimeries réunies, 1890), xv.

127. Steven Adams, *The Barbizon School and the Origins of Impressionism* (London: Phaidon, 1994), 45–46.

128. Parsons and McWilliam, "'Le Paysan de Paris,'" 54.

129. Parsons and McWilliam, "'Le Paysan de Paris,'" 39. See too Nicholas Green, *The Spectacle of Nature: Landscape and Bourgeois Culture in Nineteenth-Century France* (Manchester: Manchester University Press, 1990).

130. Alfred Sensier, *La Vie et l'oeuvre de J.-F. Millet, manuscrit publié par Paul Mantz* (Paris: A. Quantin, 1881).

131. Parsons and McWilliam, "'Le Paysan de Paris,'" 49.

132. Parsons and McWilliam, "'Le Paysan de Paris,'" 49.

133. See Robert Sayre and Michael Löwy, "Figures of Romantic Anti-Capitalism," *New German Critique* 32 (Spring-Summer 1984): 42–92.

134. Sensier, *Étude*, 3.

135. Sensier, *Étude*, 5.

136. Sensier, *Étude*, 13.

137. I am grateful to Sophie Matthiesson for sharing with me her important unpublished research work on imprisoned artists during the Revolution, including Michel. Matthiesson suggests that Michel's Babouvism may have influenced his representations of common land and collective production in his landscapes.

138. Sensier, *Étude*, 22.

139. Parsons and McWilliam, "'Le Paysan de Paris,'" 38.

140. Baron de Girardot, "Nécrologie d'amateurs d'autographes. Sensier (J.-B.-Théod.)," *L'Amateur d'autographes*, vol. 40 (August 16, 1863), 241–242.

141. *Rapport de M. Sensier, ancien notaire, commissaire du 2e arrondissement, chargé de constater le nombre des victimes et les faits mémorables des glorieuses journées des 27, 28 et 29 juillet 1830* (Paris: Firmin Didot, 1830).

142. Girardot, "Nécrologie d'amateurs d'autographes," 242. Auguste Théodore, baron de Girardot, himself authored a book on the festivals of the revolutionary period, *Les Fêtes de la Révolution, 1790—An VIII* (Nantes: Veuve Camille Mellinet, 1858). On postrevolutionary collecting, see Tom Stammers, *The Purchase of the Past: Collecting Culture in Post-Revolutionary Paris, c. 1790–1890* (Cambridge: Cambridge University Press, 2020).

143. Paul Mantz, "Alfred Sensier," foreword to Sensier, *La Vie et l'oeuvre de J.-F. Millet*, ii.

144. John Ruskin, *The Storm-Cloud of the Nineteenth Century* (New York: John Wiley and Sons, 1884), 32.

145. Ruskin, *Storm-Cloud*, 33–34.

146. Ruskin, *Storm-Cloud*, 34.

147. Judith Stoddart, *Ruskin's Culture Wars: Fors Clavigera and the Crisis of Victorian Liberalism* (Charlottesville: University of Virginia Press, 1998), 89.

148. Practically a ruin by the early nineteenth century, Saint-Pierre was restored in 1838–1845 (the tower, built for Chappe's telegraph, was removed in 1840) and then again in 1874, although in 1896 its future was still not secure, and it was closed that year. Between 1900 and 1905 it was restored definitively by Louis Sauvageot.

Chapter 5

1. Ludwig Börne, quoted in Peter Fritzsche, *Stranded in the Present: Modern Time and the Melancholy of History* (Cambridge, MA: Harvard University Press, 2004), 51. See also Inge Rippmann and Wolfgang Labuhn,

eds., *"Die Kunst—eine Tochter der Zeit": Neue Studien zu Ludwig Börne* (Bielefeld: Aisthesis Verlag, 1988), 140.

2. See Jennifer L. Roberts, "Post-Telegraphic Pictures: Asher B. Durand and the Nonconducting Image," *Grey Room* 48 (Summer 2012): 12–35.

3. Nadar, *Quand j'étais photographe* (Paris: Flammarion, 1900), 5–6.

4. Honoré de Balzac, *La Comédie humaine*, vol. 15, part 2 (Paris: A. Houssiaux, 1853), 253–254, cited in Gianni Guastella, *Word of Mouth: Fama and Its Personifications in Art and Literature from Ancient Rome to the Middle Ages* (Oxford: Oxford University Press, 2017), 41–42.

5. Richard Cobb, *Paris and Its Provinces, 1792–1802* (Oxford: Oxford University Press, 1975), 104–105.

6. On art and electricity in postrevolutionary France, see Stephanie O'Rourke, "Girodet's Galvanized Bodies," *Art History* 41, no. 5 (November 2018): 868–893.

7. Maxime du Camp, *Paris, ses organes, ses fonctions et sa vie dans la seconde moitié du XIXe siècle*, vol. 1 (Paris: Hachette, 1875).

8. On the emergence of these machines, see Patrice A. Carré, "From the Telegraph to the Telex: A History of Technology, Early Networks, and Issues in France in the Nineteenth and Twentieth Centuries," *Flux* 11 (1993): 17–31.

9. Du Camp wrote extensively about revolutionary history, analyzing the events of 1789, 1848, and 1871. See Maxime du Camp, *Histoire et critique* (Paris: Hachette, 1877); Maxime du Camp, *Souvenirs de l'année 1848* (Paris: Hachette, 1876); and Maxime du Camp, *Les Convulsions de Paris* (Paris: Hachette, 1878). On du Camp's negative assessment of the Paris Commune, see Collette E. Wilson, *Paris and the Commune, 1871–1878: The Politics of Forgetting* (Manchester: Manchester University Press, 2007), 92–130.

10. Du Camp, *Paris*, 141.

11. Du Camp, *Paris*, 136.

12. Du Camp, *Paris*, 139.

13. For broader context on this issue, see Susan Bachrach, *Dames Employées: The Feminization of Postal Work in Nineteenth-Century France* (New York: Institute for Research in History, 1984), and *Artisans d'hier des communications d'aujourd'hui, 1850–1950*, exh. cat., Archives Nationales, Hôtel de Rohan, April 9–June 8, 1981 (Paris: Archives Nationales, 1981), 50. The April 1894 edition of the American magazine *Popular Science Monthly* also commented on this aspect of the *bureau*'s organization: "The employées (mark, the word with two final *e*'s is feminine) while at work are all dressed in black blouses to preserve their dresses from the oil stains liable to result from close contact with the apparatus. They are allowed to do, when their post is free, a little work in crocheting or in tapestry." The image chosen to accompany this description, however, shows a room populated entirely by men. *Popular Science Monthly* 44 (April 1894): 799.

14. Du Camp, *Paris*, 140.

15. Du Camp, *Paris*, 140.

16. On this topic, see, among others, Jessica Riskin, "Eighteenth-Century Wetware," *Representations* 83 (Summer 2003): 97–125; Jessica Riskin, "The Defecating Duck, Or, the Ambiguous Origins of Artificial Life," *Critical Inquiry* 29, no. 4 (Summer 2003): 599–633; John Tresch, "The Machine Awakens: The Science and Politics of the Fantastic Automaton," *French Historical Studies* 34, no. 1 (Winter 2011): 87–123; and Adelheid Voskuhl, *Androids in the Enlightenment: Mechanics, Artisans, and Cultures of the Self* (Chicago: University of Chicago Press, 2013).

17. Julien Offray de La Mettrie, *L'Homme machine* (Leiden: Elie Luzac fils, 1748), 14. A connection between La Mettrie's work and the electrical telegraph, "which brought to the fore a new medial understanding of the semiotics of the body," is made in Philipp Sarasin, "The Body as Medium: Nineteenth-Century European Hygiene Discourse," trans. Brian Hanrahan, *Grey Room* 29 (Winter 2008): 54.

18. La Mettrie, *L'Homme machine*, 85.

19. La Mettrie, *L'Homme machine*, 92–93.

20. Sabine Arnaud, "Speaking Machines, the Trial of Articulation, and Deaf Education in Modern France," *Modern Intellectual History* 20, no. 3 (September 2022): 719.

21. See the classic account of industrial timekeeping in E. P. Thompson, "Time, Work-Discipline, and Industrial Capitalism," *Past and Present* 38 (December 1967): 56–97.

22. Du Camp, *Paris*, 148.

23. As du Camp himself wrote, "I am convinced that nine tenths of the commercial and financial operations that the provinces make on the Paris market are ordered by telegrams that mean something completely different from what they seem to say." Du Camp, *Paris*, 143.

24. On Caselli's early experiments with telegraphy, see Biblioteca comunale degli

Intronati, Siena, BCI P.1/2, 1, 2, 3; BCI GDS Ritratti Porri. 2332; and R.VI.37. Also, with reference to Caselli's patents, see Julia Zons, *Casellis Pantelegraph: Geschichte eines vergessenen Mediums* (Bielefeld: transcript Verlag, 2015).

25. "Télégraphe de société, de M. Darel," *La Mode, revue des modes, galerie des moeurs, album des salons* 1, no. 13, October 1829, 3–4.

26. Du Camp, *Paris*, 148.

27. On the reproduction of images via pantographic drawing in nineteenth-century France, see Katie Hornstein, "Le diagraphe de Charles Gavard et l'âge de la reproduction mécanique visuelle en France," *Histoire de l'art* 70 (July 2012): 73–82.

28. Frédéric Dillaye, *Les Nouveautés photographiques; 4e complément annuel à la théorie, la pratique, et l'art en photographie* (Paris: Librairie Illustré, 1896), 208–210.

29. On the quest for simultaneous time in the late nineteenth and early twentieth centuries, in which telegraphy plays a key role, see Peter Galison, *Einstein's Clocks, Poincaré's Maps: Empires of Time* (New York: W. W. Norton, 2003).

30. Emilio Pucci, "La transmission par fac-similé: Invention et premiers applications," *Réseaux* 63 (1994): 125–139. A 1997 exhibition in Caselli's hometown of Siena encouraged this connection by presenting a homage to the inventor in the form of "faxart" interventions by a range of contemporary artists. Enrico Crispolti and Marco Pierini, eds., *Omaggio a Giovanni Caselli. Londra, Parigi, Siena, Yamagata. Ottobre 1997* (Siena: Comune di Siena, 1997).

31. Lisa Gitelman, *Scripts, Grooves, and Writing Machines: Representing Technology in the Edison Era* (Stanford: Stanford University Press, 1999), 1–20.

32. Gitelman, *Scripts, Grooves, and Writing Machines*, 1.

33. Amir D. Aczel, *Pendulum: Léon Foucault and the Triumph of Science* (New York: Washington Square Press, 2003), 53.

34. Anton A. Huurdeman, *The Worldwide History of Telecommunications* (Hoboken: Wiley-Interscience, 2003), 150.

35. Nicola Nosengo, *L'Extinction des technosaures: histoires de technologies oubliées*, trans. Sophie Lem (Paris: Belin, 2010), 120–138.

36. Nosengo, *L'Extinction des technosaures*, 126. A single pantelegraphic message cost six francs to send. See also Alex Preda, "Les hommes de la bourse et leurs instruments merveilleux: Technologies de transmission des cours et origines de l'organisation des marchés modernes," *Reseaux* 122 (2003): 137–165.

37. On the demise and afterlife of the pantelegraph, see Julien Feydy, "Le pantélégraphe de Caselli," *Revue du musée des Arts et Métiers* 11 (June 1995): 50–57; on Chinese interest in the device see Renata Vinci, "A Telegraph for China: The Attempted Application of Caselli's Pantelegraph to Transmit Chinese Characters, 1856–87," *Interactions: Studies in Commmunication and Culture* 7, no. 3 (December 2016): 247–263.

38. "Photographic telegraphy, invented during the last century by Professor Giovanni Caselli of Florence, permitted transmission of the facsimile of any form of writing or illustration, whether manuscript or print, and letters of credit or contracts could now be signed at a distance of five thousand leagues." Jules Verne, *Paris in the Twentieth Century*, trans. Richard Howard (New York: Random House, 1996), 53.

39. On the relationship between nineteenth-century spiritualism and modern communications technologies, see Jill Galvan, *The Sympathetic Medium: Feminine Channeling, the Occult, and Communication Technologies, 1859–1919* (Ithaca: Cornell University Press, 2010); John Durham Peters, *Speaking into the Air: A History of the Idea of Communication* (Chicago: University of Chicago Press, 1999); and Jeffrey Sconce, *Haunted Media: Electronic Presence from Telegraphy to Television* (Durham: Duke University Press, 2000). On this topic, see also Jann Matlock, "Ghostly Politics," *Diacritics* 30, no. 3 (Fall 2000): 53–71.

40. See, for instance, L. Alph. Cahagnet, *The Celestial Telegraph; or, Secrets of the Life to Come Revealed through Magnetism* (London: George Pierce, 1851) or *The Spiritual Telegraph*, a leading newspaper trading in spiritual communication, published in New York City from 1852 to 1860.

41. Gabriel Delanne, *Apparitions matérialisées des vivants et des morts*, vol. 1: *Les fantômes des vivants* (Paris: Librairie Spirite, 1909), 373–375.

42. Delanne, *Apparitions matérialisées*, 375.

43. Delanne, *Apparitions matérialisées*, 375–376.

44. See Emily Eastgate Brink, "A Portrait in Pieces: Félix Nadar's *La Main du banquier D.*," *Word and Image* 37, no. 1 (January-March 2021): 88–98. Also in this year, Nadar made his photographs of the Paris catacombs, followed a couple

of years later by those of the sewers. See Shao-Chien Tseng, "Nadar's Photography of Subterranean Paris: Mapping the Urban Body," *History of Photography* 38, no. 3 (August 2014): 233–254.

45. La Gavinie, "Chronique," *La Lumière. Revue de la photographie, beaux-arts, héliographie, sciences*, September 4, 1858, 143. On the relationship between painting and photography as it appeared in *La Lumière*, the first journal dedicated to photography, see Ève Lepaon, "'L'art ne ferais pas mieux': Corrélations entre photographie et peinture dans *La Lumière*," *Études photographiques* 31 (Spring 2014).

46. La Gavinie, "Chronique," 143.

47. La Gavinie, "Chronique," 143.

48. Simone Natale, "Photography and Communication Media in the Nineteenth Century," *History of Photography* 36, no. 4 (2012): 452–454. On the confluence of photography, electric telegraphy, and computing, with reference to the telegraphic transmission of photographs (although with no mention of Caselli), see Geoffrey Batchen, "Electricity Made Visible," in *New Media, Old Media: A History and Theory Reader*, ed. Wendy Hui Kyong Chun and Thomas Keenan (New York: Routledge, 2006), 35–40. On phototelegraphy's early years, see T. Thorne Baker, *The Telegraphic Transmission of Photographs* (New York: D. Van Nostrand, 1910). See also Jonathan Dentler, "Broken Bridges and Breathless Images: Circulating Wirephotos in Midcentury America," *History of Photography* 45, no. 1 (2021): 78–91. A helpful survey of these techniques is given in R. W. Burns, *Television: An International History of the Formative Years* (London: Institution of Electrical Engineers, 1998), 3–34 (on Caselli, see 26–32).

49. Dillaye, *Les Nouveautés photographiques*, 210–214.

50. On Nadar's enthusiasm for flying machines and his 1863 photographs of helicopters (taken the same year that the pantelegraph went live), see Emily Doucet, "'The Idea Was in the Air': Nadar's Aerial Media," *Grey Room* 83 (2021): 112–137. In 1840, Cham's *Croquis en l'air* had brought the Chappe optical telegraph system into visual conversation with flying machines and other futuristic devices. Cham, *Croquis en l'air: à propos de ballons, de la chasse, du télégraphe sous-marin, de l'ambassade du Népaul, de la nouvelle Comète-Mauvais, et d'une foule d'autres événements drolatiques mais non politiques* (Paris: Au Bureau du Journal Le Charivari, 1840), 2.

51. Rosalind Krauss, "Tracing Nadar," *October* 5 (Summer 1978): 29–47.

52. Nadar, *Quand j'étais photographe*, 3–4.

53. Nadar, *Les Ballons en 1870. Ce qu'on aurait pu faire, ce qu'on a fait* (Paris: E. Chatelain, 1870), 11.

54. Nadar, *Quand j'étais photographe*, 9. Nadar returns to this phrase ("de passage en notre ville") several times in his account.

55. Nadar, *Quand j'étais photographe*, 9.

56. Nadar, *Quand j'étais photographe*, 11.

57. Nadar, *Quand j'étais photographe*, 12.

58. Eduardo Cadava, "Nadar's Photographopolis," *Grey Room* 48 (Summer 2012): 65.

59. Nadar, *Quand j'étais photographe*, 16–17.

60. Nadar, *Quand j'étais photographe*, 19–20.

61. Nadar, *Quand j'étais photographe*, 20.

62. Nadar, *Quand j'étais photographe*, 20–21.

63. Nadar, *Quand j'étais photographe*, 21. Cadava notes that Deuil also means "mourning" in French, and suggests that in taking a photograph of mourning, the young man "takes a photograph of photography itself." Cadava, "Nadar's Photographopolis," 67.

64. Nadar, *Quand j'étais photographe*, 23–24.

65. Stephen Bann, "'When I Was a Photographer': Nadar and History," *History and Theory* 48 (December 2009): 107.

66. *Journal de Toulouse: Politique et littéraire*, no. 295 (October 26, 1863): 3.

67. Markus Krajewski, *The Server: A Media History from the Present to the Baroque*, trans. Ilinca Iurascu (New Haven: Yale University Press, 2018), 246, 13.

68. Alexandre Dumas, *Count of Monte Cristo* (New York: Thomas Y. Crowell, 1894), 78.

69. "Le ministre, à long flots, verse la circulaire; / Partout le télégraphe, obscur vocabulaire, / Aux préfets attentifs apportant un sens clair, / Comme un serpent coupé, se tortille dans l'air." Barthélemy, *Némésis. Satire hebdomadaire* (Paris: Perrotin, 1833), 68.

70. Aby Warburg, "A Lecture on Serpent Ritual," trans. W. F. Mainland, *Journal of the Warburg Institute* 2, no. 4 (April 1939): 292.

71. Warburg, "A Lecture on Serpent Ritual," 292.

72. Roland Barthes, "The Eiffel Tower," trans. Richard Howard and Julie Rose, *AA Files* 64 (2012): 131.

73. On natural and artificial technological metaphors, see

Dolf Sternberger, *Panorama of the Nineteenth Century*, trans. Joachim Neugroschel (Oxford: Basil Blackwell, 1977), 17–38.

74. John Tresch, "The Prophet and the Pendulum: Sensational Science and Audio-visual Phantasmagoria around 1848," *Grey Room* 43 (Spring 2011): 16–41; also, John Tresch, *The Romantic Machine: Utopian Science and Technology after Napoleon* (Chicago: University of Chicago Press, 2012).

75. Tresch, "The Prophet and the Pendulum," 23. For a relevant reading of Meyerbeer, see Mary Ann Smart, *Mimomania: Music and Gesture in Nineteenth-Century Opera* (Berkeley: University of California Press, 2004), especially 101–131.

76. See Jean-François Sudre, *Langue musicale inventée par François Sudre* (Paris: la veuve de l'auteur and G. Flaxland, 1866).

77. Nadar's and Baudelaire's caricatures of Jongkind, Barthet, and Clésigner, executed in the 1850s on the same sheet of paper, might also be considered under this rubric. Bibliothèque nationale de France, Paris, RÉSERVE BOITE FOL-NA-88.

78. Nadar, *Quand j'étais photographe*, 25.

79. See Allan Sekula, "The Body and the Archive," *October* 39 (1986): 3–64.

80. Nadar, *Quand j'étais photographe*, 35. As Adam Begley notes, Nadar's final public act before his death in 1910 was to send a telegram to Louis Blériot, subsequently reprinted in *La Presse*, congratulating him on his crossing the English Channel. Adam Begley, *The Great Nadar: The Man behind the Camera* (New York: Tim Duggan Books, 2017), 193.

81. Krauss, "Tracing Nadar," 33. See also Cadava, "Nadar's Photographopolis," 56–77.

82. Bann, "'When I Was a Photographer,'" 95–111. Also see Stephen Bann, *Distinguished Images: Prints in the Visual Economy of Nineteenth-Century France* (New Haven: Yale University Press, 2013), 87–120. On Nadar's practice as an industrial photographer, see Elizabeth Anne McCauley, *Industrial Madness: Commercial Photography in Paris, 1848–1871* (New Haven: Yale University Press, 1994), 105–148.

83. Bann, "'When I Was a Photographer,'" 111.

84. Bann, "'When I Was a Photographer,'" 111.

85. Nadar, *Charles Baudelaire intime. Le poète vierge* (Paris: A. Blaizot, 1911), 36.

86. Nadar, *Charles Baudelaire*, 38–39. Recalling Nadar's description of Baudelaire's "jerky gait," Walter Benjamin noted the relationship between the poet's alarming appearance and the "shock experience" at the core of his art. Walter Benjamin, "On Some Motifs in Baudelaire," in *Selected Writings*, vol. 4, 1938–1940, ed. Howard Eiland and Michael W. Jennings (Cambridge, MA: Harvard University Press, 2003), 319.

87. Nadar's self-description appeared in *Journal pour rire*, January 24, 1852. As recounted in Jillian Lerner, "Nadar's Signatures: Caricature, Self-Portrait, Publicity," *History of Photography* 41, no. 2 (2017): 111.

88. Ségolène Le Men, in *Daumier, 1808–1879*, exh. cat. (Ottawa: National Gallery of Canada, 1999), 382.

89. Archives Nationales, F/90/1427–1428 B 38. On Caselli's experiment, see *Le Peuple: petit journal du soir quotidien*, no. 23 (October 14, 1863): 3.

90. Jean-Henri Fabre, *Souvenirs entomologiques: études sur l'instinct et les moeurs des insectes*, IX série (1905; repr., Paris: Delagrave, 1923), 119–128.

91. Jules Michelet, *L'Insecte* (Paris: Hachette, 1858), 208.

92. Max Kozloff, "Nadar and the Republic of Mind," *Artforum* 15, no. 1 (September 1976): 29.

Chapter 6

1. Fiona Benson, "Mayfly (*Ephemera danica*)," in *Ephemeron* (London: Jonathan Cape, 2022), 18.

2. De Contencin, "Port de Quinconces," in Noël-Marie Paymal Lerebours, *Excursions daguerriennes: vues et monuments les plus remarquables du globe*, vol. 2 (Paris: Goupil et Vibert, 1842), F 15v.

3. De Contencin, "Port des Quinconces," F 15v.

4. Gustave Nadaud, "Le Vieux Télégraphe," in *Chansons de Gustave Nadaud* (Paris: Henri Plon, 1870), 255; *Le Figaro*, August 29, 1839, quoted in Anne-Claude Ambroise-Rendu, "Le télégraphe: Considerations sur l'ubiquité et l'instantanéité de la communication moderne," *Le Temps des Médias* 18 (2012): 281.

5. Victor Hugo, quoted in Yves Lecouturier, ed., *Le Patrimoine des télécommunications françaises* (Paris: Flohic, 2002), 84.

6. The openwork spire that now tops the medieval tower of Saint-Michel was reconstructed by Paul Abadie between 1861 and 1869, above the height of the telegraph's placement there.

7. François Gautier, *L'Œuvre de Claude Chappe, créateur de l'administration française des*

télégraphes et inventeur du télégraphe aérien établi sous les auspices de la Convention Nationale (Paris: Librairie de l'Armée, 1893), 12.

8. Archives départementales des Yvelines, 6P 38. On the construction of the optical telegraph in this region, see also Archives départementales des Yvelines, 2 L/Mantes 27.

9. Quoted in Susan Sontag, *Under the Sign of Saturn* (London: Penguin Books, 2009), 111.

10. Sontag, *Under the Sign*, 118.

11. Sontag, *Under the Sign*, 116–117.

12. Ernest Jacquez, *Claude Chappe: notice biographique* (Paris: Picard A. et fils, 1893), 71.

13. Abraham Chappe in *Journal de Paris*, 13 Pluviôse Year XIII (February 2, 1805), 933.

14. *Journal de Paris*, 9 Pluviôse Year XIII (January 29, 1805).

15. Georges Didi-Huberman, *The Invention of Hysteria: Charcot and the Photographic Iconography of the Salpêtrière*, trans. Alisa Hartz (Cambridge, MA: MIT Press, 2003), 9.

16. Philippe Pinel, *Traité medico-philosophique de l'aliénation mentale*, 2nd ed. (Paris: J. Ant. Brosson, 1809), 457.

17. Jean-Martin Charcot, *Leçons du mardi à la Salpêtrière. Professeur Charcot. Polyclinique de 1888–1889. Notes de cours de MM. Blin, Charcot, Henri Colin* (Paris: E. Lecrosnier et Babé, 1889), 77 and 460.

18. Didi-Huberman recognizes the aptness of the Salpêtrière's name ("gunpowder factory") in this context (*Invention of Hysteria*, 208).

19. Jules Bernard Luys, *Leçons cliniques sur les principaux phénomènes de l'hypnotisme: dans leurs rapports avec la pathologie mentale* (Paris: Georges Carré, 1890), described at 287.

20. Seven years before the appearance of Luys's photograph, none other than Nadar published a novel titled *Le Miroir aux alouettes*, set in a clockmaker's shop; Nadar, *Le Miroir aux alouettes* (Paris: E. Dentu, 1883).

21. See Mike Jay, "Illustrations of Madness: James Tilly Matthews and the Air Loom," *Public Domain Review*, November 12, 2014, https://publicdomainreview.org/essay/illustrations-of-madness-james-tilly-matthews-and-the-air-loom#p-1-2. For a complete account, see Jay, *The Influencing Machine: James Tilly Matthews and the Air Loom* (London: Strange Attractor Press, 2012).

22. Jay, "Illustrations of Madness."

23. Jay, "Illustrations of Madness."

24. *Charges parisiennes: croquades, pochades, bêtises. Par MM. Cham, E. Lorsay, Vernier, et autres* (Paris: Aubert et Cie, n.d. [c.1844]). Plates 11–13, comprising twelve vignettes by caricaturist Cham [Amédée Charles Noé], are titled *Signes télégraphiques.*

25. Jacques Lacan, *The Psychoses: The Seminar of Jacques Lacan: Book III, 1955–1956*, ed. Jacques-Alain Miller, trans. Russell Grigg (London: Routledge, 1993), 37.

26. See Erin-Marie Legacey, *Making Space for the Dead: Catacombs, Cemeteries, and the Reimagining of Paris, 1780–1830* (Ithaca: Cornell University Press, 2019).

27. Archives Nationales, F/90/19103.

28. Louis Réau, *Les Monuments détruits de l'art français*, vol. 2 (Paris: Librairie Hachette, 1959), 258. See too Maurice Agulhon, "La 'statuomanie' et l'histoire," *Ethnologie française* 8, no. 2–3 (1978): 145–172.

29. See June Hargrove, "Shaping the National Image: The Cult of Statues to Great Men in the Third Republic," *Source: Notes in the History of Art* 29 (1991): 48–63.

30. *Journal des débats politiques et littéraires*, March 16, 1893, 1.

31. Archives Nationales, F/90/19103.

32. *La Lanterne*, July 15, 1893, 2.

33. Louis Aragon, *Paris Peasant*, trans. Simon Watson Taylor (Cambridge, MA: Exact Change, 1994), 153; Simon Baker, *Surrealism, History and Revolution* (Oxford: Peter Lang, 2007).

34. Robert Desnos, "Pygmalion et le Sphinx," *Documents* 2 (1930): 33–40.

35. Simon Baker, "Base-Metal Materialism," introduction to a translation of Desnos's essay, *Papers of Surrealism* 7 (2007), https://www.research.manchester.ac.uk/portal/files/63517391/surrealism_issue_7.pdf.

36. "Sur certaines possibilités d'embellissement irrationnel d'une ville," *Surréalisme au service de la Révolution* 5/6 (May 1933): 18–19. Also discussed in Baker, *Surrealism, History and Revolution*, 213.

37. Gaston de Pawlowski, "Inventions nouvelles et dernières nouveautés," *Les Annales politiques et littéraires* 2315, 1 August 1928, 128.

38. Hargrove, "Shaping the National Image," 61.

39. Réau, *Les Monuments détruits*, vol. 2, 258.

40. Réau, *Les Monuments détruits*, vol. 1, 320.

41. Raymond Queneau, *Courir les rues; Battre la campagne; Fendre les flots* (Paris: Gallimard, 1981), 64.

42. Jean Cocteau and Pierre Jahan, *La Mort et les statues* (1946; repr., Paris: Seghers, 1977), 32.

43. E. M. Forster, *The Machine Stops* (London: Penguin, 2001), 40.

44. Forster, *The Machine Stops*, 40–41.

45. George Sand to Gustave Flaubert, 24 February 1869, in *The George Sand–Gustave Flaubert Letters*, trans. A. L. McKenzie (London: Duckworth, 1922), 124.

46. Jonathan Sterne, "Compression: A Loose History," in *Signal Traffic: Critical Studies of Media Infrastructures*, ed. Lisa Parks and Nicole Starosielski (Urbana: University of Illinois Press, 2015), 48. In another shot from the same series, Beaumaine adjusts her position to reveal a revolutionary *bonnet rouge* and a royalist fleur-de-lys, embroidered on her dress. Philadelphia Museum of Art, 2001-62-918.

47. On the family dynamics of the Nadar photographic dynasty, see Sylvie Aubenas and Anne Lacoste, eds., *Les Nadar: Une légende photographique* (Paris: Bibliothèque nationale de France, 2018).

48. Louise Michel, *Le Siècle rouge*, Institut international d'histoire sociale, Amsterdam, Louise Michel Papers (fonds Lucien Descaves), ms. 646. Reproduced in *Trois romans: Les Microbes humains, Le Monde nouveau, Le Claque-dents*, ed. Claude Rétat and Stéphane Zékian (Lyon: Presses universitaires de Lyon, 2013), 587.

49. Victor Hugo, *Le Rhin. Lettres a un ami*, vol. 1 (Paris: Hachette, 1858), 38.

50. Laurent Benoît, *Commune de 1871. Les postes, les ballons, le télégraphe, d'après les documents et des souvenirs inédits* (Paris: Lucien Dorbon, 1934).

51. Michel, *Le Siècle rouge*, 587.

Index

Page numbers in italics indicate illustrations.

The MIT Press would like to thank the anonymous peer reviewers who provided comments on drafts of this book. The generous work of academic experts is essential for establishing the authority and quality of our publications. We acknowledge with gratitude the contributions of these otherwise uncredited readers.

This book was set in Haultin by Jen Jackowitz. Printed and bound in China.

Library of Congress Cataloging-in-Publication Data

Names: Taws, Richard, 1977- author.
Title: Time machines : telegraphic images in nineteenth-century France / Richard Taws.
Description: Cambridge, Massachusetts : The MIT Press, 2024. | Includes bibliographical references and index.
Identifiers: LCCN 2023052596 (print) | LCCN 2023052597 (ebook) | ISBN 9780262049184 (hardcover) | ISBN 9780262379694 (epub) | ISBN 9780262379687 (pdf)
Subjects: LCSH: Optical telegraph in art. | Optical telegraph. | Art and society—France—History—19th century. | Art and technology—France—History—19th century.
Classification: LCC N8251.T325 T39 2024 (print) | LCC N8251.T325 (ebook) | DDC 701/.05094409034—dc23/eng/20240513
LC record available at https://lccn.loc.gov/2023052596
LC ebook record available at https://lccn.loc.gov/2023052597

10 9 8 7 6 5 4 3 2 1